Homogeneous Relativistic Cosmologies

Princeton Series in Physics

edited by Arthur S. Wightman
and John J. Hopfield

Quantum Mechanics for Hamiltonians Defined as Quadratic Forms *by Barry Simon*

Lectures on Current Algebra and Its Applications *by Sam B. Treiman, Roman Jackiw, and David J. Gross*

Physical Cosmology *by P. J. E. Peebles*

The Many-Worlds Interpretation of Quantum Mechanics *edited by B. S. DeWitt and N. Graham*

The $P(\Phi)_2$ Euclidean (Quantum) Field Theory *by Barry Simon*

Homogeneous Relativistic Cosmologies *by Michael P. Ryan, Jr., and Lawrence C. Shepley*

Homogeneous Relativistic Cosmologies

by

Michael P. Ryan, Jr.
and
Lawrence C. Shepley

Princeton Series in Physics

Princeton University Press
Princeton, New Jersey 1975

Published by Princeton University Press, Princeton and London

Library of Congress Cataloging in Publication data will be
found on the last printed page of this book

Printed in the United States of America
by Princeton University Press
Princeton, New Jersey

ΑΓΕΩΜΕΤΡΗΤΟΣ ΜΗΔΕΙΣ ΕΙΣΙΤΩ

To G. D. – Whosoever loveth me loveth my hound

– THOMAS MORE

and

To Phyllis – Those were the days

– G. RASKIN

PREFACE

In the past decade cosmology, both theoretical and observational, has had a dramatic renascence. This book not only reflects this renewed interest but attempts to spur further theoretical research in this most majestic of fields. Here we do not treat observational cosmology, so well covered by Peebles' *Physical Cosmology*. Instead we expand on one field within general relativity.

This book is aimed at one who already knows a bit of relativity (say Track 1 of Misner, Thorne, and Wheeler's *Gravitation* – we use their sign conventions). The book, however, is self-contained, emphasizing a modern tensor analysis approach to relativistic cosmology. This modern approach should be attractive to the beginning graduate student as well as to the expert who wishes to extend his knowledge of cosmology.

The most spectacular results of the highly mathematical approach have been the singularity theorems of 1965-68. The thrust of the first half of the book is toward an introduction of these theorems. The second half delves into specific cosmological problems, and includes an introduction to the insights gained by the application of Hamiltonian techniques. A book of this type should include several features, which we have incorporated. We have outlined the book by means of flow charts for specific chapters and for the entire book. The last chapter tries to point the direction future mathematical and observational cosmology research should take. We have given a graded set of exercises, from simple calculations to deep questions worthy a Ph.D. thesis. Our bibliography includes nearly 500 important references in all aspects of mathematical cosmology.

A quick paragraph of prejudices: We are general relativists; hence we have shied away from the Brans-Dicke and other alternative theories. The expansion of the universe showed Einstein that his cosmological constant was unnecessary, and we follow him in discarding it. Finally, we are prejudiced against the apotheosizing of any cosmological principle. We shall let the real universe behave as it will.

Both of us want to thank John Wheeler and Charles Misner:

> If my slight Muse do please these curious days,
> The pain be mine, but thine shall be the praise.

Both of us have received financial support at all stages from the National Science Foundation, for which we are greatly indebted. Both of us appreciate the atmosphere provided by the Center for Relativity Theory of the University of Texas at Austin. One of us (M.R.) thanks the Department of Applied Mathematics and Theoretical Physics, Cambridge, and Dennis Sciama at the Department of Astrophysics, Oxford, for their hospitality, and the Science Research Council of Great Britain for financial support. Our friends and colleagues are too numerous; we thank them all. We have especially entertained the comments and ideas of Richard Matzner, George Ellis, Barry Collins, and Ray Sachs. Princeton University Press has been more patient than we ever could have imagined.

The Preface is the last part of a book to be written. As Prescott did we end ... "with feelings not unlike those of the traveller who, having long journeyed among the dreary forests and dangerous defiles of the mountains, at length emerges on some pleasant landscape smiling in tranquility and peace."

West Lake Hills, Texas
April, 1974

MICHAEL P. RYAN, JR.
LAWRENCE C. SHEPLEY

TABLE OF CONTENTS

Homogeneous Relativistic Cosmologies

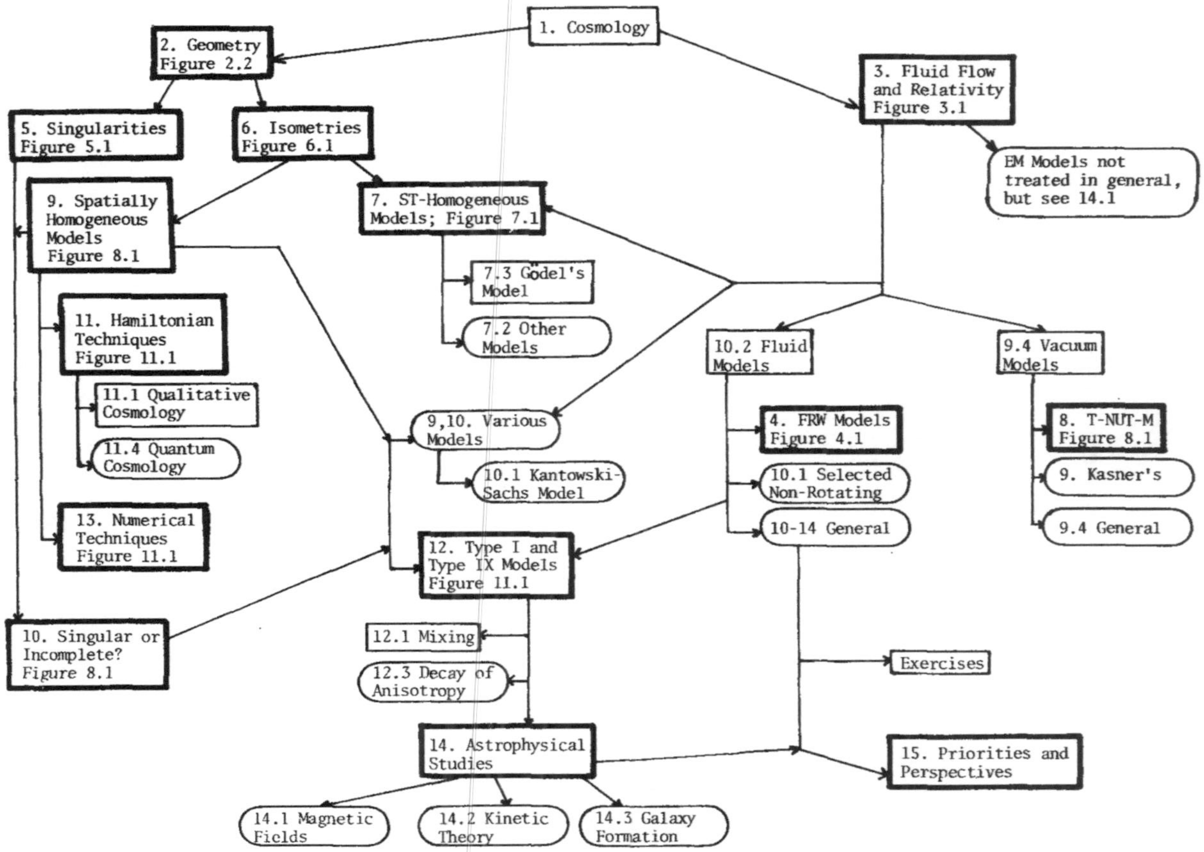
1. Cosmology
2. Geometry Figure 2.2
3. Fluid Flow and Relativity Figure 3.1
EM Models not treated in general, but see 14.1
5. Singularities Figure 5.1
6. Isometries Figure 6.1
9. Spatially Homogeneous Models Figure 8.1
7. ST-Homogeneous Models; Figure 7.1
7.3 Gödel's Model
7.2 Other Models
10.2 Fluid Models
9.4 Vacuum Models
11. Hamiltonian Techniques Figure 11.1
11.1 Qualitative Cosmology
11.4 Quantum Cosmology
9,10. Various Models
10.1 Kantowski-Sachs Model
4. FRW Models Figure 4.1
10.1 Selected Non-Rotating
10-14 General
8. T-NUT-M Figure 8.1
9. Kasner's
9.4 General
13. Numerical Techniques Figure 11.1
12. Type I and Type IX Models Figure 11.1
10. Singular or Incomplete? Figure 8.1
12.1 Mixing
12.3 Decay of Anisotropy
Exercises
14. Astrophysical Studies
15. Priorities and Perspectives
14.1 Magnetic Fields
14.2 Kinetic Theory
14.3 Galaxy Formation

1. COSMOLOGY: THE STUDY OF UNIVERSES

Who that well his warke beginneth,
The rather a good ende he winneth
– JOHN GOWER

It is difficult to explain to a layman the fact that the universe expands. How, he asks, can the universe expand? Into what is it expanding? We tell him that he does not have the correct picture in mind. Indeed the universe can "expand" in spite of the fact that it is everything and does not in any sense develop into unoccupied space. Instead, we tell him that distances between astronomical objects are becoming larger and larger as time moves forward.

Our layman will then turn to the title of this chapter and ask: If our universe is everything, what need is there to study other universes? And again the answer is very straightforward. We must simplify the study of cosmology; we must attempt to model the real universe in a mathematically tractable structure. In cosmology we study aspects of the real universe and possible aspects of the real universe. It is in fact often instructive to study features which we know not to be present in the real universe. No better example occurs than the study of anisotropic cosmological models, for so far as we can tell the physical universe is completely isotropic. However, an anisotropic model gives an extremely important example of the type of structure which may dominate the very early stages of cosmic expansion.

Our concern is not with the full range of physical and astrophysical cosmology, but rather with the geometrical and mathematical principles of general relativity as applied to cosmology. We shall treat expansion; we shall treat singularities; in short we shall discuss the arena of astrophysics and the boundary of that arena. Our study is limited to the simplest structure – homogeneous cosmologies – for the simplicity evoked by the homogeneity symmetry nonetheless allows very complex models.

The Problem of Fall

The most serious problem of modern theoretical cosmology is the existence of the initial singularity or "big bang." In all cosmological models this singularity appears. It is a region of infinitely dense matter, of infinitely strong gravitational forces. It is the beginning of spacetime, the boundary where our theories of space and time must be false. In the early days of gravitational physics, a problem of similar moment was the problem of fall (Koyré, 1955). The close analogy between the problem of singularity and the problem of fall is instructive.

Galileo established laws of inertia and some properties of gravitation. The question then arose: What is the path of a body falling under the influence of gravity? This question was not trivial, for the detailed law of gravitation had not yet been formulated. Several complicating features arose to slow the solution to the problem of fall: The mathematics (geometry) was not up to the problem; physicists found the mathematical reasoning difficult to apply. The new laws of physics were not well understood; the best mathematicians often had wrong ideas of the concepts involved. Finally, the center of attraction, the earth, is rotating; this rotation complicated the investigations of many researchers.

In the early seventeenth century, before the mathematical and physical ideas of Newton had appeared, the problem of fall was not solved. To introduce this problem in modern terms, imagine that an observer is stationed in the gravitational field of a point particle and throws out a small projectile. The path this projectile will follow can be discovered through step by step calculation making use of the fact that the projectile is always accelerated toward the central attractive point. A compact formula for the path was not available in the seventeenth century, but one special case is easily solved. A requirement of symmetry greatly simplifies the calculation. The initial velocity of the projectile is directly toward the central massive point, this being a naturally symmetric direction. The result is obvious: The projectile will hit the massive center in a finite time. (See Figure 1.1.)

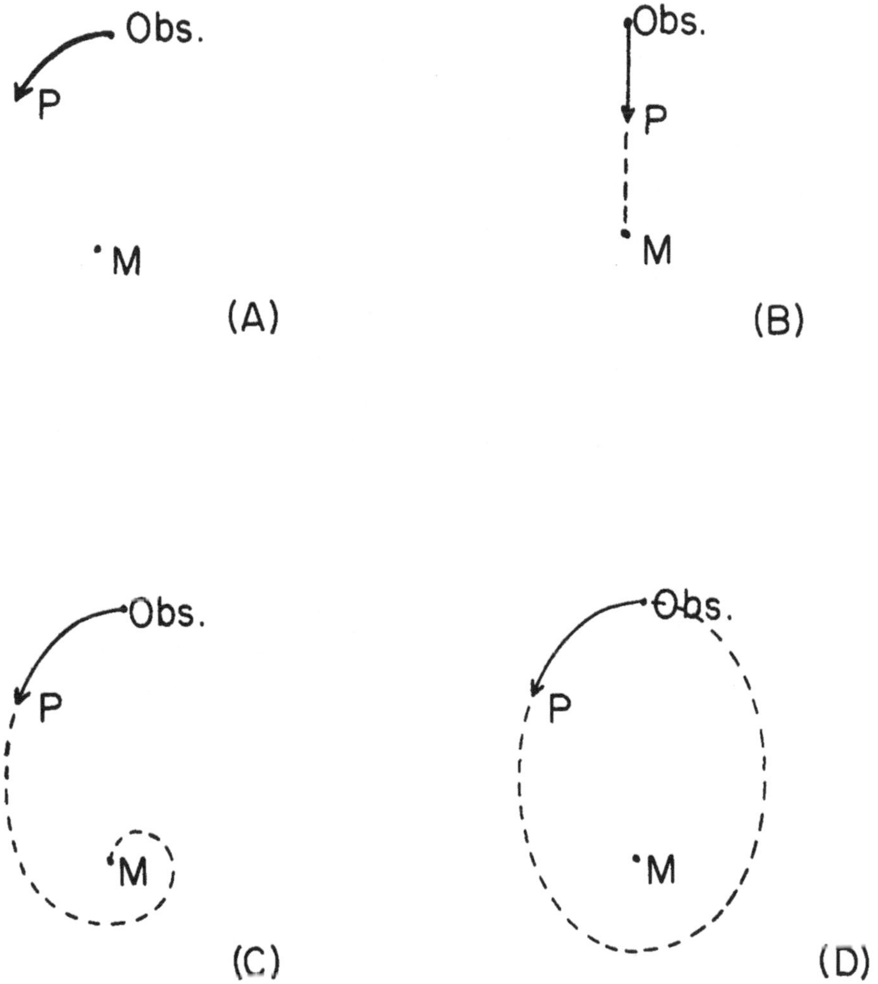

Fig. 1.1. The Problem of Fall in Pre-Newtonian Gravitation. (A) Numerical integration gives the orbit of a projectile P fired from a given Observer in a given direction. M is a central attractive point particle. (B) The path of P is found even more easily when the initial velocity is directly toward M. P hits M after a finite time. (C) In the more general case (non-radial), does the projectile P spiral in to hit M ... (D) or does P "bounce" back, missing M and remaining in a stable orbit?

The problem that remained unsolved before Newton was the fate of any other projectile. Will it spiral in to hit the central point? Is there a built-in end-point to the problem of orbit tracing? In this case no planet could have a stable orbit unless non-gravitational forces operated.

The central problem of modern cosmology bears a striking resemblance to the Renaissance problem of fall. Just as one particularly "symmetric" projectile path ends in disaster, so cosmological models chosen by the strongest symmetry requirements exhibit singular beginning points.

The cosmological models dealt with in this book, for example, are required to be symmetric under a three-dimensional group of isometries so that computation of their properties will be simple. Unlike the special projectile path that ended in disaster, however, these cosmological models are very general. Many properties, such as matter rotation, can be arbitrarily chosen (up to seven adjustable parameters). In spite of this generality, every cosmological model evolves a singularity of some sort.

The problem of fall was solved by recognizing that only very special initial conditions lead to destruction of the orbiting particle. This method of escaping trouble does not work in gravitational collapse of stars. Penrose, Hawking, Geroch and others have shown that whatever shape or state of motion a star is in when it starts to collapse, it will unavoidably reach a singular state. In cosmology, also, the relaxation of symmetry does not prevent the singularity. In cosmology this singularity appears as the beginning of the universe, although in some models a second singularity appears as the universe recollapses on itself. It is the beginning singularity that is disturbing, for it is an effect without a cause.

The Beginnings of Modern Cosmology

The problem of fall was an important chapter in the history of gravitation. At that time, too, appeared the very first applications of gravitation to cosmology. To understand the principles, techniques, and problems of modern cosmology it is worthwhile to look at selected incidents over the past several hundred years which have culminated in the theory of general relativistic cosmology.

Whether a static, matter-filled universe is reconcilable with Newton's theory of gravity was the subject of perhaps the earliest correspondence which could be termed modern cosmology. In four letters to Richard Bentley, Newton explored the possibility that matter might be spread evenly throughout an infinite space. It was Bentley's suggestion that this even distribution might be stable, but Newton felt the matter would tend to coagulate into large massive bodies. However, he apparently also thought that these massive bodies themselves could be stably spread throughout all space. Modern Newtonian cosmology, which is remarkably similar to relativistic cosmology, shows that Newton was wrong about this stability. The many other discussions of cosmology in the time of Newton were more of a theological or descriptive nature and do not have what modern researchers feel is the correct outlook: the explanation of cosmic features by use of terrestrial physical laws.

This situation, that the laws of physics known on earth were not applied to cosmological problems, continued with one notable exception. This exception was the effect commonly known as Olbers' Paradox (Jaki, 1969). Olbers' Paradox is the problem of why the night sky is dark. About a century before Olbers both Halley and Chéseaux realized that although the light from a star diminishes as the square of the distance to the star, the number of stars in a spherical shell increases as the square of the radius of the shell. The accumulated effect of the light intensity sould make the night sky as bright as the surface of the sun. In all fairness to Olbers the general scientific community was not able to appreciate fully the work of Halley and Chéseaux because of an insufficient sense of cosmic infinity. The problem is an especially important one, however, in that it depends critically on whether the universe is finite in space, finite in time, or static. The modern resolution of the paradox is that the universe is expanding and that the matter in the universe has a finite age. In particular, it is not a good enough explanation merely to postulate finite space sections in the universe.

The history of contemporary cosmology really began with Einstein's application of the principles of general relativity. Curiously Newtonian gravitation was not employed in cosmological theory until some time after relativistic cosmology was founded. Seeliger (1895) and Neumann (1896) had prepared the way for a cosmological constant in the 1890's, however.

Einstein's cosmology is an example of a slight lack of confidence in general relativity. So strong was his belief that the universe was static that he introduced a cosmological term to modify his original theory. Had he anticipated the work of Friedmann and shown that general relativity required an expanding universe, most researchers of the period would have found his theory unacceptable. Einstein himself at first refused to believe the results of Friedmann and only reluctantly accepted the idea that a non-static cosmological model is possible.

When Hubble completed his redshift survey, the results were astounding. They showed that there is a systematic redshift of light from distant galaxies, increasing as the distance to these galaxies. There have been many attempts to explain this redshift, some of which are quite ingenious. But the most widely accepted explanation, the explanation which now fits several other pieces of data, is that the universe is expanding. When most people accepted the notion of a non-static universe, an enormous number of cosmological models, both relativistic and non-relativistic, appeared. The best expression of the spherically symmetric relativistic cosmological models was presented by Robertson (1929), the mathematical niceties being a product both of Robertson (1935, 1936) and Walker (1936). We will call these models FRW universes to acknowledge the pioneering work of Friedmann. More general cosmologies, including models which result from the application of Newtonian gravitation, were presented by Milne (1934) and Milne and McCrea (1934). It is interesting to note that static cosmological models are not possible in Newtonian theory without the addition of a term much like the cosmological constant of Einstein.

Relativistic cosmologies and Newtonian cosmologies share an important feature. The expansion typically begins with a bang: There is a finite time in the past of any given observer where the density is infinite. This infinite density epoch represents the most fundamental problem of contemporary cosmology. The description of a singularity, whether of infinite density or of another type, and the proof in general of its existence, has only been carried through in recent years. Penrose, Hawking, Geroch, Misner – these are the names of the people who, with many others, have shown that singularities are a common feature of cosmological models. At present the detailed physical interpretation of these singularities remains an unresolved problem.

Although general relativity cosmology was the first cosmological theory, others have come and gone, sometimes without adequate reason for giving them up. Milne's theory, Hoyle's steady state theory, and others have at times been actively investigated but at present are no longer of interest. In the case of the steady state theory there is good, but by no means conclusive evidence that the theory does not meet observation. In particular, the discovery of the 3 K black body background radio emission is accepted by most people as evidence that the universe was significantly different in the past, in violation of the principle of the steady state universe.

Nowhere in this volume do we mention the cosmological principle except here. This principle is a fancy name for a simplifying set of assumptions. Its application results in homogeneous cosmologies, and to some, the principle also implies either a steady state situation or a static model. We prefer to leave the terminology of "principle" to the past, where it was a guide and a solace to researchers. Where we make symmetry assumptions, we state them as mere assumptions to aid in the solution of equations. We do not accept a cosmological principle as a Procrustean law, but leave to the observer the question of whether the universe has chosen to obey any of these assumptions.

Outline

We treat in this book a general consortium of cosmic problems. As a map to our treatment, we have drawn up a flow chart printed on the end papers of this volume. The flow starts with the general mathematical and physical foundations of general relativity and proceeds to relativistic hydrodynamics. The mathematics proceeds to the detailed theory of symmetry and singularity. Detailed applications to homogeneous cosmologies follow. We end with Hamiltonian techniques and with some remarks on astrophysical studies. Our conclusion is "A Call to Arms," for we feel that there are many interesting cosmological problems to be solved.

2. GEOMETRY IN THE LANGUAGE OF FORMS

So if a man's wit be wandering,
let him study the mathematics
– FRANCIS BACON

2.1. Points, Manifolds, and Geometrical Objects

Nothing is so vital to general relativity as the physical reality of an "event," or point, in spacetime, completely separate from coordinate systems used to describe it. On the surface of the earth Moscow is Moscow no matter what latitude or longitude we assign to it. Modern mathematics recognizes this separateness in the concept of a *manifold*, the set of points on which is placed the geometry of spacetime.

In general relativity the manifold is spacetime. A point of the manifold is identified with a physical event. A sample event is shown in Figure 2.1. As a point in a manifold it is independent of any coordinate system.

Fig. 2.1. Motion picture of a region in spacetime surrounding the event of a ball colliding with the ground. The event is that location in space and time when the ball just touches the ground. No matter the speed of film travel, the magnification of the lens, or the orientation of the camera – these are coordinate effects and do not affect the nature of the event itself.

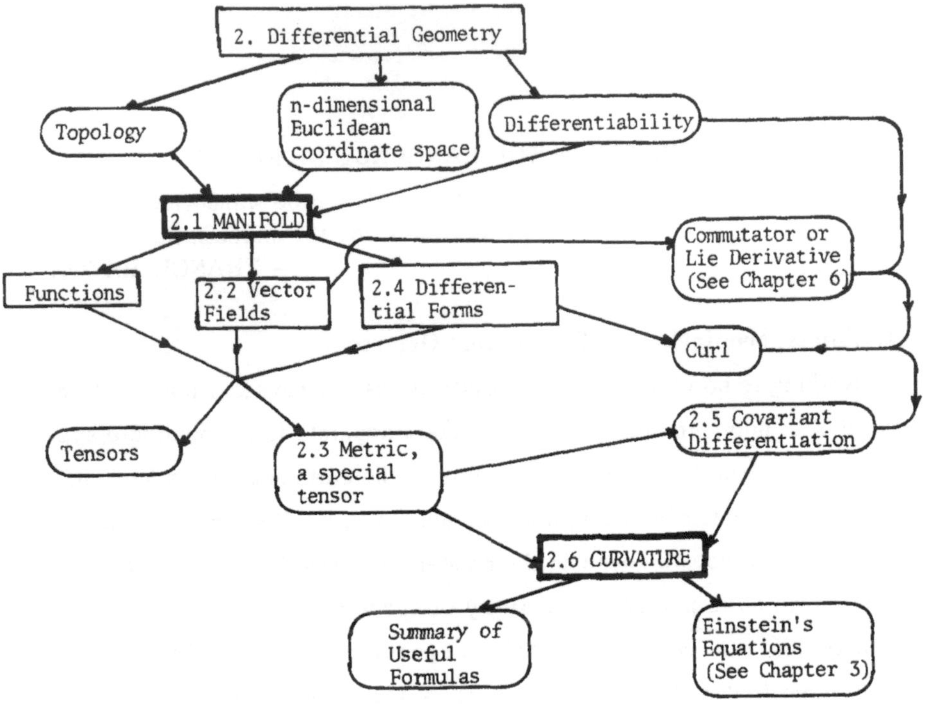

Fig. 2.2. Logical relations among some of the ideas of Chapter 2. Chapter 2 includes examples and other minor ideas not shown here.

This identification of an event with a mathematical point was a daring step due to Newton. Recently, the alternate concept has developed of a sponge-like construct which gives the appearance of smoothness only down to the scale of quantum fluctuations $(\sim (\hbar G/c^3)^{\frac{1}{2}} = 1.6 \times 10^{-33}\text{cm})$. At that scale the smoothness vanishes (Weyl, 1949; Wheeler, 1962b; Penrose, 1966). We shall not adopt this alternate viewpoint. Instead we will stay with the classical view of spacetime as a continuous and differentiable manifold.

On this manifold we shall place "geometrical objects" (Veblen and Whitehead, 1932; Schouten, 1954), the simplest of which are *function*, *vector field*, *metric*, and *differential form*. We shall develop these concepts in this chapter to the extent needed for the analysis of homogeneous cosmologies (see the flow diagram in Figure 2.2) in a form independent of coordinate systems. Definitions will be short and many proofs will be omitted (see standard works on differential geometry, e.g., Helgason, 1962, or Hicks, 1965). We shall end the chapter with a description of the three-sphere, a manifold we shall meet later on in the Friedmann-Robertson-Walker (FRW) universes. We shall also give a short table of useful formulas which have been developed in the chapter.

Definition of Manifold – Topology and Differentiability

A manifold is a set of points, basic subsets of which are labeled *open sets*. The open sets obey the property that any union of open sets is open (the set $0 < x < 1$ is an open set in the real line). Some of the subsets of the manifold will be *closed* (the complement of an open set; example $0 \leq x \leq 1$) and some neither open nor closed $(0 < x \leq 1)$. A set such as a manifold upon which open and closed subsets are defined is a *topological space*. An open set containing the point P is a *neighborhood* of P.

The defining properties of a *manifold* M are: (1) M is a topological space; (2) about every point P in M there is at least one neighborhood (open set) N in which a coordinate system (a local homeomorphism between points in N and the points in the n-dimensional space R^n of real numbers) may be set up. An open set N together with a coordinate system in N is called a *coordinate patch* or *coordinate neighborhood*. The number of coordinate patches needed to cover a manifold may be greater than one. Figure 2.3 shows an ordinary sphere (S^2), which needs two coordinate patches.

In order for a manifold to be useful in physics it must have a structure which distinguishes between differentiable and non-differentiable functions. An allowed coordinate system is one in which the coordinate functions are all differentiable. A *differentiable manifold* is one covered by a collection of allowed coordinate patches with the property that wherever coordinate patches overlap, one system is given in terms of the other by

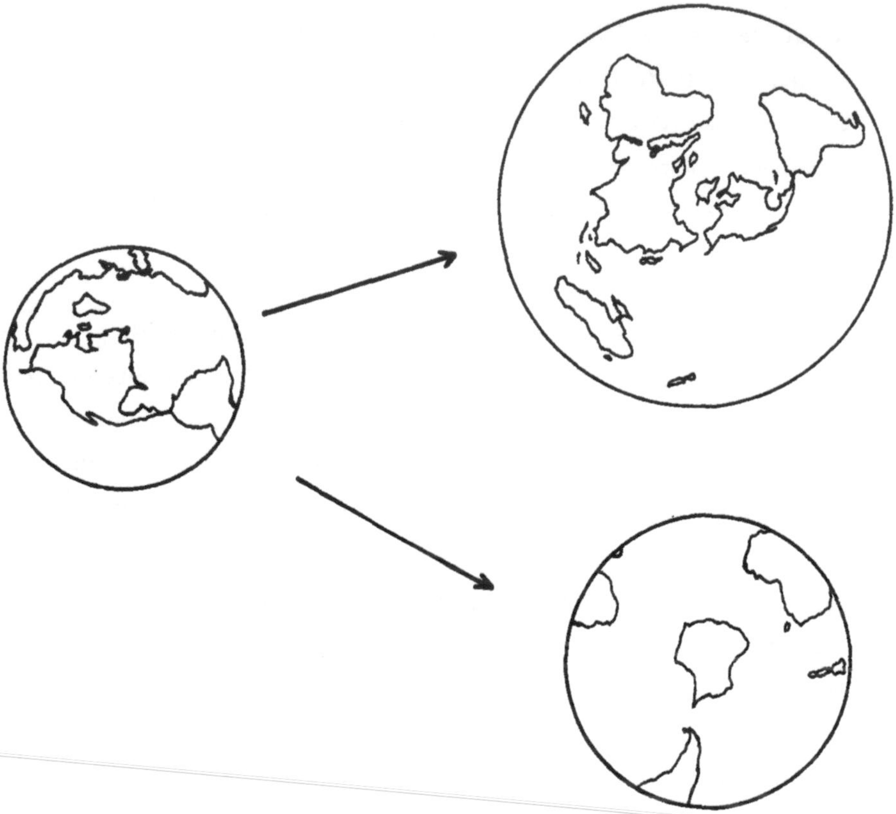

Fig. 2.3. A Spherical Surface S^2. The left hand drawing represents the Earth as photographed from space. Two coordinate patches are drawn on the right as flat maps. The upper one especially is distorted near the edge — this distortion is a well-known coordinate effect.

infinitely differentiable (C^∞) functions (weaker differentiability requirements are sometimes useful; Munkres, 1963).

A function on M is *differentiable* when it is a differentiable function of these coordinates. Another coordinate system on an open set N is *allowed* if it consists of n (the dimension of M) differentiable functions x^σ which uniquely specify all points in N. Any covering of M by allowed coordinate patches defines its differentiable structure. It is interesting to note that the differentiable structure need not be unique: Milnor

(1956) gives an example of a manifold which admits two, non-equivalent differentiable structures.

2.2. Vector Field: A Derivation

In mathematics the concept of vector field is closely tied to the concept of differentiability: A vector field is a differential operator. In order to connect this concept with the usual physical concept of a vector as an arrow connecting two points (P and Q), consider a function f on the manifold M. The change in f between the points P and Q depends on the vector $\overrightarrow{PQ}$ and the function itself. If P and Q are in the same coordinate patch (Δx^σ the difference of their coordinates),

$$f(Q) - f(P) \equiv \Delta f \approx \Delta x^\sigma(\partial f/\partial x^\sigma) = \text{vectorial derivative}. \qquad (2.1)$$

The dependence of Δf on displacement is contained in the linear differential operator

$$\Delta x^\sigma(\partial/\partial x^\sigma) \equiv \Delta x^\sigma \partial_\sigma ,$$

to be thought of as the vector $\overrightarrow{PQ}$.

Modern differential geometry refines this idea of a vector as follows: (1) Take the limit as $\Delta x^\sigma \to 0$ to define a local concept (*tangent vector*) which preserves the directional properties of $\overrightarrow{PQ}$. (2) Insure that this concept is independent of coordinates. (3) Define the concept of *vector field*, consisting of a tangent vector at each point of the manifold.

The resulting coordinate-independent concept of a vector field is that of a differential operator on M, an operator V which carries differentiable functions on M into other differentiable functions. V must be:

(i) linear $\qquad V[f(P)+g(P)] = Vf(P) + Vg(P)$

(ii) a derivative operation $\qquad V(fg) = gVf + fVg$.

Bases

An important example of a vector field is the one obtained by differentiation with respect to a coordinate. Consider a coordinate neighborhood N with n functions $x^\mu(\mu=1,\cdots,n)$ whose values $x^\mu(P)$ are the coordinates of the point P. As an example, we define the operator ∂_2 by

$$\partial_2 f = \frac{\partial f(x^1, x^2, \cdots, x^n)}{\partial x^2} .$$

This differential operator is the vector tangent to lines defined by $x^\mu =$ const, $\mu \neq 2$. In the approximation of (2.1) $\partial_2 f$ gives the change in f between $P \equiv (x^1, x^2, \cdots)$ and $Q \equiv (x^1, x^2+1, \cdots)$. The operator ∂_2 is considered to lie in spaces *tangent* to the manifold at each point P (see Hicks, 1965; Helgason, 1962). This ∂_2 is portrayed as an arrow pointing along the x^2 coordinate direction at P and at every other point of M (Figure 2.4).

In a similar fashion $\partial_1, \partial_3, \cdots, \partial_n$ are defined. The n operators ∂_μ are *base vectors*. The base vectors are linearly independent; that is, every vector field in N may be expressed uniquely as a linear combination of the ∂_μ with coefficients which are differentiable functions in N:

$$V = v^\sigma \partial_\sigma .$$

The n functions v^σ are components of the tangent vector field V, or the *contravariant components* of the vector field V.

At the point P the vector field V has the value $V(P) = v^\sigma(P)\partial_\sigma$. The set of all vectors V at P is the *tangent space* M_P of M at P. M_P is clearly n-dimensional.

V has existence independent of coordinate systems. Let us therefore consider new coordinate functions $\bar{x}^\mu$ defining the basis $\bar{\partial}^\mu$. The components of V may change, but not V itself:

$$V = v^\mu \partial_\mu = \bar{v}^\sigma \bar{\partial}_\sigma .$$

If we let V act on the function x^μ, we find

$$v^\mu = (\partial x^\mu / \partial \bar{x}^\sigma) \bar{v}^\sigma . \tag{2.2}$$

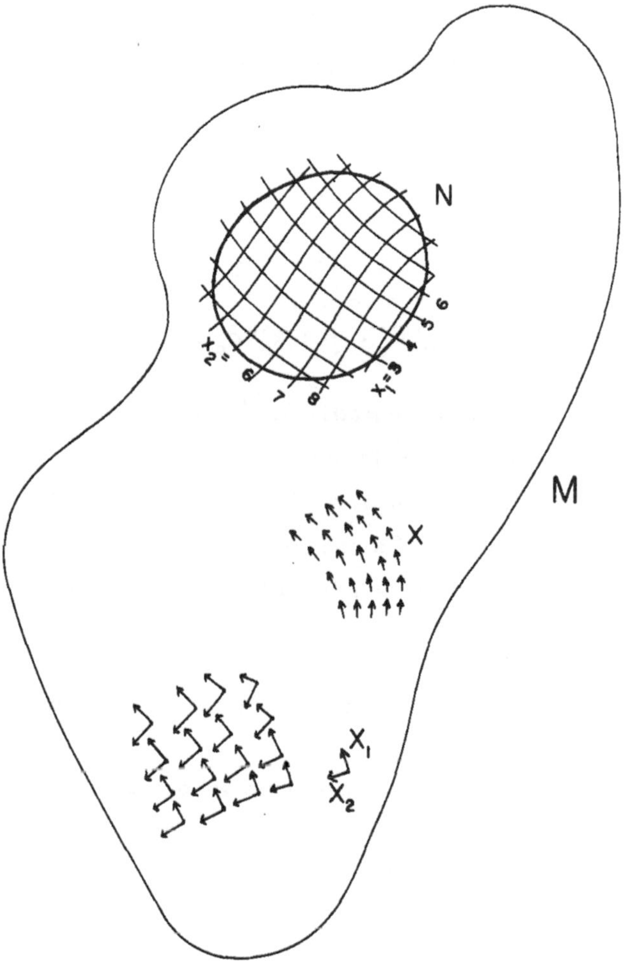

Fig. 2.4. A Two-Dimensional Manifold M. The coordinate patch N is an open set in which the coordinates x_1 and x_2 uniquely describe points. The basis vectors ∂_1 and ∂_2 are parallel to the grid lines in N. X is a vector field, and two linearly independent vector fields, such as X_1, X_2 form a basis for the set of vector fields in M.

We shall call a set of base vectors derived from a set of coordinates a *coordinated basis*.

Not every basis is a coordinated basis. Any n vector fields linearly independent in an open set of M may be used as a basis in that open set.

To determine if a given set $X_1, \cdots, X_n$ of vector fields is a basis, first write each vector in some convenient coordinated basis,

$$X_1 = x_1{}^1 \partial_1 + \cdots + x_1{}^n \partial_n = x_1{}^\sigma \partial_\sigma$$

$$\vdots$$

$$X_n = x_n{}^1 \partial_1 + \cdots + x_n{}^n \partial_n = x_n{}^\sigma \partial_\sigma .$$

The set $\{X_i\}$ is a basis if the matrix $(x_i{}^\sigma(P))$ has non-zero determinant at every point P. Other terms for a set of basis vectors are repére mobile or ennuple in general, triad when $n = 3$, and tetrad or vierbein when $n = 4$.

To arrive at an example of a non-coordinated basis consider ordinary spherical coordinates. In the coordinated basis $\{\partial_r, \partial_\theta, \partial_\phi\}$ the velocity of a particle is

$$V = v^r \frac{\partial}{\partial r} + v^\theta \frac{\partial}{\partial \theta} + v^\phi \frac{\partial}{\partial \phi}, \quad \text{where} \quad v^r = \frac{dr}{dt}, \; v^\theta = \frac{d\theta}{dt}, \; v^\phi = \frac{d\phi}{dt}.$$

Usually, however, the velocity components are defined to be:

$$\hat{v}^r = \frac{dr}{dt}, \quad \hat{v}^\theta = r\frac{d\theta}{dt}, \quad \hat{v}^\phi = r \sin\theta \frac{d\phi}{dt} .$$

This definition uses the formula

$$V = \hat{v}^r X_r + \hat{v}^\theta X_\theta + \hat{v}^\phi X_\phi ,$$

where

$$X_r = \frac{\partial}{\partial r}, \quad X_\theta = \frac{1}{r}\frac{\partial}{\partial \theta}, \quad X_\phi = \frac{1}{r \sin\theta}\frac{\partial}{\partial \phi} .$$

The vectors X_r, X_θ, X_ϕ form a non-coordinated basis.

Commutators

A coordinated basis has the property

$$(\partial_\mu \partial_\nu - \partial_\nu \partial_\mu) f = 0$$

for any function f. For a more general basis $X_\mu X_\nu - X_\nu X_\mu$ is not necessarily zero. We shall find non-commuting bases useful in describing the general homogeneous manifold.

We make the notion of *commutator* precise by defining the commutator of two vector fields U, V to be the operator

$$\begin{aligned}[U, V] &\equiv UV - VU , \\ [U, V]f &= U(Vf) - V(Uf) .\end{aligned} \tag{2.3}$$

The commutator has the derivative property

$$[U, V]fg = f[U, V]g + g[U, V]f$$

and is therefore a vector field. The commutator satisfies the Jacobi identity

$$[U,[V,W]] + [V,[W,U]] + [W,[U,V]] = 0 \text{ for all vector fields } U, V, W. \tag{2.4}$$

The commutator [U, V] is also called the *Lie derivative* (Chapter 6) of V with respect to U. That is,

$$\mathcal{L}_U V \equiv [U, V] . \tag{2.5}$$

In a coordinated basis (coordinates x^μ), if $U = u^\sigma \partial_\sigma$, $V = v^\sigma \partial_\sigma$ then

$$\mathcal{L}_U V = [U, V] = (u^\sigma \partial_\sigma v^\tau - v^\sigma \partial_\sigma u^\tau)\partial_\tau . \tag{2.6}$$

In a non-coordinated basis, $U = \hat{u}^\mu X_\mu$, and $V = \hat{v}^\mu X_\mu$, so that

$$[U, V] = \{\hat{u}^\mu (X_\mu \hat{v}^\lambda) - \hat{v}^\mu (X_\mu \hat{u}^\lambda) + \hat{u}^\mu \hat{v}^\nu C^\lambda_{\mu\nu}\} X_\lambda . \tag{2.7}$$

The functions $C^\lambda_{\mu\nu}$ arise from the expansion of the vector field $[X_\mu, X_\nu]$ in the basis $\{X_\mu\}$ as

$$[X_\mu, X_\nu] = C^\lambda_{\mu\nu} X_\lambda . \tag{2.8}$$

The functions $C^\lambda_{\mu\nu}$ are the *structure coefficients* of the basis X_μ. In a general basis the $C^\lambda_{\mu\nu}$ do not vanish. In fact, it can be shown that if $C^{\lambda}_{\mu\nu} = 0$ the basis is a coordinated basis (Hicks, 1965). If the manifold admits a group of isometries (Chapter 6), the most convenient basis is one with the $C^{\lambda}_{\mu\nu}$ determined by the group structure.

2.3. The Metric

In relativity a gravitational field is described as curvature in the spacetime manifold. The bending of particle paths is due jointly to: (1) non-gravitational forces, and (2) the curvature of the manifold itself. A path (geodesic; see Chapter 3) which is bent only because of the curvature of the manifold is the world line of a test particle moving in a gravitational field. In relativity the metric of the manifold determines its curvature. The relation between curvature and metric is reminescent of that between the electromagnetic four-potentials and the electromagnetic field (Wheeler, 1962b).

A cosmological model is not only a manifold, but a manifold-with-metric, a *pseudo-Riemannian manifold.* (Riemannian manifold is reserved by us for a manifold with a positive-definite metric.) Because the metric plays such an important role in relativity we shall discuss it before we consider other geometrical objects.

Metric: Distance Measure and Operator

To many physicists, a *metric* is a structure which determines the distance between two nearby points and the angle between two lines. More precisely, a *metric* is a bilinear, non-singular function on the set of pairs of vector fields. We shall use this second definition. In the next section we shall show that a metric is a second-rank covariant tensor field.

Actually we may connect the two definitions of metric quite naturally. Consider two nearby points P and Q in a pseudo-Riemannian manifold M. We suppose P and Q to be in the same coordinate patch, the differences in their coordinates being Δx^{μ}. The square of the distance between them be written

$$\Delta s^2 \approx g_{\mu\nu} \Delta x^\mu \Delta x^\nu .$$

The $g_{\mu\nu}$ are the *components* of the metric. The matrix $g_{\mu\nu}$ is symmetric and a function of position on M. We use Δs^2 by convention, even though in relativity Δs^2 may be negative. We now define an *operator* dx^μ which acts on the differential operator $\partial/\partial x^\nu$ to produce δ^μ_ν:

$$dx^\mu \frac{\partial}{\partial x^\nu} = \delta^\mu_\nu = \begin{cases} 1, & \mu = \nu \\ 0, & \mu \neq \nu \end{cases} .$$

On writing the vector $\overrightarrow{PQ}$ as $\Delta x^\sigma \partial/\partial x^\sigma$, we can identify Δs^2 as the "dot product" of $\overrightarrow{PQ}$ with itself:

$$\Delta s^2 = g_{\mu\nu} \Delta x^\sigma \Delta x^\tau [dx^\mu(\partial/\partial x^\sigma)] [dx^\nu(\partial/\partial x^\tau)] \equiv g(X, X) . \tag{2.9}$$

The operator $g_{\mu\nu} dx^\mu dx^\nu$ is a bilinear, non-singular function on pairs of vector fields.

Mathematically a *metric* is an operator which acts on two vector fields U and V to produce a function (written $g(U, V)$ or $U \cdot V$). Such a metric is required to be:

(i) bilinear

$$U \cdot (V + W) = U \cdot V + U \cdot W, \quad (U + W) \cdot V = U \cdot V + W \cdot V ,$$

(ii) symmetric

$$U \cdot V = V \cdot U ,$$

and

(iii) non-singular

$$\text{If } U \cdot V = 0 \text{ for all } U, \text{ then } V = 0 .$$

In a general basis we define n^2 functions $g_{\mu\nu}$ by

$$g_{\mu\nu} \equiv X_\mu \cdot X_\nu = g(X_\mu, X_\nu) . \tag{2.10}$$

These functions are the *components* of the metric in the basis $\{X_\mu\}$. At a point P the function $g_{\mu\nu}$ has the value $g_{\mu\nu}(P)$. By (ii) and (iii), respectively, the matrix $g_{\mu\nu}(P)$ is symmetric and non-singular (det $g_{\mu\nu}(P) \neq 0$). In general relativity we also require that M be four-dimensional and $g_{\mu\nu}(P)$ have signature $(-+++)$ everywhere.

A Non-Singular Metric

In Chapter 5 we will refer to "non-singular metrics," points of "singularity" and so forth, and will define these concepts more precisely there. Briefly, a metric is *non-singular* in an open set N if in N the metric obeys (i), (ii), and (iii) above. If in following a metric about a manifold M we come to a point P where either (i), (ii), or (iii) breaks down we shall call the operator at that point a *metric which is singular.* Strictly this is an abuse of language, since at P we have left the subset M′ of M which has a metric on it.

One cannot use a particular set of functions $g_{\mu\nu} \equiv X_\mu \cdot X_\nu$ to determine the truly singular points of the metric. At the edge of a coordinate patch, for example, det $g_{\mu\nu}(P)$ may vanish. This zero may mean that the X_μ have become degenerate at P or that the X_μ are no longer differentiable vector fields (so that $C^\mu_{\alpha\beta}$ are not finite, differentiable functions near P). To recognize a *true* from an *apparent* singularity we must attempt to find another basis $\{\overline{X}_\mu\}$ such that $\overline{C}^\mu_{\alpha\beta}$ and $\overline{g}_{\mu\nu}$ are finite, differentiable functions at P($\overline{g}_{\mu\nu}(P)$ must also be non-singular). Such a basis always exists if the singularity is not a true one.

2.4. Differential Forms

The geometrical language of *forms* is especially useful in describing antisymmetric, covariant tensor fields. In modern differential geometry we define this concept without reference to coordinates, using the concepts of operation. We define a *differential form* (of first degree), also called a *one-form,* as a linear operator on vector fields. That is, if ω is a one-form and U a vector, $\omega(U)$ is a function, so that $\omega(U)(P)$ is a real number.

Dual Bases

If $\{X_\mu\}$ is a basis we define a set of one-forms $\{\omega^\mu\}$ by

$$\omega^\mu(X_\nu) = \delta^\mu_\nu \ . \tag{2.11}$$

The functions $\omega^\mu(X_\nu)$ are the constant functions δ^μ_ν. These ω^μ are called the *duals* of X_μ or the *basis dual* to $\{X_\mu\}$. The most general one-form ω can be written as a linear combination of the dual basis forms ω^μ

$$\omega = b_\sigma \omega^\sigma \ .$$

The duals of a coordinated basis $\{\partial_\mu\}$ are written dx^μ. The form dx^μ is *not* a component of a vector but one of a set of n linear operators. These forms will be called a *coordinated basis of forms.* In this basis

$$\omega = a_\sigma dx^\sigma \ .$$

As with vectors we require ω to be unchanged under change of coordinates from x^μ to $\bar{x}^\mu$:

$$\omega = \bar{a}_\sigma d\bar{x}^\sigma = a_\tau dx^\tau \ .$$

It is not difficult to show that:

$$\bar{a}_\tau = a_\sigma (\partial x^\sigma / \partial \bar{x}^\tau) \ . \tag{2.12}$$

Note that the a_σ transform like the "covariant components of a vector field" of the older literature.

If $\omega = b_\sigma \omega^\sigma$ and $U = u^\sigma X_\sigma$ then

$$\omega(U) = b_\sigma u^\sigma \ . \tag{2.13}$$

This expression is called the *contraction* of ω with U.

Multiplication of Forms — Tensors — The Metric

In older literature the complicated geometrical objects known as tensors were defined by the transformation properties of their components.

In even older literature these objects were defined by the tensor multiplication of covariant and contravariant vectors which were combined to produce a general tensor. Modern differential geometry has returned to the earlier method. We shall define *tensor multiplication* $\otimes$ on one-forms. Helgason (1962) shows how to extend this to vector fields, and so on to build up tensors.

The *tensor product* $\omega \otimes \sigma$ of two one-forms is a bilinear operator acting on pairs of vector fields (U, V):

$$(\omega \otimes \sigma)(U, V) = \omega(U)\sigma(V) . \tag{2.14}$$

In a coordinated basis $(\omega = a_\sigma dx^\sigma,\ \sigma = b_\tau dx^\tau,\ U = u^\gamma \partial_\gamma,\ V = v^\delta \partial_\delta)$:

$$(\omega \otimes \sigma)(U, V) = (a_\alpha u^\alpha)(b_\beta v^\beta) , \tag{2.15}$$

and $\omega \otimes \sigma$ may be written as

$$\omega \otimes \sigma = a_\mu b_\nu dx^\mu \otimes dx^\nu .$$

The tensor product on forms and vectors is used to build up tensors of arbitrary rank. The product of r forms and s vectors is a tensor of covariant rank r and contravariant rank s. The general tensor is the sum of such elementary products. The typical tensor can be written as a linear combination of basis elements:

$$T^{\alpha\cdots\beta}{}_{\mu\cdots\nu} X_\alpha \otimes \cdots \otimes X_\beta \omega^\mu \otimes \cdots \otimes \omega^\nu . \tag{2.16}$$

The functions $T^{\alpha\cdots\beta}{}_{\mu\cdots\nu}$ are the *components* of the tensor. Contraction is an operation which lowers the covariant and contravariant ranks each by one by the operation illustrated in (2.13).

If $\{\omega^\mu\}$ is a basis for one-forms, $\omega^\mu \otimes \omega^\nu$ is a basis for all covariant tensors of rank two, i.e., for all bilinear operators which act on pairs of vector fields. The bilinear operator $\cdot$ or g (the metric) can be expressed as

$$g = \bar{g}_{\mu\nu} \omega^\mu \otimes \omega^\nu$$

or

$$U \cdot V = g(U, V) = (\bar{g}_{\mu\nu} \omega^\mu \otimes \omega^\nu)(U, V) .$$

Because of the non-singularity of g, g cannot be expressed as a simple product of one-forms $\omega \otimes \sigma$, but g must be a linear combination of at least n such elementary products. It is easy to show

$$X_\mu \cdot X_\nu \equiv g_{\mu\nu} = \bar{g}_{\mu\nu} .$$

It is customary to write

$$\omega^\mu \omega^\nu = \frac{1}{2} (\omega^\mu \otimes \omega^\nu + \omega^\nu \otimes \omega^\mu) , \qquad (2.17)$$

and, since $g_{\mu\nu} = g_{\nu\mu}$,

$$g = ds^2 = g_{\mu\nu} \omega^\mu \omega^\nu . \qquad (2.18)$$

We write the bilinear operator g as ds^2 in remembrance of (2.9).

In general covariant and contravariant vectors are distinct objects. Given a metric g, however, an equivalence relation is set up between contravariant vectors X and covariant vectors ω. X is said to be the contravariant image of ω if

$$g(X, Y) = \omega(Y) \text{ for all vectors } Y . \qquad (2.19)$$

Because g is non-singular, X is uniquely defined by this relation if ω is given. Conversely, ω is uniquely determined by X. In component form (in a basis), the relation between the components a^μ of X, b_μ of ω (where $X = a^\mu X_\mu$, $\omega = b_\mu \omega^\mu$, and $\{X_\mu\}$ and $\{\omega^\mu\}$ are dual bases) is

$$g_{\mu\sigma} a^\sigma = b_\mu . \qquad (2.20)$$

Often b_μ is written as a_μ.

The *contravariant metric tensor* $\hat{g}$ is the second-rank, symmetric, contravariant tensor whose components $g^{\mu\nu}$ are the components of the matrix inverse of $(g_{\mu\nu})$. The tensor $\hat{g}$ acts on a pair of covariant vector fields ω, σ to yield a function. If ω, σ are the covariant images of the contravariant vector fields X, Y, then $\hat{g}$ is uniquely given in a basis-independent manner by the expression

$$\hat{g}(\omega, \sigma) = g(X, Y) \, . \tag{2.21}$$

In a basis, we write

$$\left.\begin{array}{ll} \omega = a_\mu \omega^\mu, & \sigma = c_\mu \omega^\mu \\ X = a^\mu X_\mu, & Y = c^\mu X_\mu \end{array}\right\} g_{\mu\nu} a^\alpha = a_\mu, \; g_{\mu\alpha} c^\alpha = c_\mu$$

$$\hat{g} = g^{\mu\nu} X_\mu X_\nu, \quad g = g_{\mu\nu} \omega^\mu \omega^\nu \tag{2.22}$$

$$(g^{\mu\nu}) = (g_{\mu\nu})^{-1}$$

$$g^{\mu\nu} a_\mu c_\nu = g_{\mu\nu} a^\mu c^\nu \, .$$

The metric g and its contravariant form $\hat{g}$ may be used to contract a tensor T on two contravariant or two covariant indices. If the components of T in some basis are, for example, $T_{\alpha\beta}{}^{\gamma}{}_{\delta}$, then the *contraction with* $g^{\alpha\beta}$ on the first two indices is the tensor whose components are $g^{\alpha\beta} T_{\alpha\beta}{}^{\gamma}{}_{\delta}$.

Just as we singled out the symmetric part of $\omega^\mu \otimes \omega^\nu$ we can write the antisymmetric part

$$\omega \wedge \sigma \equiv \frac{1}{2} (\omega \otimes \sigma - \sigma \otimes \omega) \, , \tag{2.23}$$

the *wedge product* of ω and σ. The wedge products of basis forms $\omega^\mu \wedge \omega^\nu$ are a basis for the space of two-forms on M. The generic two-form is $F = f_{\mu\nu} \omega^\mu \wedge \omega^\nu$. The components $f_{\mu\nu}$ are an antisymmetric matrix of functions. The wedge product of a two-form and a one-form is a three-form. This process may be carried out to any rank, n products defining n-forms (functions are zero-forms).

Exterior Differentiation or Curl; Structure Coefficients

Every physicist is familiar with quantity called the "differential" of a function, $df(t) = \frac{df}{dt} dt$. This concept is refined in modern differential geometry by use of an operator d, called the *curl, gradient* or *exterior derivative* operator, operating on r-forms. We shall first define d on functions, then proceed to forms of higher rank.

The operator d on a function f is defined by

$$df = (X_\mu f)\omega^\mu \tag{2.24}$$

in a basis $\{X_\mu\}$ whose dual basis is $\{\omega^\mu\}$. A basis-free definition of d is that d is a linear operator carrying a function f into the unique one-form df defined by $df(U) = Uf$, where U is any vector field. In a coordinated basis we have the familiar expression

$$df = (\partial_\mu f)dx^\mu . \tag{2.25}$$

We uniquely extend d to forms of higher rank by the requirements: i) d converts an r-form into an (r+1)-form; ii) $d(d\omega) = 0$ for any ω; iii) $d(\omega \wedge \sigma) = d\omega \wedge \sigma + (-1)^r \omega \wedge d\sigma$ if ω is an r-form (remember, functions are zero-forms).

It is from df in a coordinated basis that we get the notation dx^μ for basis one-forms. If we operate on the n coordinate functions x^μ with d we get dx^μ, n one-forms which can be easily shown to be identical to the duals dx^μ of the vectors ∂_μ.

From ii) above we have $d(dx^\mu) = 0$. We shall find that this expression is equivalent to the statement that the ∂_μ all commute (their structure coefficients are zero). Let $\{\omega^\mu\}$ be a basis of one-forms dual to a basis $\{X_\mu\}$ which has non-zero commutators. The curl of any ω^μ is a two-form $d\omega^\mu$ and hence a linear combination of the basis of two-forms $\{\omega^\mu \wedge \omega^\nu\}$:

$$d\omega^\mu = D^\mu_{\alpha\beta}\,\omega^\alpha \wedge \omega^\beta .$$

It can be shown that the $D^\mu_{\alpha\beta}$ are related to the structure coefficients $C^\mu_{\alpha\beta}$ of (2.8) by

$$D^\mu_{\alpha\beta} = -\tfrac{1}{2}\, C^\mu_{\alpha\beta} . \tag{2.26}$$

A Picture of a One-Form – Closed and Exact Forms

We shall try to give an intuitive picture of a "form" which we have defined in an abstract way. We shall use the property of d that: If $d\omega = 0$, then $\omega = d\alpha$ for some form α, at least in a limited region (Spivak, 1965).

A closed form ω is one which has $d\omega = 0$. An *exact* form $\bar{\omega}$ is one which can be written $\bar{\omega} = d\alpha$ for some α. As we stated above each closed form ω is exact, at least over limited regions. Whether $d\omega = 0$ implies $\omega = d\alpha$ everywhere on M depends on the topology of M.

We can obtain a picture of a general exact one-form by constructing a set of $n-1$-dimensional hypersurfaces in an n-dimensional manifold M

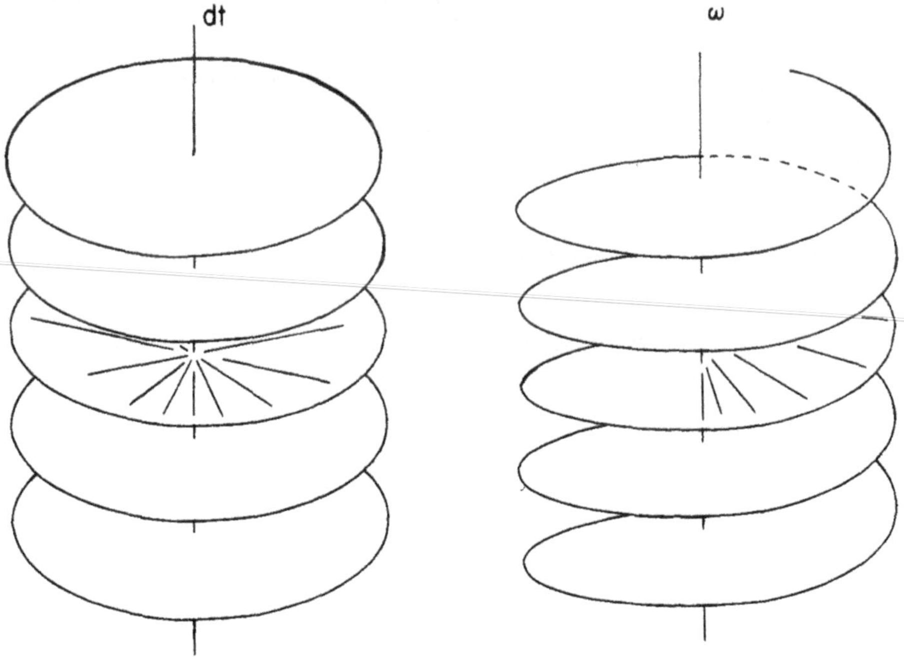

Fig. 2.5. One-Forms in 3-d Space, Schematically Indicated. Because dt is closed, or curl-free, so that $d(dt)=0$, dt locally determines a set of 2-dimensional surfaces ($t=\text{const}$). Here t, x, and y are coordinates in the space in which dt and ω exist. The one-form ω shown is $\omega=\cos t\,dx+\sin t\,dy$, and so $d\omega \wedge \omega \neq 0$. The structure determined by ω is a spiral or screw arrangement of 2-dimensional surface elements defined as the locus of endpoints of vectors X such that $\omega(X) = 0$.

labeled by a parameter t. The parameter t is a function which is constant on these hypersurfaces. We may identify dt, the curl of t, with the hypersurfaces themselves (see Figure 2.5) and portray dt as an arrow perpendicular to the hypersurfaces.

If we want to portray an arbitrary one-form ω, consider ω at a point P. At P ω may be represented by a small "chip" of a hypersurface which contains all the vectors U such that $\omega(U) = 0$. If $d\omega \wedge \omega = 0$, the form is said to be *hypersurface orthogonal* and the chips can be sewn together to form a surface. If ω is curl-free, $d\omega = 0$, the chips can be sewn together without stretching. If $d\omega \wedge \omega \neq 0$, the chips cannot, even with distortion, be combined to form a surface.

To portray an r-form we may extend the method just outlined. For an r-form σ, (n−r)-dimensional chips at P are determined by the vectors U such that $\sigma(U) = 0$. Again the condition that we may sew these chips together (possibly by distorting them) in a finite region to make an (n−r)-dimensional hypersurface is $d\sigma \wedge \sigma = 0$. This condition is obviously satisfied when $d\sigma = 0$, in which case σ is the curl of an (r−1)-form, and the sewing together involves no stretching.

2.5. Covariant Differentiation

In tensor analysis covariant differentiation ∇ is an operation which is a type of differentiation which (a) reduces to partial differentiation on functions; (b) converts a tensor T to one of higher covariant rank; (c) applied to the metric gives zero; (d) contains Christoffel symbols which are symmetric in their lower two indices. We shall define an operator with these four properties expressed in a coordinate independent manner. There will be two important new directions, however.

First, we shall concentrate on ∇_U, covariant differentiation with respect to the vector field U rather than on ∇. The operation ∇T on a tensor T produces $\overline{T}$, where $\overline{T}$ is a tensor of one higher covariant rank than T, but $\nabla_U T$ for a vector field U is a tensor of the same rank as T. Second, our definition will be invariant under a change of basis. In non-coordinated bases we shall find that the analogues of the Christoffel symbols need not be symmetric.

∇_U *as a Differentiation Operator*

We shall list four defining properties of ∇_U. The first two are:

(I) ∇_U is an operation which carries a tensor field into another tensor field $\nabla_U(T)$ of the same rank and which is linear in U:

$$\nabla_{(fU + gV)} = f\nabla_U + g\nabla_V, \quad f, g \text{ functions; } U, V \text{ vector fields.} \tag{2.27}$$

This property is equivalent to property (b) above.

(II) The operator ∇_U applied to a function gives

$$\nabla_U f = Uf \; . \tag{2.28}$$

When applied to tensor products, ∇_U is a derivative operator:

$$\nabla_U(S \otimes T) = \nabla_U(S) \otimes T + S \otimes \nabla_U(T) \; . \tag{2.29}$$

These properties are the analogues of (a). Also, ∇_U commutes with contraction C in the sense that the result of applying ∇_U and C to a given tensor field is independent of the order of application.

Using (I) and (II) we can show that (for any function f and vectors U, V):

$$\nabla_U(fV) = f\nabla_U(V) + (Uf)V \; .$$

If we write $U = u^\sigma X_\sigma$, $V = v^\sigma X_\sigma$ then $\nabla_U(fV)$ may be computed if we know $\nabla_{X_\mu}(X_\nu)$ for all μ, ν. This derivative defines the *connection coefficients* $\Gamma^\alpha_{\beta\gamma}$ by

$$\nabla_{X_\mu}(X_\nu) = \Gamma^\sigma_{\nu\mu} X_\sigma \; . \tag{2.30}$$

We know that $\nabla_{X_\mu}(X_\nu)$ is a vector field; the $\Gamma^\sigma_{\nu\mu}$ are just its components in the basis $\{X_\mu\}$.

If we let $U = u^\mu X_\nu$, $V = v^\nu X_\mu$, then our definitions give rise to a useful formula:

$$\nabla_U(V) = u^\sigma \nabla_{X_\sigma}(v^\tau X_\tau) = u^\sigma [X_\sigma v^\tau + v^\alpha \Gamma^\tau_{\alpha\sigma}] X_\tau \ . \tag{2.31}$$

The quantity $\nabla_{X_\mu}(V)$ is a vector field, so we can write $\nabla_{X_\mu}(V) = v^\sigma{}_{;\mu} X_\sigma$, where the $v^\sigma{}_{;\mu}$ are its set of components. Writing $X_\mu[v^\sigma] \equiv v^\sigma{}_{,\mu}$ we have

$$\nabla_{X_\mu}(Y) = (v^\sigma{}_{;\mu}) X_\sigma = (v^\sigma{}_{,\mu} + \Gamma^\sigma_{\tau\mu} v^\tau) X_\sigma \ , \tag{2.32}$$

and

$$\nabla_U(V) = (v^\mu{}_{;\sigma} u^\sigma) X_\mu \ . \tag{2.33}$$

These expressions are common in relativity, but here $v^\mu{}_{,\sigma}$ is not $\partial_\sigma v^\mu$ and $\Gamma^\mu_{\alpha\beta} \neq \Gamma^\mu_{\beta\alpha}$ in general.

We may also compute $\nabla_U(\omega)$ where ω is a one-form, with $\omega = a_\sigma \omega^\sigma$. The result is

$$\begin{aligned} \nabla_{X_\mu}(\omega) = \nabla_{X_\mu}(a_\sigma \omega^\sigma) &= (X_\mu a_\sigma - \Gamma^\tau_{\sigma\mu} a_\tau)\omega^\sigma \\ &= (a_{\sigma,\mu} - \Gamma^\tau_{\sigma\mu} a_\tau)\omega^\sigma \ . \end{aligned} \tag{2.34}$$

With these formulas we can compute $\nabla_U(T)$ for any tensor T. For instance, the second rank tensor $g_{\mu\nu}\omega^\mu \otimes \omega^\nu$ has the covariant derivative

$$\nabla_{X_\kappa}(g_{\mu\nu}\omega^\mu \otimes \omega^\nu) = (X_\kappa g_{\mu\nu} - \Gamma^\tau_{\nu\kappa} g_{\mu\tau} - \Gamma^\tau_{\mu\kappa} g_{\tau\nu})\omega^\mu \otimes \omega^\nu \ . \tag{2.35}$$

The Vanishing of $\nabla_U(g)$ and of Torsion – The First Cartan Equation

In order to make the $\Gamma^\mu_{\alpha\beta}$ unique it is necessary to specify two additional properties of ∇_U:

(III) The covariant derivative of the metric tensor g vanishes:

$$\nabla_U(g) = 0 \ .$$

Because of (2.35) we can then write

$$g_{\mu\nu,\alpha}\omega^\alpha = (\Gamma^\sigma_{\nu\alpha} g_{\mu\sigma} + \Gamma^\sigma_{\mu\alpha} g_{\sigma\nu})\omega^\alpha \ , \tag{2.36}$$

where $g_{\mu\nu,a} = X_a g_{\mu\nu}$. (Note that the curl of the function $g_{\mu\nu}$ is $dg_{\mu\nu} = g_{\mu\nu,a}\omega^a$.) Let us define the *connection forms*

$$\omega^\mu_\nu \equiv \Gamma^\mu_{\nu\sigma}\omega^\sigma . \tag{2.37}$$

Property (III) then can be written

$$dg_{\mu\nu} = \omega_{\mu\nu} + \omega_{\nu\mu}, \qquad \omega_{\mu\nu} \equiv g_{\mu\sigma}\omega^\sigma_\nu . \tag{2.38}$$

(IV) The final property we shall demand of ∇_U is *zero torsion* (see Helgason, 1962). That is:

$$\nabla_U(V) - \nabla_V(U) = [U, V] = \mathcal{L}_U(V) .$$

Because of their limited usefulness in general relativity (although see Edelen, 1962, and Einstein, 1955) we shall not consider covariant derivatives with non-zero torsion.

It can be shown that in a basis $\{X_\mu\}$ ($C^\mu_{\alpha\beta}$ defined above)

$$C^\mu_{\alpha\beta} = \Gamma^\mu_{\beta\alpha} - \Gamma^\mu_{\alpha\beta} . \tag{2.39}$$

This relation implies

$$d\omega^\mu = -\tfrac{1}{2} C^\mu_{\alpha\beta}\omega^\alpha \wedge \omega^\beta = -\Gamma^\mu_{\beta\alpha}\omega^\alpha \wedge \omega^\beta ,$$

or

$$d\omega^\mu = -\omega^\mu_\sigma \wedge \omega^\sigma . \tag{2.40}$$

This is, for zero torsion, the *first Cartan equation.*

We may use (2.37) and (2.40) to compute $\Gamma^\mu_{\alpha\beta}$ for a general basis. If we define a basis $\{X_\mu\}$ such that $g_{\mu\nu}$ is a constant in that basis (2.38) reduces to $\omega_{\mu\nu} + \omega_{\nu\mu} = 0$. This special case is often useful in relativity. In general, however, we have

$$\begin{aligned} \Gamma^\mu_{\alpha\beta} &= \tfrac{1}{2} g^{\mu\sigma}(g_{\sigma\alpha,\beta} + g_{\sigma\beta,\alpha} - g_{\alpha\beta,\sigma}) \\ &\quad + \tfrac{1}{2}(-C^\mu_{\alpha\beta} + g_{\alpha\sigma}g^{\mu\tau}C^\sigma_{\tau\beta} + g_{\sigma\beta}g^{\mu\tau}C^\sigma_{\tau\alpha}) \end{aligned} \tag{2.41}$$

where $g^{\mu\sigma} g_{\sigma\nu} \equiv \delta^{\mu}_{\nu}$ and $g_{\mu\nu,\alpha} \equiv X_{\alpha} g_{\mu\nu}$. A case of special interest is $C^{\mu}_{\alpha\beta} = 0$ (a coordinated basis). In this case (2.41) reduces to a well-known form: the Christoffel symbols.

2.6. Curvature – The Second Cartan Equation

Let us define (without motivation for the moment) the *curvature operation* on two general vector fields U and V to be

$$R(U,V) = \nabla_U \nabla_V - \nabla_V \nabla_U - \nabla_{[U,V]} \ . \tag{2.42}$$

Eventually we shall use this operation to define the gravitational field. The operation $R(U,V)$ is a tensor of one covariant rank and one contravariant rank, since it operates on a vector field to produce another vector field, and since for any function f

$$R(U,V)(fW) = fR(U,V)(W) \ .$$

The proof of this property is straightforward, but tedious.

The Riemann Curvature Tensor

Using $R(U,V)$ we can define a tensor field of three *covariant* ranks and one *contravariant* rank which operates on three vector fields U, V, W and one differential form ω to produce a function:

$$R(U,V,W,\omega) = f \ . \tag{2.43}$$

This tensor is the *Riemann curvature tensor*. The definition of R is

$$R(U,V,W,\omega) = \omega[R(U,V)W] \ . \tag{2.44}$$

($R(U,V)W$ is a vector field, so $R(U,V,W,\omega)$ is a function.)

To be a tensor, R must be linear in all its entries. It is obviously linear in ω and we know that $R(U,V)$ is a linear operator. We shall leave it to the reader to prove linearity in U and V.

For a basis $\{X_\mu\}$, $R(X_\mu, X_\nu)(X_\alpha)$ is the vector field

$$R(X_\mu, X_\nu)(X_\alpha) = R^{\sigma}_{\alpha\mu\nu} X_\sigma \ .$$

The $R^{\sigma}_{\alpha\mu\nu}$ are a set of functions called the components of the Riemann curvature tensor.

In a coordinated basis $(U = u^{\sigma}\partial_{\sigma}, [\partial_{\mu}, \partial_{\nu}] = 0)$

$$[R(\partial_{\mu}, \partial_{\nu})(U)]\partial_{\sigma} = (u^{\sigma}{}_{;\nu\mu} - u^{\sigma}{}_{;\mu\nu})\partial_{\sigma} \tag{2.45}$$

which implies the familiar equation:

$$u^{\sigma}{}_{;\nu\mu} - u^{\sigma}{}_{;\mu\nu} = R^{\sigma}_{\alpha\mu\nu} u^{\alpha} . \tag{2.46}$$

In a general basis, a long computation shows

$$R^{\sigma}_{\alpha\mu\nu} = \Gamma^{\sigma}_{\alpha\nu,\mu} - \Gamma^{\sigma}_{\alpha\mu,\nu} + \Gamma^{\tau}_{\alpha\nu}\Gamma^{\sigma}_{\tau\mu} - \Gamma^{\tau}_{\alpha\mu}\Gamma^{\sigma}_{\tau\nu} - C^{\tau}_{\mu\nu}\Gamma^{\sigma}_{\alpha\tau} , \tag{2.47}$$

where $\Gamma^{\sigma}_{\alpha\nu,\mu} \equiv X_{\mu}\Gamma^{\sigma}_{\alpha\nu}$ and $C^{\tau}_{\mu\nu} = \Gamma^{\tau}_{\nu\mu} - \Gamma^{\tau}_{\mu\nu}$. In a coordinated basis this reduces to the usual expression for $R^{\sigma}_{\alpha\mu\nu}$.

The Second Cartan Equation

Using (2.47) we can show that the *curvature forms*

$$\theta^{\mu}_{\nu} \equiv d\omega^{\mu}_{\nu} + \omega^{\mu}_{\alpha} \wedge \omega^{\alpha}_{\nu} \tag{2.48}$$

have the useful property (the *second Cartan equation*)

$$\theta^{\mu}_{\nu} = \frac{1}{2} R^{\mu}_{\nu\sigma\tau}\, \omega^{\sigma} \wedge \omega^{\tau} . \tag{2.49}$$

Equations (2.38), (2.40) and (2.49) are all we need to compute the connection coefficients and the Riemann curvature tensor for a metric in any basis. Taken together, (2.38) and (2.40) may be solved for ω^{μ}_{ν} (this is usually much simpler then computing $\Gamma^{\alpha}_{\mu\nu}$ in a coordinated basis). A straightforward computation gives the θ^{μ}_{ν} and (2.48) allows us to read off the components $R^{\mu}_{\nu\sigma\tau}$ directly.

The Ricci tensor components $R_{\mu\nu}$ and the scalar curvature R are:

$$R_{\mu\nu} \equiv R^{\sigma}_{\mu\sigma\nu} ; \qquad R \equiv g^{\mu\nu} R_{\mu\nu} . \tag{2.50}$$

The Einstein field equations are

$$R_{\mu\nu} - \frac{1}{2} g_{\mu\nu} R = T_{\mu\nu} \,. \tag{2.51}$$

($T_{\mu\nu}$ are the components of the stress-energy tensor in terms of the ω^μ and in a system of units where $8\pi G/c^4 = 1$.)

2.7. The Three-Sphere as an Example of a Differentiable Manifold: Metric, Vector Fields, and Structure Coefficients

The three-sphere S^3 is the set of all points in four-dimensional Euclidean space R^4 with coordinates x^1, x^2, x^3, x^4 such that

$$\sum_{\mu=1}^{4} (x^\mu)^2 = 1 \,. \tag{2.52}$$

In R^4 we also use the Euclidean metric $g_{\mu\nu} = \delta_{\mu\nu}$ $(\mu, \nu = 1, 2, 3, 4)$.

We should like to construct a set of basis vectors for the manifold S^3. We need three linearly independent vectors X_1, X_2, X_3 at every point of S^3. Because every point of S^3 is also a point of R^4 we can write any vector at a point of S^3 as $X_i = a_i{}^\mu \partial_\mu$, where the $a_i{}^\mu$ are functions of x^1, x^2, x^3, x^4. For any function f on S^3, $X_i(f)$ must be a function on S^3, so the $a_i{}^\mu$ need be defined only for points in S^3. Consider the three vectors in R^4:

$$\begin{aligned} X_1 &= x^2\partial_1 - x^1\partial_2 + x^4\partial_3 - x^3\partial_4 \\ X_2 &= x^3\partial_1 - x^4\partial_2 - x^1\partial_3 + x^2\partial_4 \\ X_3 &= x^4\partial_1 + x^3\partial_2 - x^2\partial_3 - x^1\partial_4 \,. \end{aligned} \tag{2.53}$$

Each of these three vectors is a vector in S^3 when x^1, x^2, x^3, x^4 satisfy (2.52). The X_i are linearly independent and it can be shown that

$$X_i \cdot X_j = \delta_{ij}, \qquad i, j = 1, 2, 3 \,. \tag{2.54}$$

Thus the X_i form an orthonormal basis for S^3.

Computing the commutators of the vectors (2.53) we find

$$[X_1, X_2] = 2X_3 \qquad \textit{et cyc}$$

so that the structure coefficients are given by

$$C^i{}_{st} = 2\varepsilon_{ist} \,. \tag{2.55}$$

Thus the X_i are a non-degenerate basis everywhere on S^3, and the metric whose components are defined by (2.54) is non-singular everywhere on S^3.

Coordinates on S^3

No coordinate system will cover all of S^3, but consider the coordinates $\bar{x}^i \equiv x^i$, $i = 1,2,3$ at every point where $x^4 > 0$. The $\bar{x}^i$ are good coordinates everywhere in this region. The three coordinated basis vectors $\bar{\partial}_i$ can be written in terms of the ∂_μ of R^4 as

$$\bar{\partial}_i = \partial_i - \frac{x^i}{x^4}\partial_4 \,, \tag{2.56}$$

where $x^4 \equiv [1-(x^1)^2 - (x^2)^2 - (x^3)^2]^{\frac{1}{2}}$.

The metric g has components in this basis:

$$\bar{g}_{ij} = \bar{\partial}_i \cdot \bar{\partial}_j = \delta_{ij} - \bar{x}^i \bar{x}^j \left(1 - \sum (\bar{x}^s)^2\right)^{-1} .$$

We see that g_{ij} has a singularity at $x^4 = 0$, but this singularity is spurious, as it is due to the breakdown of the coordinates $\bar{x}^i$ at $x^4 = 0$. We know that the metric is actually non-singular because there exists a basis $\{X_i\}$ in which its components are regular everywhere: (2.53). In terms of the $\bar{\partial}_i$ we may write

$$\begin{aligned} X_1 &= \bar{x}^2\bar{\partial}_1 - \bar{x}^1\bar{\partial}_2 + x^4\bar{\partial}_3 \\ X_2 &= \bar{x}^3\bar{\partial}_1 - x^4\bar{\partial}_2 - \bar{x}^1\bar{\partial}_3 \\ X_3 &= x^4\bar{\partial}_1 + \bar{x}^3\bar{\partial}_2 - \bar{x}^2\bar{\partial}_3 \,. \end{aligned}$$

Let us now consider the three-sphere of radius b, that is the points in R^4 such that $\sum (x^i)^2 = b^2$. We define $Y_i = aX_i$ where the X_i are given by (2.53) (with the new restriction $\sum (x^i)^2 = b^2$). For the moment we choose $a = -\frac{1}{2}$, so that

$$Y_i \cdot Y_j = \frac{1}{4} b^2 \delta_{ij}$$
$$[Y_i, Y_j] = -\varepsilon_{sij} Y_s . \tag{2.57}$$

Let us now change to a new basis $\{\overline{Y}_i\}$ where

$$\overline{Y}_i = a_i{}^s Y_s ,$$

the $a_i{}^s$ being a matrix of constants such that $\sum_i a_i{}^s a_i{}^t = \delta_{st}$ ($a_i{}^s$ is a 3×3 orthogonal matrix). Equation (2.57) remains valid for the $\overline{Y}_i$, if the determinant of $a_i{}^j$ is one. The $\{\overline{Y}_i\}$ and $\{Y_i\}$ bases will be useful in the discussion of spatially homogeneous cosmologies based on the three-dimensional orthogonal group.

Table 2.1. Summary of Useful Formulas

VECTOR FIELDS:

$V(f+g) = V(f) + V(g)$
$V(fg) = g\,V(F) + f\,V(g)$

Coordinated basis:

$V = v^\mu \partial_\mu$ where $\partial_\mu = \dfrac{\partial^\mu}{\partial x^\mu}$

General basis:

$X_\mu = \xi_\mu{}^\sigma \partial_\sigma$, $V = \hat{v}^\sigma X_\sigma$
$V(f) = f_{,\sigma} \hat{v}^\sigma$, where $f_{,\sigma} = X_\sigma(f)$
(in a coordinated basis $X_\mu(f) = \partial_\mu f$)

Changing coordinated bases:

$V = v^\mu \partial_\mu = \bar{v}^\mu \bar{\partial}_\mu$, $\bar{v}^\mu = \dfrac{\partial x^\mu}{\partial \bar{x}^\nu} v^\nu$

ONE FORMS:

$dx^\mu(\partial_\nu) = \delta^\mu{}_\nu$ where $\partial_\mu \equiv \dfrac{\partial}{\partial x^\nu}$

A general form:

$\omega = a_\mu dx^\mu$
dx^μ is a coordinated basis of forms

A general basis:

$\omega^\mu = \xi^\mu{}_\sigma dx^\sigma$, $\omega = \hat{a}_\upsilon \omega^\sigma$
A basis $\{\omega^\sigma\}$ is dual to a basis of vectors $\{X_\sigma\}$ if $\omega^\sigma(X_\tau) = \delta^\sigma{}_\tau$

If $\omega = a_\mu \omega^\mu$, $U = u^\mu X_\mu$, $\omega(U) = a_\mu u^\mu$

Table 2.1. Summary of Useful Formulas
(Continued)

VECTOR FIELDS:

Changing general bases:

$V=v^{\mu}X_{\mu}$, and if $\bar{X}_{\mu}=A_{\mu}{}^{\nu}X_{\nu}$,
$\bar{v}^{\mu}=(A^{-1})^{\mu}{}_{\nu}v^{\nu}$ and $V=\bar{v}^{\mu}\bar{X}_{\mu}$

Commutators:

$[U,V]\equiv UV-VU$
$\mathcal{L}_{U}V\equiv[U,V]$

In a basis X_{μ}: $U=u^{\mu}X_{\mu}$, $V=v^{\mu}X_{\mu}$
$[U,V]=\{u^{\mu}(X_{\mu}v^{\sigma})-v^{\mu}(X_{\mu}u^{\sigma})+u^{\mu}v^{\nu}C^{\sigma}{}_{\mu\nu}\}X_{\sigma}$ where
$[X_{\mu},X_{\nu}]\equiv C^{\sigma}{}_{\mu\nu}X_{\sigma}$

Tensor Product of Vectors: $U\otimes V$

ONE FORMS:

Changing bases:

Coordinated: $\omega=a_{\mu}dx^{\mu}=\bar{a}_{\mu}d\bar{x}^{\mu}$,
$$\bar{a}_{\nu}=\frac{\partial x^{\mu}}{\partial\bar{x}^{\nu}}a_{\mu}$$

Non-coordinated: $\bar{\omega}^{\mu}=A^{\mu}{}_{\nu}\omega^{\nu}$,
$\bar{a}_{\mu}=(A^{-1})_{\mu}{}^{\nu}a_{\nu}$ and $\omega=\bar{a}_{\mu}\bar{\omega}^{\mu}$

Tensor Product of Forms: $\sigma\otimes\omega$

Wedge Product: $\omega\wedge\sigma=\frac{1}{2}(\omega\otimes\sigma-\sigma\otimes\omega)$
$\omega\sigma=\frac{1}{2}(\omega\otimes\sigma+\sigma\otimes\omega)$

Exterior Derivative:

On functions: $df=\partial_{\mu}f\,dx^{\mu}$

On general forms: $d^{2}=0$

$d(\omega\wedge\sigma)=d\omega\wedge\sigma+(-1)^{r}\omega\wedge d\sigma$
(r the order of ω)

On a basis: $d\omega^{\mu}=-\frac{1}{2}C^{\mu}{}_{\sigma\tau}\omega^{\sigma}\wedge\omega^{\tau}$.

Metric Tensor:

$g=g_{\mu\nu}\omega^{\mu}\otimes\omega^{\nu}$, $g(U,V)\equiv U\cdot V$, $g(X_{\mu},X_{\nu})=g_{\mu\nu}$.

The one form $u=u_{\mu}\omega^{\mu}$ is the image of the vector $U=u^{\mu}X_{\mu}$ if $\{\omega^{\mu}\}$ is dual to $\{X_{\mu}\}$ and $u_{\mu}=g_{\mu\nu}u^{\nu}$.

Covariant Derivative:

On functions: $\nabla_{U}f=Uf$.

On the tensor product of two vectors S,T: $\nabla_{U}(S\otimes T)=\nabla_{U}(S)\otimes T+S\otimes\nabla_{U}(T)$

$\nabla_{U}(fV)=f\nabla_{U}(V)+[U(f)]V$

If $U=u^{\mu}X_{\mu}$, $V=v^{\mu}X_{\mu}$, $\nabla_{U}(V)=u^{\sigma}[v^{\mu}{}_{,\sigma}+\Gamma^{\mu}{}_{\sigma\tau}v^{\tau}]X_{\mu}\equiv u^{\sigma}(v^{\mu}{}_{;\sigma})X_{\mu}$

where $\nabla_{U}X_{\mu}\equiv u^{\nu}\Gamma^{\sigma}{}_{\nu\mu}X_{\sigma}$ and $v^{\mu}{}_{,\nu}\equiv X_{\nu}(v^{\mu})$.

On one-forms: $\nabla_{X_{\mu}}(\omega)=(a_{\nu,\mu}-\Gamma^{\sigma}{}_{\nu\mu}a_{\sigma})\omega^{\nu}$, if $\omega=a_{\mu}\omega^{\mu}$, where $a_{\mu,\nu}\equiv X_{\nu}(a_{\mu})$.

Table 2.1. Summary of Useful Formulas
(Continued)

Connection Forms:

$\omega^\mu{}_\nu \equiv \Gamma^\mu{}_{\nu\alpha}\omega^\alpha$

The metric has zero covariant derivative: $dg_{\mu\nu} = \omega_{\mu\nu} + \omega_{\nu\mu}$, $\omega_{\mu\nu} \equiv g_{\mu\alpha}\omega^\alpha{}_\nu$.

The condition: $[U,V] = \mathcal{L}_U(V) = \nabla_U V - \nabla_V U$ is the condition for zero torsion.

It implies $C^\mu_{\alpha\beta} = \Gamma^\mu{}_{\beta\alpha} - \Gamma^\mu{}_{\alpha\beta}$

and also $d\omega^\mu = -\omega^\mu{}_\nu \wedge \omega^\nu$ (The first Cartan equation).

In general: $\Gamma^\mu{}_{\alpha\beta} = \frac{1}{2} g^{\mu\sigma}(g_{\sigma\alpha,\beta} + g_{\sigma\beta,\alpha} - g_{\alpha\beta,\sigma})$

$$+\frac{1}{2}(-C^\mu_{\alpha\beta} + g_{\tau\alpha} g^{\mu\sigma} C^\tau_{\sigma\beta} + g_{\tau\beta} g^{\mu\sigma} C^\tau_{\sigma\alpha}).$$

The Curvature Tensor:

$R(U,V) \equiv \nabla_U \nabla_V - \nabla_V \nabla_U - \nabla_{[U,V]}$, $R(U,V,W,\omega) \equiv \omega[R(U,V)W]$

$R(X_\alpha, X_\beta)(X_\mu) \equiv R^\sigma_{\mu\alpha\beta} X_\sigma$ R is the Riemann curvature tensor

In general: $R^\sigma_{\mu\alpha\beta} = \Gamma^\sigma{}_{\mu\beta,\alpha} - \Gamma^\sigma{}_{\mu\alpha,\beta} + \Gamma^\tau{}_{\mu\beta}\Gamma^\sigma{}_{\tau\alpha} - \Gamma^\tau{}_{\mu\alpha}\Gamma^\sigma{}_{\tau\beta} - C^\tau_{\alpha\beta}\Gamma^\sigma{}_{\mu\tau}$.

The components of the Ricci tensor are: $R_{\alpha\beta} \equiv R^\sigma_{\alpha\sigma\beta}$, the Ricci scalar is $R \equiv R^\alpha_\alpha$.

Einstein's field equations are $R_{\mu\nu} - \frac{1}{2} R g_{\mu\nu} = T_{\mu\nu}$ if $8\pi G/c^4 = 1$ ($T_{\mu\nu}$ stress-energy tensor).

The Curvature Forms:

$\theta^\mu{}_\nu \equiv d\omega^\mu{}_\nu + \omega^\mu{}_\alpha \wedge \omega^\alpha{}_\nu$

$\theta^\mu{}_\nu = \frac{1}{2} R^\mu{}_{\nu\alpha\beta}\omega^\alpha \wedge \omega^\beta$

Antisymmetric Second Derivative of Components of a Vector Field in a Coordinated Basis:

$u^\alpha{}_{;\nu\mu} - u^\alpha{}_{;\mu\nu} = R^\alpha{}_{\sigma\mu\nu} u^\sigma$

3. SPACETIME AND FLUID FLOW

I came like water and like wind I go
— OMAR KHAYYAM

3.1. Relativity and Hydrodynamics

In this chapter we shall concentrate on the description of a fluid in general relativity and fluid-filled cosmological models. Emphasis in this chapter will be on the use of the coordinate free language of Chapter 2 for relativistic hydrodynamics. Figure 3.1 is an outline of the chapter.

Hydrodynamics

In the theory of relativity we study the behavior of a four-dimensional manifold M on which there is a metric of signature (−+++). The path of any particle in this manifold is affected by the curvature of the manifold. This matter in turn determines the geometry through Einstein's field equations

$$R_{\mu\nu} - \frac{1}{2} R g_{\mu\nu} = k\, T_{\mu\nu} , \tag{3.1}$$

with $R_{\mu\nu}$ being the components of the Ricci tensor; R the Ricci scalar, $R = g^{\alpha\beta} R_{\alpha\beta}$; and $T_{\mu\nu}$ the components of the stress-energy tensor. We shall choose units such that the Einstein gravitational constant $k = 8\pi G/c^4$ equals one.

We shall usually fill our model universes with a smooth, perfect (isotropic pressure) fluid, as is customary in cosmology. DeVaucouleurs (1970), Yu and Peebles (1969), and Misner (1967c, 1967d, 1968) have described cosmological models in which the matter cannot be described by such a perfect fluid. Birkhoff (1960) has pointed out that many well-known non-smooth hydrodynamic phenomena such as shock waves,

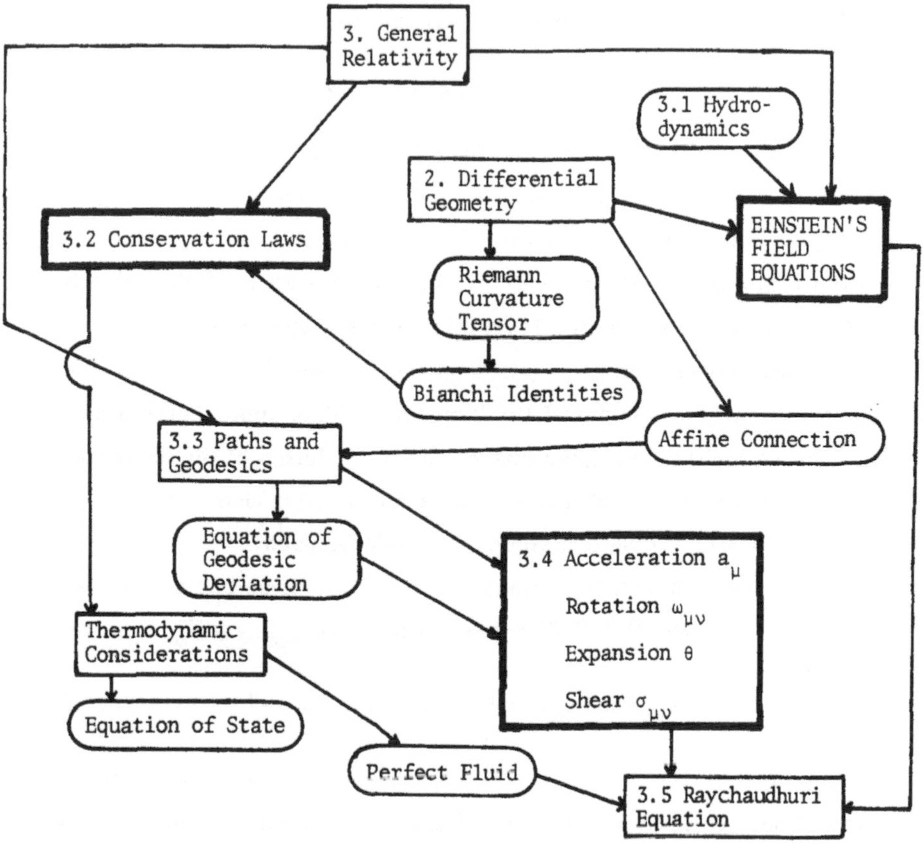

Fig. 3.1. Flow Chart of Chapter 3.

cavitation, and turbulence should not be ignored. For simplicity, however, we shall follow the usual practice and discuss perfect-fluid models.

Fluid Stress-Energy Tensor

The stress-energy tensor T appears as the source term in Einstein's field equations. Although T may be defined in several different ways, we will use the definition most practical for perfect-fluid cosmologies. This method (Eisenhart, 1924) concentrates on the algebraic structure of T, a second-rank, symmetric tensor. We choose units so that $G = (8\pi)^{-1}$, $c = 1$.

In coordinate free language, T has the form

$$T = (w+p)\, u \otimes u + pg \tag{3.2}$$

where u is a timelike, unit-magnitude differential form (covariant vector field), g is the metric field, and w and p are scalars, the energy density and pressure respectively. The fluid velocity is most often written in the contravariant form U, that is, as a vector field. The field u is the covariant image of U (indices lowered by use of g).

For most cosmological purposes $p \approx 0$, so we shall often use the approximation $p = 0$ ("dust-filled" models). This approximation breaks down during highly condensed phases of the evolution of the universe where even the concept of a fluid breaks down and a kinetic theory (Bichteler, 1967) or quantum approach should be used.

Let us now consider the components of T in some basis. In a vector field basis $\{X_\mu\}$ $(\mu = 0,1,2,3)$ g has components $g_{\mu\nu} = g(X_\mu, X_\nu)$, and u has components

$$u_\mu = u(X_\mu), \text{ where } g^{\mu\nu} u_\mu u_\nu = -1 . \tag{3.3}$$

The contravariant vector field U is $U = u^\mu X_\mu$, where $u_\mu = g_{\mu\sigma} u^\sigma$. We have defined U to be a unit vector field, $g(U,U) = u_\mu u^\mu = -1$. In the basis $\{X_\mu\}$, (3.2) becomes

$$T_{\mu\nu} = (w+p) u_\mu u_\nu + p g_{\mu\nu} . \tag{3.4}$$

The energy density w is equal to $\rho(1+\varepsilon)$, where ε is the internal energy, and p is the rest-mass density (see Taub, 1967). The function ρ is found by multiplying the number of particles per unit volume by the rest mass of each, and consequently obeys the "continuity law"

$$(\rho u^\sigma)_{;\sigma} = 0 . \tag{3.5}$$

The *entropy* S of the fluid is related to ε, ρ and p by the thermodynamic equation $\theta dS = d\varepsilon + p d(1/\rho)$ where θ is the temperature.

The scalars w and p are the timelike and spacelike eigenvalues of T respectively, there being three spacelike eigenvalues, each p. This degeneracy of the spatial eigenvalues is due to the isotropy of the pressure – the defining characteristic of a perfect fluid (Eisenhart, 1924).

3.2. Thermodynamics, The Bianchi Identity, and Conservation Laws

We assume equilibrium thermodynamics for cosmological models (see Taub, 1959), that is, the constancy of entropy along the fluid flow lines: $\nabla_U S = S_{,\mu} u^\mu = 0$. The equilibrium thermodynamics hypothesis implies

$$w_{,\mu} u^\mu = \nabla_u w = -(w+p) u^\mu{}_{;\mu} \,. \tag{3.6}$$

From quite general, physical considerations it can be shown that T obeys the *conservation law*

$$T^{\mu\nu}{}_{;\nu} = 0 \,. \tag{3.7}$$

In the case of a perfect fluid this law separates into two equations, (3.6) and

$$u_{\mu;\sigma} u^\sigma = -(w+p) p_{,\sigma} (u^\sigma u_\mu - \delta_\mu{}^\sigma) \,, \tag{3.8}$$

This latter equation is completely analogous to the conservation of momentum equation of Newtonian hydrodynamics (Euler's equations of motion, see Birkhoff, 1960).

The Bianchi Identity

Equations (3.7) and (3.1) imply

$$\left(R^\mu{}_\nu - \frac{1}{2} R \delta^\mu{}_\nu\right)_{;\mu} = 0 \tag{3.9}$$

if Einstein's field equations hold. This latter equation is actually a geometric identity. If we take the curl of the curvature forms $\theta^\mu{}_\nu$, we find $d\theta^\mu{}_\nu = d\omega^\mu{}_\sigma \wedge \omega^\sigma{}_\nu - \omega^\mu{}_\sigma \wedge d\omega^\sigma{}_\nu$. We insert $\theta^\mu{}_\nu = \frac{1}{2} R^\mu{}_{\nu\alpha\beta} \omega^\alpha \wedge \omega^\beta$ and $d\omega^\mu = -\omega^\mu{}_\nu \wedge \omega^\nu$ to find

$$R^\alpha{}_{\beta\rho\sigma;\tau} \omega^\rho \wedge \omega^\sigma \wedge \omega^\tau = 0 \,. \tag{3.10}$$

This identity is called the Bianchi identity. Equation (3.8) is the twice-contracted version of the Bianchi identity.

The Equation of State

In a basis (not necessarily a coordinated basis) the full Einstein field equations for a perfect fluid read

$$R_{\mu\nu} - \frac{1}{2} R g_{\mu\nu} = (w+p) u_\mu u_\nu + p g_{\mu\nu} \; . \tag{3.11}$$

We contract with $g^{\mu\nu}$ to find $R = w - 3p$ (since $u^\mu u_\mu = -1$). We may therefore rewrite (3.11) as

$$R_{\mu\nu} = (w+p) u_\mu u_\nu + \frac{1}{2} (w-p) g_{\mu\nu} \; . \tag{3.12}$$

We now have ten partial differential equations for the $g_{\mu\nu}$, plus (3.3), (3.6) and (3.8) relating w, p and u^μ. Notice that we lack one equation, as the above are fifteen equations for sixteen unknowns.

This type of indeterminacy occurs in Newtonian hydrodynamics (Courant and Friedrichs, 1948) and as in that case, supplementary thermodynamic conditions lead to a well-defined problem. In principle we shall give a set of equations

$$\varepsilon = f(\rho, S) ; \quad p = f(\rho, S) \; , \tag{3.13}$$

where ε, ρ, S were defined above. These, with the equation $S_{,\mu} u^\mu = 0$, give seventeen equations for seventeen unknowns. In fact, in cosmology, we usually assume S to be a group invariant (constant in space) and therefore $S = \textit{const}$. A fluid with constant S is called *isentropic*, and its pressure obeys an equation of the form

$$p = p(w) \; , \tag{3.14}$$

which is called an *equation of state*. We shall always assume the existence of such an equation of state (for a more physical treatment of equations of state see Harrison, Thorne, Wakano, and Wheeler, 1965).

The equations of state we shall generally use are

1) *vacuum*: $T_{\mu\nu} = 0$, or $w = p = 0$.
2) *dust*: $p = 0$.

   This is closest to the present universe, and it serves as a very good model for the general case.
3) *gamma-law*: $p = \gamma w$ where γ is a constant.

   The *radiation* or *photon gas* is a special case of this for $\gamma = \frac{1}{3}$. This is a good model for a highly condensed cosmology.
4) *polytropic*: $p = \kappa w^{\left(1+\frac{1}{n}\right)}$. κ = const.; n = const. (the polytropic index).

3.3. Geodesics and Clouds of Particles

In the dust case Equation (3.8) becomes

$$u_{\alpha;\sigma}u^{\sigma} = 0 \tag{3.15}$$

which is the equation for a *geodesic*. A test particle, a small particle which reacts to gravitational forces only, also follows a geodesic path (Bergmann, 1942). Because of these two cases we will consider geodesics in general.

Geodesic, a Parametrized, Self-Parallel Path

A geodesic is, first of all, a parametrized *path*, a map from a segment of the real line R into a manifold M:

$$p : R \to M$$

(for every t in R, p(t) is a point in M). The numbers in R are the *domain* of parameters, while the set of points p in M is the *image* or *range* of p in M. The one-dimensional subset of M mapped out by p is a path or world-line. The vector U tangent to the path is the operator which acts on functions f restricted to the image of p, yielding

$$Uf = \frac{df(p(t))}{dt} . \tag{3.16}$$

The vector U is a map from R to the tangent space of p(t) for each t.

The path p is said to be differentiable or "smooth" if in an allowed coordinate system $\{x^\mu\}$, p is represented by n (= dimension of M) differentiable functions of t, $x^\mu(t)$. Similarly a vector field W on p is smooth if the components of W are differentiable functions of t in any allowed coordinate system. The components of the tangent vector U are

$$u^\mu = \frac{dx^\mu(t)}{dt} \qquad \text{i.e.} \qquad U = u^\mu \partial_\mu .$$

Paths whose tangent U obeys $U \cdot U < 0$; $U \cdot U = 0$; $U \cdot U > 0$ are called *timelike, null,* and *spacelike,* respectively. In relativity all massive particles travel along timelike paths, while photons travel along null paths. Tachyons, hypothetical particles with spacelike world-lines, have not yet been found in nature.

It is easy to see that $\nabla_U W$ for any vector W defined on p is well-defined, including $\nabla_U U$. A *geodesic path* p is a path whose tangent vector obeys

$$\nabla_U U = aU \tag{3.17}$$

where a is some function. This equation says that the transport of U along U (by the symbolic use of ∇_U) is parallel to U (see Figure 3.2).

Reparametrization (a change of coordinates in the parameter segment R) changes the function a. There is always a parametrization of a geodesic path for which $a = 0$, that is

$$\nabla_U U = 0 \quad \text{or, in terms of components} \quad u^\mu{}_{;\sigma} u^\sigma = 0 . \tag{3.18}$$

A parameter τ which makes (3.18) true is called an *affine* parameter. We may construct another affine parameter $\hat{\tau}$ out of τ by $\hat{\tau} = a\tau + b$, $a \neq 0$.

If we contract (3.18) with u^μ we find

$$U \cdot U = \text{const. along } p, \text{ if an affine parameter is used.}$$

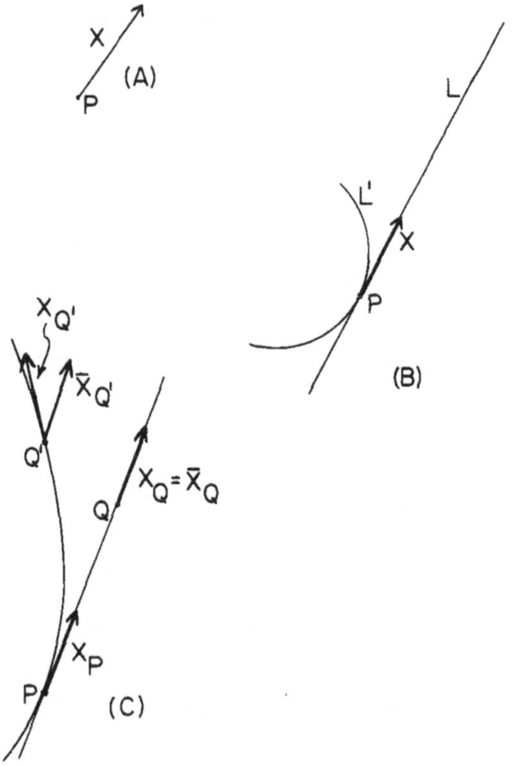

Fig. 3.2. Geodesics and Non-Geodesics. (A) X is a vector at the point P. (B) L is a geodesic through P whose tangent coincides with the vector X at P. L′ is a path, but not a geodesic, which also has as tangent at P the vector X. (C) A geodesic is characterized by the following property: If its tangent is parallely translated along the geodesic path to Q, then the parallel translate $\bar{X}_Q$ will coincide with the tangent at Q, i.e. X_Q. With a non-geodesic path, the parallel translate $\bar{X}_{Q'}$ will not coincide with the tangent $X_{Q'}$ of the path at Q′.

Without loss of generality we may choose τ such that $U \cdot U = \pm 1$ or 0. If $U \cdot U = 1$ (U spacelike) τ is called the *proper distance* along U. If $U \cdot U = -1$ (U timelike) τ is called *proper time*. We may still choose an affine parameter even if U is null, but $U \cdot U =$ const. is useless in this case as $U \cdot U = 0$ for any parameter.

It is an interesting property of a geodesic segment that it is an extreme path between its endpoints P and Q. That is, (3.18) is the Euler-Lagrange equation for a variational principle of the form

$$\delta \int g_{\mu\nu} \frac{dx^\mu}{d\tau} \frac{dx^\nu}{d\tau} d\tau = 0 \ .$$

While a geodesic is extremal it is not necessarily the "shortest" path between P and Q. Nor is it true that there must exist a geodesic between any two points, even if the manifold has no artificially set boundaries (Calabi and Markus, 1962).

Motion of a Cloud of Particles

Let us now consider a *cloud* of non-colliding particles, that is, a set of timelike paths such that one and only one path passes through any point in the manifold M. Such a cloud is represented by a timelike vector field everywhere on M. If the cloud consists of particles which interact only gravitationally, then U is a geodesic field. Whether U is geodesic or not, we shall choose parameters so that $U \cdot U = -1$.

If we join two nearby paths by an infinitesimal line segment, and remember from Chapter 2 that such a segment is equivalent to a contravariant vector, then we can speak of the vector W joining two paths (see Figure 3.3). We shall parametrize the paths so that W joins points with the same parameter. We can extend such a definition to a vector field over an open region in M.

We will now show that $\mathcal{L}_U W = 0$, where $\mathcal{L}_U W = [U, W]$ is the Lie derivative of W along p. To do this, write $\mathcal{L}_U W$ in some basis $\{X_\mu\}$ as

$$(\mathcal{L}_U W)^\mu = w^\mu{}_{;\sigma} u^\sigma - u^\mu{}_{;\sigma} w^\sigma \ . \qquad (3.19)$$

In a coordinated system ($\Gamma^a_{\mu\nu}$ symmetric on μ and ν) the covariant derivatives may be replaced by ordinary derivatives. Now choose a coordinate system $\{\tau, y^i\}$ where y^i is a three-parameter set labelling particles

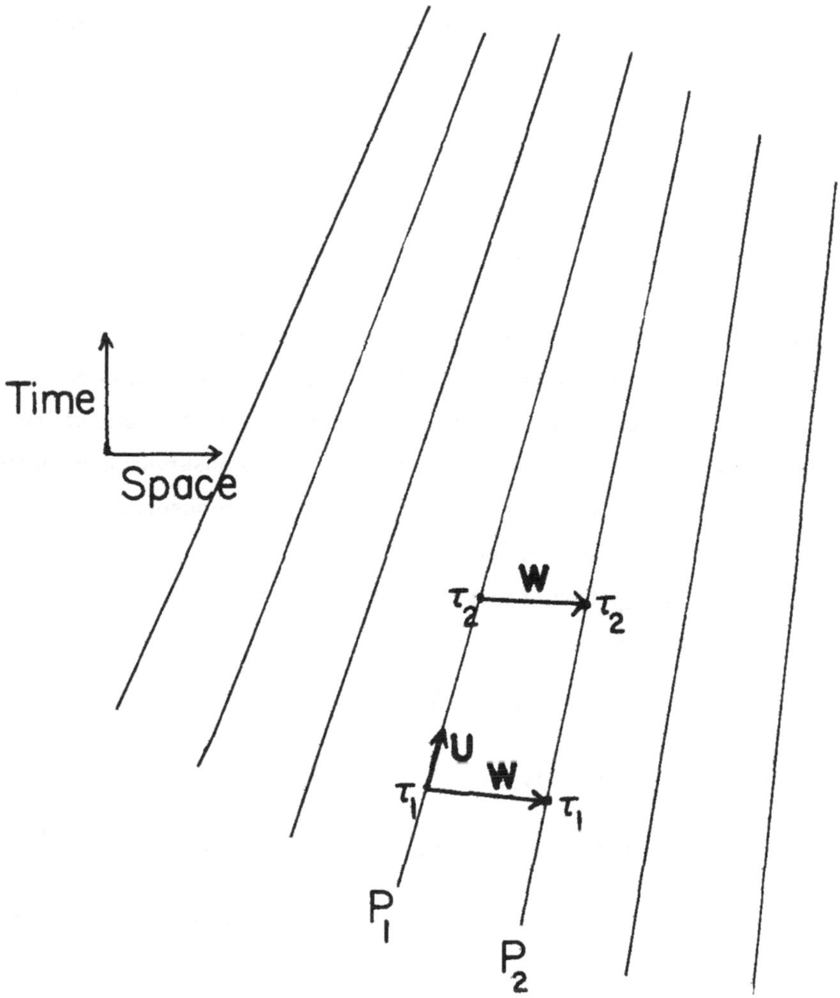

Fig. 3.3. Coherent Motion of a Cloud of Test Particles. Each particle is represented by a path with tangent vector field U, parametrized by proper time τ. The vector field W connects the path P_1 with another typical particle path P_2, the connection being between points at the same proper time. W has the property that $\mathcal{L}_U W = 0$.

and τ is the affine parameter along particle paths. In this coordinate system $u^\mu{}_{,\sigma} = 0$ (since $u^\mu = \delta^\mu{}_0$). Since W joins points with the same affine parameter, the contravariant components of W are independent of τ

along the path. Thus $w^{\mu}{}_{,\sigma}u^{\sigma} = \partial w^{\mu}/\partial\tau$ vanishes. Because we are using a coordinated basis,

$$\mathcal{L}_U W = 0 \ . \tag{3.20}$$

This is a basis independent statement, so it is true in all bases.

Reaction of a Cloud of Test Particles to Curvature – Equation of Geodesic Deviation

When U is a geodesic vector field and W any vector field such that $\mathcal{L}_U W = 0$, then the second derivative of W provides information on the effect of curvature on the cloud. This information takes the form of the *equation of geodesic deviation*. We shall not go into observational topics here, but the equation of geodesic deviation does have observational application (see Kristian and Sachs, 1966).

To derive the equation of geodesic deviation, we first rewrite (3.19) as (this uses zero torsion, Section 2.5):

$$\nabla_U W - \nabla_W U = 0 \ .$$

On taking the covariant derivative of this equation with respect to U:

$$\nabla_U \nabla_U W = \nabla_U \nabla_W U - \nabla_W \nabla_U U = R(U, W)U \ , \tag{3.21}$$

we have the equation of geodesic deviation. In a coordinated basis (see, for example, Weber, 1961):

$$w^{\mu}{}_{;\sigma\tau}\, u^{\sigma} u^{\tau} = R^{\mu}{}_{\sigma\tau\rho}\, u^{\sigma} u^{\tau} w^{\rho} \ . \tag{3.22}$$

3.4. Acceleration, Rotation, Shear, and Expansion

Many properties of a cloud of particles do not depend on the Einstein field equations. A cloud is represented by a timelike vector field U, and we parametrize U by proper time (whether U is geodesic or not) so that $U \cdot U = -1$. The cloud may be thought of as a continuous fluid. In a basis $\{X_{\mu}\}$, U has the form $U = u^{\mu} X_{\mu}$. The covariant components of U are $u_{\mu} = g_{\mu\sigma} u^{\sigma}$.

The first covariant derivative of U may be written in the form

$$u_{\mu;\nu} = -a_\mu u_\nu + \sigma_{\mu\nu} + \omega_{\mu\nu} + \frac{1}{3} h_{\mu\nu} , \tag{3.23}$$

where

$$\begin{aligned}
\sigma_{\mu\nu} &= \left[\frac{1}{2}(u_{\sigma;\tau}+u_{\tau;\sigma}) - \frac{1}{3} u^{\alpha}{}_{;\alpha} h_{\sigma\tau}\right] h^{\sigma}{}_{\mu} h^{\tau}{}_{\nu} , \\
\omega_{\mu\nu} &= \frac{1}{2}(u_{\sigma;\tau} - u_{\tau;\sigma}) h^{\sigma}{}_{\mu} h^{\tau}{}_{\nu} , \\
\theta &= u^{\sigma}{}_{;\sigma} , \\
a_\mu &= u_{\mu;\sigma} u^{\sigma} , \\
h_{\mu\nu} &= g_{\mu\nu} + u_\mu u_\nu .
\end{aligned} \tag{3.24}$$

The tensor h whose components are $h_{\mu\nu}$ is the *projection operator* onto the set of vectors perpendicular to U. The properties of h are $h_{\mu\sigma} u^{\sigma} = 0$, $h_{\mu\sigma} h^{\sigma}{}_{\nu} = h_{\mu\nu}$, $h^{\sigma}{}_{\sigma} = 3$, and $h^{\mu}{}_{\sigma} w^{\sigma} = w^{\mu}$ if $w^{\tau} u_{\tau} = 0$. Ehlers (1961) has given the above quantities the following names (we label terms by their components) a_μ – *acceleration*, $\omega_{\mu\nu}$ – *rotation tensor* ($\omega \equiv [\omega^{\sigma\tau}\omega_{\sigma\tau}]^{\frac{1}{2}}$ is the *rotation*), $\sigma_{\mu\nu}$ – *shear tensor* ($\sigma \equiv [\sigma^{\mu\tau}\sigma_{\mu\tau}]^{\frac{1}{2}}$ – the *shear*), θ – (volume) *expansion*. Notice that $\omega_{\mu\nu}$ is antisymmetric, $\sigma_{\mu\nu}$ is symmetric and traceless, and $a_\mu u^{\mu} = \omega_{\mu\sigma} u^{\sigma} = \sigma_{\mu\sigma} u^{\sigma} = 0$.

Fermi Transport and Fluid-Flow Parameters

What is the physical signicance of acceleration, rotation, shear, expansion? The acceleration measures the response of a particle to non-gravitational fields (its departure from geodesic motion). In a perfect fluid the acceleration is determined by the pressure gradient (see 3.8).

The other quantities defined by (3.24), rotation, expansion, and shear, measure the rate at which the cloud deforms with respect to a *Fermi-transported basis* (Synge, 1960). This basis is a set of three vectors $\{X_i\}$, each of which is orthogonal to U and each of which obeys

$$\nabla_U X_i = ((\nabla_U U)\cdot X_i)U, \quad \text{and} \quad X_i \cdot U = 0 \tag{3.25}$$

If the three vectors are chosen orthonormal at a point P then (3.25) preserves not only their orthonormality, but also the fact that $X_i \cdot U = 0$ *all* along the path generated by U from P. Fermi-transport has the following properties: 1) If U is geodesic (3.25) reduces to the natural requirement that $\nabla_U X = 0$. 2) A point gyroscopic (for example, an electron) which is accelerated will precess according to (3.25) (Synge, 1960).

To measure the cloud's deformation, we consider the vector W connecting two nearby particles. We have shown that $\mathcal{L}_U W = [U, W] = 0$, and by proper choice of the affine parameters for different particles we can set $W \cdot U = 0$. The *components of* W *in the* Fermi-transported basis $W_i = X_i \cdot W$ are physically measurable, and changes in W_i along a path are described by $\omega_{\mu\nu}$, $\sigma_{\mu\nu}$, and θ. We use (3.20) and (3.25) to show:

$$\nabla_U(W_i) = \nabla_U(W\cdot X_i) = (\nabla_W U)\cdot X_i + (W\cdot U)(\nabla_U U)\cdot X_i = (\nabla_W U)\cdot X_i\,. \tag{3.26}$$

In a comoving basis ($u^a = \delta^a{}_0$, X_i has components $X_i{}^a = \delta_i{}^a$ (i = 1, 2, 3)), $\omega_{0\mu} = \sigma_{0\mu} = h_{0\mu} = 0$ and also, $h_{ij} = g_{ij}$. The equation for $\nabla_U W_i$ shows that in a time $\delta\tau$ in this basis the change in W, δW, is

$$\delta W_i = \left(\omega_{is} + \sigma_{is} + \frac{1}{3}\,\theta h_{is}\right) W^s\,\delta\tau\,. \tag{3.27}$$

If W is allowed to trace out a surface S at $\tau = 0$, (3.27) shows how S will deform in a time $\delta\tau$: S will *expand* by a relative volume $\theta\delta\tau$, and S will *rotate* and *shear* by amounts given by $\omega_{ij}\delta\tau$ and $\sigma_{ij}\delta\tau$. Thus (3.26), or more precisely (3.27), justifies the names given the quantities defined in (3.24).

Invariant Definition of Rotation

Of the quantities defined in (3.24), rotation plays a special role in the collapse of a cosmological model (see next section). We shall therefore examine it in some detail. It is interesting to note that the rotation tensor has an especially simple definition in terms of differential forms and is equivalent to a vector Ω which is orthogonal to U.

Consider the covariant fluid velocity u. At a point P_0 in spacetime we take coordinates so that $g_{\mu\nu} = \eta_{\mu\nu}$, $g_{\mu\nu,\lambda} = 0$, and $u^\mu = (1, 0, 0, 0)$ (this can always be done at *one* point). As we move away from P_0 in the spacelike hypersurface $t = x^0 = \text{const.}$, the particle velocities acquire spacelike components unless u is orthogonal to the hypersurface. The existence of these spacelike components indicate rotation with respect to an inertial frame.

The covariant vector u need not be the gradient of the function which defines the hypersurface in order to be normal to the hypersurface. However, u must be of the form $u = r df$, where r and f are functions. We shall take this form to be the defining characterization for vanishing rotation. It can be shown that $u = r df$ can only be satisfied if (u a covariant vector field)

$$du \wedge u = 0 . \tag{3.28}$$

The differential three-form $du \wedge u$ is a completely antisymmetric covariant tensor of rank three. It has components $\tilde{\Omega}_{\alpha\beta\gamma}$ in a coordinated basis, and there are at most four independent components. We now define a *vector field*, Ω, the "dual" (in the sense of differential forms) of $\tilde{\Omega}$, $*du \wedge u$, where

$$\Omega^\mu = (*du \wedge u) = \frac{|g|^{-\frac{1}{2}}}{3!} \varepsilon^{\mu\alpha\beta\gamma} \tilde{\Omega}_{\alpha\beta\gamma} , \tag{3.29}$$

(with $|g|$ the absolute value of the determinant of $g_{\mu\nu}$, and $\varepsilon^{\mu\alpha\beta\gamma}$ the completely antisymmetric array with $e^{0123} = +1$). The vector field Ω is always orthogonal to $u (\Omega^\mu u_\mu = 0)$ and is called the *rotation vector* (Gödel, 1950; Taub, 1959). In a comoving basis $(u^\mu = (1, 0, 0, 0))$ $\Omega_0 = 0$, and the three spacelike components Ω_i are equivalent to the three independent components of $\omega_{\mu\nu}$.

3.5. The Raychaudhuri Equation and the Conservation of Rotation

If U is geodesic (a^μ vanishes) an interesting formula due to Raychaudhuri (1955b) describes the rate of change of θ, the expansion. Raychaudhuri's equation in the case of vanishing pressure (or, as in the

Friedmann-Robertson-Walker models, if the fluid velocity is geodesic in spite of $p \neq 0$) is equivalent to the (00) component of the field equations $R_{\mu\nu} = (w+p)u_\mu u_\nu + \frac{1}{2}(w-p)g_{\mu\nu}$. In any spatially homogeneous model the spacelike homogeneous ($t = const.$) hypersurfaces may be described by a timelike geodesic field tangent to the t-axis. In this case θ is the relative expansion rate of the $t = \text{const.}$ hypersurface and Raychaudhuri's equation governs θ.

The general form of Raychaudhuri's equation is

$$\begin{pmatrix}\text{Expansion}\\ \text{derivative}\end{pmatrix} = -\begin{pmatrix}\text{energy density}\\ \text{term}\end{pmatrix} - \begin{pmatrix}\text{shear}\\ \text{term}\end{pmatrix} - \begin{pmatrix}\text{expansion}\\ \text{term}\end{pmatrix} + \begin{pmatrix}\text{rotation}\\ \text{term}\end{pmatrix}$$

Rotation enters with a sign opposite to the rest of the terms, so it is especially significant.

Raychaudhuri's equation follows from the definition of the Riemann tensor in terms of the commutator of covariant differentiation (see Section 2.6). For a vector field u (components u_μ) we have, in a coordinated basis,

$$u_{\mu;\alpha\beta} - u_{\mu;\beta\alpha} = R^{\sigma}{}_{\mu\alpha\beta}\, u_\sigma \,.$$

We raise the μ index, contract on μ and α, and contract again with u^β. We find:

$$\theta_{,\sigma} u^\sigma + \frac{1}{3}\theta^2 = -R_{\sigma\tau} u^\sigma u^\tau - \sigma^2 + \omega^2 \,. \qquad (3.30)$$

This is the *Raychaudhuri equation.*

For a fluid stress-energy tensor, $R_{\sigma\tau} u^\sigma u^\tau = \frac{1}{2}(w+3p)$ which is greater than or equal to zero if $p \geq -w/3$. Even if the u^σ of (3.30) is not the local fluid velocity but the velocity of a cloud of test particles ($a^\mu \neq 0$) and $\bar{u}$ is the fluid velocity field then $\bar{u}^\mu u_\mu \leq -1$ since u and $\bar{u}$ are both timelike and unit. Consequently $R_{\sigma\tau} u^\sigma u^\tau \geq \frac{1}{2}(w+3p) \geq 0$ if $p \geq -w/3$. If we only allow small negative pressures then all the terms in (3.30) contributing to $\theta_{,\sigma} u^\sigma$ are negative except for the term in ω^2.

Rotation and Gravitational Collapse – Conservation of Rotation

If rotation is zero and $U = dx^{\mu}/d\tau$ for some parameter τ then

$$\frac{d\theta}{dt} + \frac{1}{3}\theta^2 \leq 0 . \tag{3.31}$$

Hence if $\theta \neq 0$ at some time τ,

$$\frac{d}{d\tau}(1/\theta) \geq 1/3 ,$$

so $\theta \to \infty$ in a finite proper time along the particle path. This infinite value of the expansion is indicative of a singularity of sorts: It shows that particle paths cross.

When U is the cosmic fluid velocity this singularity is a true singularity as defined in Chapter 5. When U describes the motion of a non-rotating cloud of test particles the singularity at $\theta = \infty$ may or may not be physically real. (Even in Minkowski space, if we aim a cloud of particles at a point, θ will blow up.) Detailed examination is needed in this case.

If $\omega^2 \neq 0$ then it is possible that θ will never become infinite. When the cloud of particles is the source of the gravitational field (U the fluid velocity) then the possibility of θ remaining finite for all time suggests that a rotating, non-singular cosmological model may exist. Maitra (1966) has given an example of a rotating, non-singular, dust-filled model. His model is not a cosmological model, however, because of its axial symmetry, and only pecular non-singular models are known (see, for example, Collins, 1974).

Because of the special position of rotation, we shall derive the law of conservation of rotation. This law governs the behavior of ω^2 during epochs when θ and $R_{\sigma\tau}u^{\sigma}u^{\tau}$ become large. This conservation of rotation law is of practical use only in limited circumstances (notably when the shear vanishes; see Ehlers, 1961; and Ellis, 1967). In these cases, and in an approximate manner in other cases, this law can tell if ω^2 can become large enough to dominate $R_{\sigma\tau}u^{\sigma}u^{\tau}$, θ^2, and σ^2 in (3.30).

Assuming u to be a geodesic field, so that $\omega_{\mu\nu}$ are the components of du, we find from $d^2u = 0$:

$$\frac{1}{2}\,\omega^2{}_{,\lambda}\,u^\lambda + \frac{2}{3}\,\theta\omega^2 + 2\omega^{\sigma\mu}\omega_{\sigma\lambda}\,\sigma^\lambda{}_\mu = 0\,. \tag{3.32}$$

When $\omega^{\sigma\mu}\omega_{\sigma\lambda}\sigma^\lambda{}_\mu$ vanishes (3.32) may be integrated using the fact that θ is the relative rate of expansion of the volume V of a small region of fluid (Ehlers, 1961, has integrated (3.32) and has extended the result to "conformally geodesic" clouds, in which the acceleration is proportional to a gradient). We write

$$\theta = \frac{1}{V}\frac{dV}{d\tau} \tag{3.33}$$

where τ is the proper time along the path of particle p_1. Equation (3.32) becomes

$$\frac{1}{\omega}\frac{d\omega}{d\tau} + \frac{2}{3V}\frac{dV}{d\tau} = 0, \qquad \text{or} \qquad \omega = AV^{-\frac{2}{3}}\,,$$

where A is a constant on each world line of the cloud.

If the cloud is a real fluid cloud, with rest density ρ, then ρ obeys the continuity law (3.5). This law can be integrated to show

$$\rho = BV^{-1}\,, \tag{3.34}$$

where B is constant on the path p_1. Near gravitational collapse, V is small, and one might think that ω^2 dominates the effect of ρ in $R_{\sigma\tau}u^\sigma u^\tau$. This domination by ω^2 might cause one to think that collapse to a singularity would not ensue. However, the detailed effects of $R_{\mu\nu}u^\mu u^\nu$ and of $\sigma_{\mu\nu}$ must be studied before any definitive statement concerning a singularity can be made.

4. FRIEDMANN-ROBERTSON-WALKER MODELS: BEGIN WITH A BANG

> Who knows from whence this great creation sprang?
> ...The Most High Seer that is in highest heaven,
> He knows it – or perchance even He knows not –
>
> From the RIG-VEDA

4.1. Field Equations of the Closed FRW Universe

The closed Friedmann-Robertson-Walker (FRW) universe (Friedmann, 1922; Robertson, 1929; Walker, 1935) is the most provocative and important cosmological model which has been devised since Bruno. It is also one of the simplest. It is isotropic, spatially homogeneous, and fluid-filled. Each spatial section is closed (compact, yet without boundary, finite in extent and volume). Compactness of the spatial sections was considered vital by Einstein (1917) in his earliest cosmological ideas and it is still an intriguing idea, if not necessary as once postulated.

The most shocking feature of this model is its expansion: The volume of the spatial sections changes with time. This expansion leads to a singularity at a finite time in the past when the volume of a spatial section becomes zero and matter becomes infinitely dense and infinitely hot (the *Big Bang* at the beginning of the universe). This singularity and the fact that physically reasonable models have such a singularity lead to an interest in singularities and in homogeneous models as vehicles to study such singularities. Whether such singularities are obligatory is an unanswered question at the moment.

The FRW models serve as an introduction to the study of homogeneous models. This chapter describes them mathematically. We will also briefly mention cosmography. (Figure 4.1 is a flow chart.)

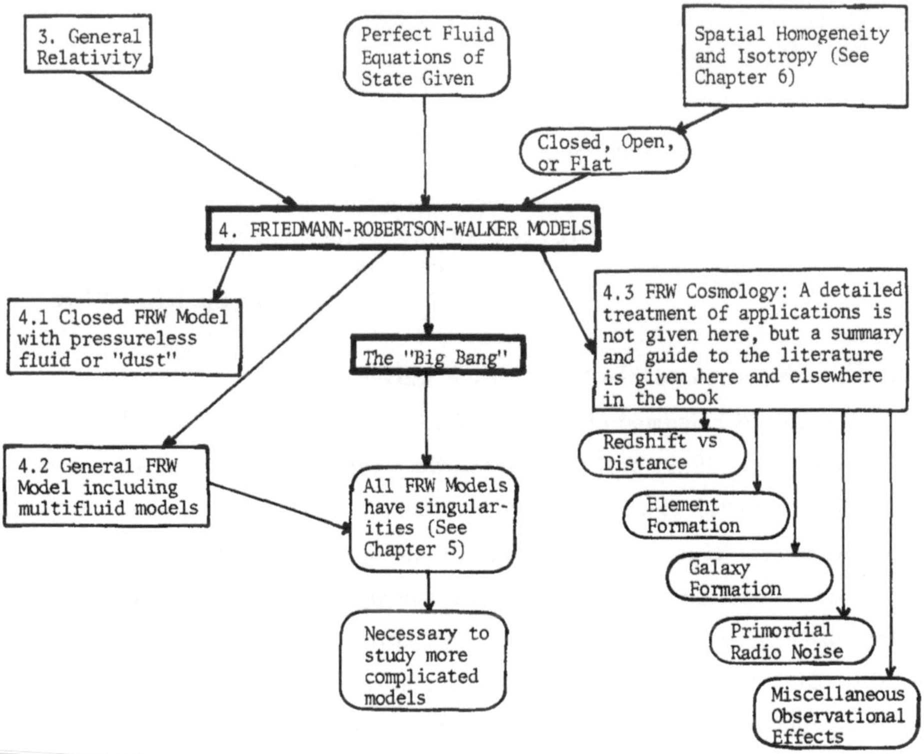

Fig. 4.1. Flow Chart of Chapter 4.

The Metric of the Closed FRW Universe

In a coordinated basis (coordinates x^0, x^1, x^2, x^3) the metric of the closed FRW universe is $(g_{\mu\nu} = \partial_\mu \cdot \partial_\nu)$ (Weber, 1961)

$$g_{11} = g_{22} = g_{33} = G^2(1+r^2)^{-2}, \quad g_{00} = -1,\ g_{\mu\nu} = 0,\ \mu \neq \nu,$$
$$\text{with } G = G(x_0), \quad r^2 = (x^1)^2 + (x^2)^2 + (x^3)^2 . \tag{4.1}$$

The manifold on which this metric is placed is the set of all points whose coordinates lie in the range $-\infty < x_i < \infty\,(i = 1, 2, 3)$ and $-T_1 < x^0 < T_2$ (for some numbers T_1, T_2).

At first glance no $x^0 = const$ slice of this manifold seems compact, and the manifold even seems singular at $r = \infty$; but let us consider the basis

$$
\begin{aligned}
X_0 &= \partial_0 \\
X_1 &= -\tfrac{1}{2}\,(x^1x^3 + x^2)\partial_1 - \tfrac{1}{2}\,(x^2x^3 - x^1)\partial_2 + \tfrac{1}{4}\,(2(x^3)^2 + 1 - r^2)\partial_3 \\
X_2 &= \tfrac{1}{2}\,(x^1x^2 - x^3)\partial_1 + \tfrac{1}{4}\,(2(x^2)^2 + 1 - r^2)\partial_2 + \tfrac{1}{2}\,(x^2x^3 + x^1)\partial_3 \\
X_3 &= -\tfrac{1}{4}\,(2(x^1)^2 + 1 - r^2)\partial_1 - \tfrac{1}{2}\,(x^1x^2 + x^3)\partial_2 - \tfrac{1}{2}\,(x^1x^2 - x^2)\partial_3\,.
\end{aligned}
\tag{4.2}
$$

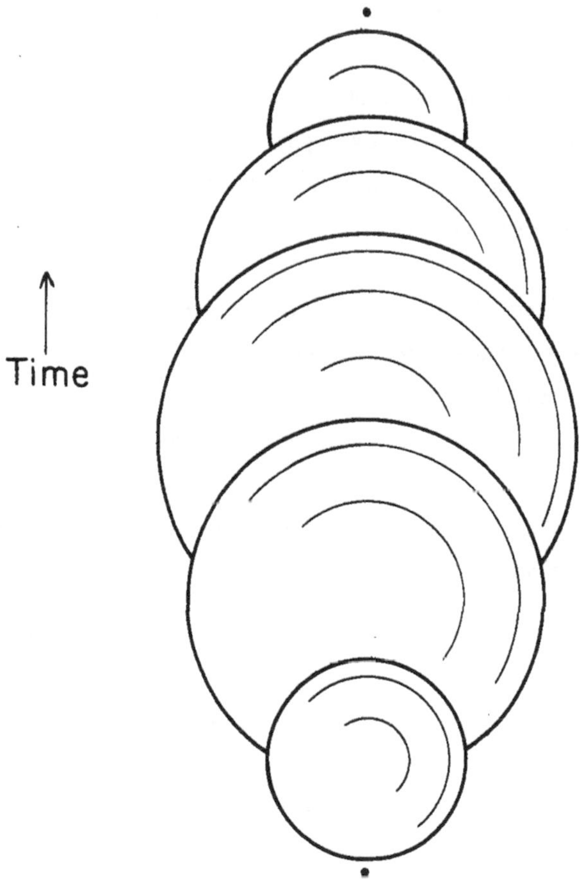

Fig. 4.2. The Closed FRW Universe. Spacelike invariant hypersurfaces are topologically three-spheres and are metrically isotropic. Consequently the homogeneous hypersurfaces in the FRW universe may be represented as expanding and contracting spheres. These spheres collapse to a point after a finite amount of time. The matter density ρ becomes infinite at the time of collapse.

A simple computation shows that $[X_i, X_j] = \varepsilon_{ijk} X_k$ (ε_{ijk} is the completely antisymmetric symbol with $\varepsilon_{123} = 1$), and $[X_0, X_i] = 0$. In this basis the metric is

$$X_0 \cdot X_0 = -1, \quad X_0 \cdot X_i = 0, \quad X_i \cdot X_j = \frac{1}{16} G^2 \delta_{ij} . \tag{4.3}$$

This metric and the commutation relations among the space vectors show that at x^0 = const. the geometry of the space section is a three-sphere (see Section 2.7) of radius $G(x^0)$.

At this point we mention a way of picturing the closed FRW universe. The equator of a three-sphere is a two-sphere. Since all equators are equivalent, we shall then picture each x^0 = const. surface as a sphere which evolves in time (Figure 4.2).

Connection Forms and Ricci Tensor

In order to give a complete description of the metric (4.3) we need only give the functional form of $G(x^0)$. In order to compute $G(x^0)$ we shall use the basis (σ^0, σ^i), where the σ^μ are the duals of the vectors $Y_0 = \partial_0$, $Y_i = \frac{4}{G} X_i$. In this basis $(g_{\mu\nu}) = (\eta_{\mu\nu}) = \text{diag}(-1,1,1,1)$, and (2.38) reads $\sigma_{\mu\nu} + \sigma_{\nu\mu} = 0$ ($\sigma^\mu{}_\nu$ being the connection forms). The computation of $d\sigma^\mu$ yields

$$d\sigma^0 = d(dx^0) = 0; \quad d\sigma^1 = \frac{\dot{G}}{G}\, \sigma^0 \wedge \sigma^1 + \frac{4}{G}\, \sigma^0 \wedge \sigma^3 \quad \textit{et cyc.} \left(\cdot \text{ means } \frac{d}{dx^0} \right). \tag{4.4}$$

From the first Cartan equation, $d\sigma^\mu = -\sigma^\mu{}_\nu \wedge \sigma^\nu$, we find

$$\sigma^0{}_\mu = \frac{\dot{G}}{G}\, \sigma^\mu; \quad \sigma^1{}_2 = \frac{2}{G}\, \sigma^3 \quad \textit{et cyc.} \tag{4.5}$$

The curvature forms, $\theta^\mu{}_\nu = d\sigma^\mu{}_\nu + \sigma^\mu{}_\alpha \wedge \sigma^\alpha{}_\nu$, can be readily computed. Two typical examples are

$$\theta^0{}_1 = \frac{\ddot{G}}{G}\, \sigma^0 \wedge \sigma^1, \quad \theta^1{}_2 = \left(\frac{\dot{G}^2}{G^2} + \frac{4}{G^2} \right) \sigma^1 \wedge \sigma^2 .$$

From $\theta^{\mu}{}_{\nu} = \frac{1}{2} R^{\mu}{}_{\nu\alpha\beta}\sigma^{\alpha} \wedge \sigma^{\beta}$ we find

$$R^{0}{}_{i0i} = \frac{\ddot{G}}{G}; \quad R^{i}{}_{jij} = \frac{\dot{G}^2}{G^2} + \frac{4}{G^2} \quad (i, j = 1, 2, 3, \ i \neq j; \ \text{no sum}) \tag{4.6}$$

with all the rest of the Riemann tensor components zero.

By summation we find the Ricci tensor, which is diagonal, to be

$$R_{00} = -3\frac{\ddot{G}}{G}; \quad R_{11} = R_{22} = R_{33} = \frac{\ddot{G}}{G} + 2\frac{\dot{G}^2}{G^2} + \frac{8}{G^2} . \tag{4.7}$$

Time-Evolution and Singularity

We now want to insert (4.7) into the field equations

$$R_{\mu\nu} = \rho u_{\mu} u_{\nu} + \frac{1}{2}\rho g_{\mu\nu} . \tag{4.8}$$

The $T_{\mu\nu}$ giving the right-hand side is that of dust (fluid matter with $p = 0$), so that $w = \rho$, the rest-mass density. We will later include pressure.

In our orthonormal frame, since $u_{\mu} u^{\mu} = -1 = -u_0{}^2 + u_1{}^2 + u_2{}^2 + u_3{}^2$ we have $|u_0| \geq 1$. The fact that $R_{\mu\nu}$ is diagonal implies $u_i = 0$, $i = 1, 2, 3$, so we take $u_{\mu} = (-1, 0, 0, 0)$. The field equations (4.8) now reduce to

$$-6\frac{\ddot{G}}{G} = \rho ; \quad 3\frac{\dot{G}^2}{G^2} + \frac{12}{G^2} = \rho . \tag{4.9}$$

If we consider the equation $T^{\mu\nu}{}_{;\nu} = 0$ we find that it reduces in the dust case to $u^{\mu}{}_{;\sigma} u^{\sigma} = 0$ (compare equation 3.8) which here is a tautology, and $(\rho u^{\mu})_{;\mu} = 0$. This latter equation reads

$$\dot{\rho}/\rho = \Gamma^{\sigma}{}_{0\sigma} ,$$

With the help of (4.5) this equation can be rewritten as

$$\dot{\rho}/\rho = -3\dot{G}/G \rightarrow \rho G^3 = M = \text{const.} , \tag{4.10}$$

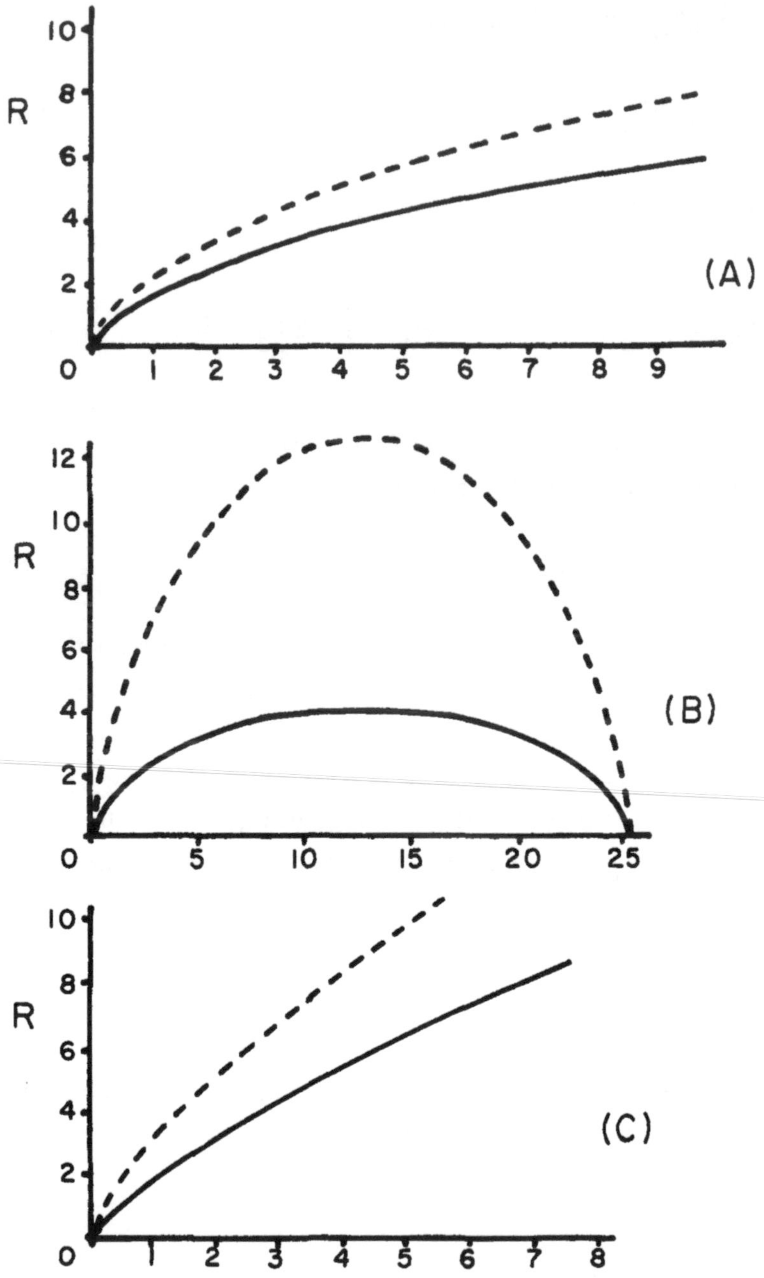

Fig. 4.3. A) R vs t for a k=0 FRW Universe. The solid line is the solution for pressureless matter, and the dashed line is for radiation with the constant Γ chosen so that the energy density of the radiation is equal to that of the dust at R=10. B) R vs t for a k=+1 FRW Universe. Again the solid line is dust and the dashed line radiation. The constants μ and Γ were chosen so that the distance between the two zeros of R would be the same. C) R vs t for a k = −1 FRW Universe. The solid line is dust and the dashed line radiation. The constants μ and Γ were chosen as in 4.3A.

The content of the field equations now reduces to the single equation

$$3\frac{\dot{G}^2}{G^2} + \frac{12}{G^2} = \frac{M}{G^3} \, . \tag{4.11}$$

The solution to (4.11) is illustrated in Figure 4.3. The details of the form of G are not as important as the fact that G becomes zero at a finite value of the time coordinate x^0: As $G \to 0$, $\rho \to \infty$. Since x^0 is the proper time measured by an observer travelling with the fluid, such an observer will run into an epoch of infinite density with a finite time. In addition he or she has emerged from an infinitely dense region a finite time in his past. These two regions are barriers – barriers beyond which the equations cannot predict the form of $G(x^0)$ or ρ. These barriers or "singularities" represent a breakdown of some aspect of the postulates which lead to the closed FRW universe.

4.2. The General FRW Model – Mathematics and Mystery

The present universe is described quite well by an FRW model even though the general FRW model has the type of singularity discussed above. The general model which is isotropic everywhere is also homogeneous (Walker, 1935). Friedmann (1922, 1924) investigated the closed and open isotropic, homogeneous models and Robertson (1929, 1933, 1935-6) shows that these plus the flat model are the only isotropic, homogeneous cosmologies.

Form of the Metric

The metric of the general FRW model may be written as

$$ds^2 = -dt^2 + R^2 d\sigma_k^2 \, , \tag{4.12}$$

where R is a function of t and $d\sigma_k{}^2$ is the metric of a three-space of constant curvature k. The three-curvature k is independent of t, and without loss of generality we may always choose $k = \pm 1, 0$. When $k = +1$, $d\sigma_1{}^2$ is the metric of a three-sphere, and the metric ds^2 includes that of the "closed" FRW model described above. If $k = 0$, $d\sigma_0{}^2 = (dx^1)^2 + (dx^2)^2 + (dx^3)^2$, and the $t = \text{const.}$ three-surfaces are flat. If $k = -1$, $d\sigma^2_{-1}$ is the metric of a hyperbolic space.

We can write the metric of each of these types of three-spaces in an orthonormal frame as $d\sigma^2_k = (\omega^1)^2 + (\omega^2)^2 + (\omega^3)^2$, with $d\omega^i = \frac{1}{2} C^i{}_{jk}\omega^j \wedge \omega^k$ for some constants $C^i{}_{jk}$. The $C^i{}_{jk}$ for $k = +1, 0, -1$ are respectively the structure constants for homogeneous spaces of Bianchi types IX, I, and V (we shall consider Bianchi-type spaces in detail in Chapter 6).

Let us define a time coordinate τ by $[\gamma(\tau)]^{\frac{1}{2}}\, d\tau = dt$, where proper time (t-time) is chosen by taking $\gamma = 1$. Another useful choice for γ yields $R(\tau) \propto e^{-\tau}$ (Misner, 1969, and Hughston, 1969, use Ω for this time variable). This second coordinate is valid only so long as $R(t)$ is a monotonic function: If a turnaround (a point where $dR/dt = 0$) occurs then a new time coordinate must be chosen.

Affine Connection and Ricci Tensor

We shall compute the connection coefficients and the Ricci tensor in an orthonormal basis $\{\sigma^\mu\}$ defined by

$$\sigma_0 = \gamma^{-\frac{1}{2}} d\tau; \quad \sigma^i = R\omega^i \;, \tag{4.13}$$

for which $ds^2 = \eta_{\mu\nu}\sigma^\mu\sigma^\nu$. Table 4.1 gives the complete Einstein equations for a stress-energy-tensor $T_{\mu\nu} = (w+p)u_\mu u_\nu + pg_{\mu\nu}$. From this table we see that $R_{0i} = 0$ for all FRW metrics, so that $u_\mu = \{-1,0,0,0\}$ in our basis.

We can solve the field equations under the assumption of an equation of state $p = p(w)$ by solving either for R or γ while taking the other to have a fixed functional form. A help in the solution process is the conservation law $T^{\mu\nu}{}_{;\nu} = 0$ which in our case reduces in content to

$$w_{,0} + 3(w+p)\frac{R_{,0}}{R} = 0 \;. \tag{4.14}$$

Table 4.1. Computation of the Christoffel Symbols, Ricci Tensor, and Einstein Equations for the FRW Models

See Chapter 6 for values of the C^{i}_{jk} for Bianchi Types IX (k=+1), I (k=0), V(k=−1).

General	FRW
$ds^2 = \eta_{\mu\nu}\sigma^{\mu}\sigma^{\nu}$	$ds^2 = \eta_{\mu\nu}\sigma^{\mu}\sigma^{\nu}$
$d\sigma^{\mu}$ Note: $\sigma^0 = \gamma^{-\frac{1}{2}} d\tau$, $R = R(\tau)$ and $' = \frac{d}{d\tau}$	$d\sigma^0 = 0;\ d\sigma^i = \frac{R'}{R}\gamma^{\frac{1}{2}}\sigma^0 \wedge \sigma^i + \frac{1}{2}R^{-1}C^{i}_{st}\sigma^s \wedge \sigma^t$
$d\sigma^{\mu} = -\sigma^{\mu}{}_{\nu} \wedge \sigma^{\nu};\ \sigma_{\mu\nu} + \sigma_{\nu\mu} = 0$	$\sigma^0_i = \frac{R'}{R}\gamma^{\frac{1}{2}}\sigma^i;\ \sigma^i_j = \frac{1}{2}R^{-1}(C^{i}_{js} - C^{j}_{is} - C^{s}_{ij})\sigma^s$
$\theta^{\mu}{}_{\nu} = d\sigma^{\mu}{}_{\nu} + \sigma^{\mu}{}_{\alpha} \wedge \sigma^{\alpha}{}_{\nu} = \frac{1}{2}R^{\mu}{}_{\nu\alpha\beta}\sigma^{\alpha} \wedge a^{\beta}$	$\theta^0{}_i = \left\{\left[\frac{R''}{R} - \left(\frac{R'}{R}\right)^2\right]\gamma + \frac{1}{2}\frac{R'}{R}\gamma'\right\}\sigma^0 \wedge \sigma^i +$ $+ \frac{1}{2}\frac{R'}{R^2}\gamma^{\frac{1}{2}}(C^{j}{}_{is} - C^{i}{}_{js} - C^{s}{}_{ji})\sigma^j \wedge \sigma^s$ $\theta^i_j = \left(\frac{R'}{R}\right)^2\gamma\sigma^i \wedge \sigma^j +$ $+ \frac{1}{4}R^{-2}(C^{i}{}_{js} - C^{j}{}_{is} - C^{s}{}_{ij})C^{s}{}_{\ell m}\sigma^{\ell} \wedge \sigma^m$
$R_{\mu\nu} = R^{\sigma}{}_{\mu\sigma\nu}$	$R_{00} = -3\gamma\frac{R''}{R} - \frac{3}{2}\gamma'\frac{R'}{R}$ $R_{0i} = 0$ $R_{ij} = \left[\gamma\frac{R''}{R} + 2\gamma\left(\frac{R'}{R}\right)^2 + \frac{1}{2}\gamma'\frac{R'}{R} + \frac{1}{2R^2}k\right]\delta_{ij}$
$R_{\mu\nu} = (w+p)u_{\mu}u_{\nu} + \frac{1}{2}(w-p)g_{\mu\nu}$ $u_{\mu} = [-1,0,0,0]$ From $T_{\mu\nu} = (w+p)u_{\mu}u_{\nu} + pg_{\mu\nu}$	$-3\gamma\frac{R''}{R} - \frac{3}{2}\gamma'\frac{R'}{R} = \frac{1}{2}(w+3p)$ $\gamma\frac{R''}{R} + 2\gamma\left(\frac{R'}{R}\right)^2 + \frac{1}{2}\gamma'\frac{R'}{R} + \frac{1}{2R^2}k = \frac{1}{2}(w-p)$

In the frame we are using, $w_{,0} = X_0(w)$, where $\sigma^0(X_0) = 1$. Thus $X_0 = \gamma^{\frac{1}{2}} \partial_0$, so the $({}_{,0})$ symbol can be replaced by $d/d\tau$ (denoted by $'$).

If we let R be a known function of τ so that $\gamma(\tau)$ becomes the function to be solved for, we find:

$$\gamma = \frac{4wR^2 - 3k}{12(R')^2} . \tag{4.15}$$

Thus, by giving $p = p(w)$ we can solve (4.14) for w, then using our form for R to find γ from (4.15) to complete the solutions.

Multifluid Solutions

If we choose as an example, $p = Aw$, $A = \text{const.}$, we find from (4.14) that

$$w = BR^{3(1+A)}; \quad \gamma = \frac{4BR^{-(1+3A)} - 3k}{12(R')^2} . \tag{4.16}$$

If we let $R = e^{-\tau}$, we find that $w = Be^{3(1+A)\tau}$, $\gamma = \frac{1}{3}Be^{3(1+A)\tau} - \frac{1}{3}ke^{2\tau}$ (see Hughston, 1969). Equation (4.16) includes dust $(A = 0)$ and radiation gas $(A = \frac{1}{3})$.

The linearity in w of (4.14) allows a solution where two or more non-interacting fluids are present. Suppose $T^{\mu\nu} = \sum_a [(w_a + p_a)u^\mu{}_a u^\nu{}_a + p_a g^{\mu\nu}]$ where each u_a has components $(1, 0, 0, 0)$. Then

$$\gamma = \frac{4\left(\sum_a w_a\right)R^2 - 3k}{12(R')^2} \tag{4.17}$$

and (4.14) goes to

$$w_a' + 3(w_a + p_a)(R'/R) = \xi_a; \quad \sum_a \xi_a = 0 . \tag{4.18}$$

(See Hughston and Shepley, 1970.) If $\xi_1 = \xi_2 = \cdots = \xi_N = 0$ and $p_a(w_a)$ is known for each fluid, $\gamma(\tau)$ and $w_a(\tau)$ may be found if $R(\tau)$ is given.

We can extend this analysis to any member of non-interacting fluids as long as $u^\mu = (1,0,0,0)$ for each of them.

If we choose $R(\tau) = e^{-\tau}$ for a specific solution, the proper time t may be found by integrating $dt/d\tau = \gamma^{-\frac{1}{2}}(\tau)$. The function $\gamma^{-\frac{1}{2}}$ is usually not integrable in terms of elementary functions when many fluids are present.

Dust and Radiation

Two very useful models of the physical universe are FRW models containing dust $(p(w)=0)$ and radiation $(p(w)=w/3)$. Dust is often used for the present universe, where pressure is effectively zero. Radiation is used to describe the early universe because extremely hot gases are often postulated to have this equation of state.

In these two cases the conservation law (4.14) can be solved to yield

$$w = MR^{-3},\ M = \text{const (dust)};\quad w = \Gamma R^{-4},\ \Gamma = \text{const (radiation)}\ . \tag{4.19}$$

With these forms for w the field equations reduce to one. The others are redundant. For t-time $(\gamma = 1,\ \cdot = d/dt)$ we find that R is given by

$$3\left(\frac{\dot{R}}{R}\right)^2 + \frac{3k}{4R^2} = \frac{M}{R^3}\ \text{(dust)} \tag{4.20a}$$

$$3\left(\frac{\dot{R}}{R}\right)^2 + \frac{3k}{4R^2} = \frac{\Gamma}{R^4}\ \text{(radiation)}\ . \tag{4.20b}$$

Note that the dust equation becomes (4.11) if $k = +1$ and we let $G = 4R$.

The general solution to (4.20) (see Figure 4.3) for all three values of k has $R = 0$ at one moment. For $k = +1$ there is a turnaround time and $R \to 0$ at two different times. For $k = 0$ or -1, R is monotonic.

4.3. The "Big Bang" and Cosmology

Why is the idea of a singularity where $R = 0$ disturbing? To answer this question we must look at the real universe.

A Picture of the Universe

At present the real universe is isotropic, expanding, and filled with a roughly uniform (however, see De Vaucouleurs, 1970) cloud of clusters of galaxies which do not interact with one another. The simplest model to fit this observation is one of the FRW models filled with a zero pressure fluid. That is, we assume we are in a "typical" position and that the universe would look roughly the same anywhere.

We might expect that the universe may have been anisotropic and inhomogeneous at earlier times and that it has settled down to the universe we see today. The FRW models do not take this possibility into account. It is usual, however, to ignore this objection and assume the FRW models are true back to $R = 0$.

While in theory we could measure $p(w)$ now, this measurement is presently beyond our observational capacities. We generally take equations of state which are known from terrestrial phenomena, and as we mentioned above, p now may be taken to be zero.

Observational Parameters

If we assume $p(w)$ to be known we can solve (4.14) if we can find initial data from observation. To illustrate how this is done it is sufficient to consider a dust-filled universe for which $p = 0$. In order to solve (4.20a) we need to know $R(t_0)$, k, and M, where t_0 is the present.

The usual measurable quantities are: a) the present matter density $w(t_0)$ (which gives us $M/[R(t_0)]^3$); b) $(\dot{R}/R)|_{t_0}$, the Hubble constant; and c) $(\ddot{R}/R)|_{t_0}$ (or some equivalent measure of the acceleration; e.g., $q \equiv (R\,\dot{R}^{-2}\ddot{R})|_{t_0}$). The first two of the quantities allow us to solve (4.14) for $R(t_0)$ if $k = \pm 1$. Since we may always rescale R at any one time by a change of coordinates if $k = 0$, we may arbitrarily set $R(t_0)$ in this case.

Observation (c) is then redundant, since $-6\ddot{R}/R = M/R^3$ (differentiate (4.20a)). This relation provides a valuable check on our observations. If the redundant observation does not agree with the other two it must be doubted whether general relativity is the proper theory to explain cosmology.

In fact, the three measurements do *not* agree, but generally there are few qualms about this disagreement. Although $(\dot{R}/R)|_{t_0}$ is well known, $w(t_0)$ and $(\ddot{R}/R)|_{t_0}$ are not. The uncertainty in $w(t_0)$ arises because only luminous matter is measurable directly and there may be "dark" matter we do not see. The uncertainty in the acceleration is due to large experimental error at the high redshifts needed to measure it. At present the acceleration data imply $k = +1$ (see Sandage, 1972-73, and Chapter 15), while the luminous-matter density implies $k = -1$. Generally one chooses between these two values on the basis of theoretical prejudice.

Cosmography

While we do not know the exact form of the metric of our universe, we can use the FRW universes as a first guess. Moreover, the behaviors of the $k = \pm 1, 0$ models are close enough for times before the present (which is why it is so hard to distinguish among them) that we may use any one of them to discuss a variety of problems. These problems include galaxy formation, and element formation, under the influence of various equations of state, $p = p(w)$.

For many purposes the universe is best approximated by an FRW model with two non-interacting fluids making up $T_{\mu\nu}$. The first of these is dust $(p = 0)$ and the second radiation $(p = \frac{1}{3}w)$. From the conservation law for non-interacting fluids (4.14) radiation obeys $w = \Gamma/R^4$, Γ a constant. The number density of photons n is proportional to R^{-3}. The temperature T may be defined as the ratio w/n, so $T \propto R^{-1}$. The equation of evolution for such a universe would be

$$3(\dot{R}/R)^2 + \frac{3k}{4R^2} = \frac{M}{R^3} + \frac{\Gamma}{R^4} . \tag{4.21}$$

The constant Γ is determined by measurements of the black-body radiation which appears to fill the universe (see Dicke et al., 1965). It is

Table 4.2. A Two-Fluid Model Contrasted with the Real Universe. The solution of equation (4.21) for $k=0$ provides the model: A universe filled with non-interacting radiation (energy density w) and dust (energy density ρ). In this model the temperature T is defined by $T \propto w/\rho$ (T is the temperature of the radiation gas). The proper horizon X_H for an observer (defined more precisely in Chapter 12) is the radius within which all matter can in principle be seen by the observer. This matter is the only cosmic matter which can affect the observer.
This model is fit to the following observations: a) the present Hubble constant $\dot{R}/R = 1.7 \times 10^{-18}\ \mathrm{sec}^{-1}$ (from equation 15.1); this value sets the time now as $\underline{t=3.8\times10^{17}}$ sec; b) the present temperature of black body radiation $\underline{T=2.7}$ K (corresponding to $\underline{w=4.4\times10^{-34}}$ $\mathrm{gm\cdot cm^{-3}}$). These numbers are underlined below. The function $R(\propto 1/T)$ is not plotted.

t (sec) time	T (°K) temperature	ρ ($\mathrm{gm\cdot cm^{-3}}$) dust	w ($\mathrm{gm\cdot cm^{-3}}$) radiation	X_H (cm) horizon	Speculations or Observations Concerning the Real Universe
0	∞	∞	∞	0	The "Big Bang," a feature of theory, not a feature of the real universe.
2.7×10^{-44}	9.5×10^{31}	4.8×10^{65}	6.0×10^{92}	1.6×10^{-33}	Horizon = $(\hbar G/c^3)^{1/2}$. Quantized gravity and geometry, ignored by model, needed here.
6.4×10^{-22}	6.0×10^{20}	1.2×10^{32}	1.1×10^{48}	3.9×10^{-11}	Horizon = $\hbar/mc$ = Compton radius of electron. A quantum field theory description of cosmic matter needed here.
2.3×10^{-4}	1.0×10^{12}	5.6×10^{5}	8.3×10^{12}	1.4×10^{7}	Copious elementary particles present. In "composite particle model," the temperature has never been significantly higher.
2.3×10^{2}	1.0×10^{9}	5.6×10^{-4}	8.3	1.4×10^{13}	Nuclei begin to form. Electron neutrinos are decoupled from other matter.

1.5×10^{11}	3.4×10^{4}	1.1×10^{-17}	1.1×10^{-17}	9.4×10^{21}	Approximate time when densities of dust and radiation are equal. Temperature falls to point where hydrogen combines. Matter and radiation henceforth decoupled (model assumes decoupled matter always). Start of present era; radiation has little dynamical effect. Galaxies start to form.
7.7×10^{12}	4.0×10^{3}	1.6×10^{-20}	2.1×10^{-21}	4.3×10^{23}	
$\underline{3.8\times10^{17}}$	$\underline{2.7}$	5.5×10^{-30}	$\underline{4.4\times10^{-34}}$	3.4×10^{28}	Present. Underlined numbers used to fix parameters of model. Visible mass density much below computed value of ρ: universe may thus be more nearly a $k=-1$ model. Observed deceleration parameter (see Chapter 15) somewhat above that computed for this model: universe may thus be more nearly a $k=+1$ model.
4.1×10^{18}	1.2	4.8×10^{-31}	1.9×10^{-35}	3.7×10^{29}	If the universe is a $k=+1$ model, recontraction should be about to begin.

found to be much smaller than M. Thus at present the M/R^3 term is dominant and describes the universe. As R becomes smaller in the past T goes up (we may use T as a measure of the stage of evolution of the universe). Eventually the term in Γ/R^4 dominates (the "radiation-dominated" epoch), so that (4.20b) describes the universe. Table 4.2 summarizes the behavior of such a two-fluid universe. It must be noted that interaction between the radiation and matter does take place, so that table is not strictly accurate. The table is for a $k = 0$ model since the assumption that the terms in k are negligible before the present is an excellent approximation. It is against the background of the universe described above that the ideas of galaxy formation, element formation, separation of matter from antimatter, anisotropy, and inhomogeneity at early times, and singularities are usually discussed.

The "Big Bang"

One feature of the general FRW universe is a singularity, $R = 0$, at some finite time t_0. We can see this singularity in (4.14) and (4.15) $(\gamma = 1)$ which can be combined to give

$$\ddot{R}/R = -\frac{1}{6}(w+3p) . \tag{4.22}$$

Since a real fluid cannot support large negative pressures we can take $p > -\frac{1}{3}w$ and write

$$\frac{d}{dt}[(\dot{R}/R)^{-1}] = 1 + \frac{R^2}{6(\dot{R})^2}(w+3p) > 0 . \tag{4.23}$$

This inequality implies that $\dot{R}/R$ must become infinite at a finite time t_0. Therefore R must go to zero at t_0 because of (4.15). As in Table 4.2 we can always choose $t_0 = 0$.

Infinite Density

We can say that $p \leq w$, because we want the speed of sound, $c_s = (dp/dw)^{\frac{1}{2}}$ to be less than the speed of light $(c = 1)$. The two limits on p we have assumed, $-\frac{1}{3}w < p < w$, may be used in (4.14) to show

$$R^{-2} < w < R^{-6} \quad (\text{so } w \to \infty \text{ as } R \to 0) .$$

This infinite density arises in every FRW model. (Remember we will always set the cosmological constant Λ zero; there are singularity-free FRW-type universes if $\Lambda \neq 0$.)

The infinity in w means that the singularity at $R = 0$ is not due to a poor choice of coordinates. A physically measurable quantity, w, becomes infinite for every observer at a finite time in his past. The FRW models fail at $t = t_0$ and no present theory can predict the behavior of the universe for $t < t_0$.

The FRW Model as Motivation for the Study of Singularities

The astrophysicist can be content that an FRW model describes our universe for a long time into the past. The relativity theorist, however, is interested in large gravitational effects. These effects occur in the FRW universes near t_0 – at the point where the model breaks down. He asks himself if there is any model which is non-singular at earlier times and approaches an FRW model at present. Evidence is mounting that there is no such model, but in studying this problem two basic avenues of research, each of interest in contexts far wider than cosmology have been explored. These subjects are the theory of symmetries and the theory of singularities. Much of the rest of this book is devoted to these two topics.

5. SINGULARITIES IN A SPACETIME

> ...I tell you that I can trace my ancestry back
> to a protoplasmal primordial atomic globule
>
> – WILLIAM SCHWENCK GILBERT

5.1. The Riemannian Manifold Contrasted with the Pseudo-Riemannian Manifold

Each FRW cosmological model is said to be singular. It has at least one region within which the density is unbounded. A freely falling observer in this region, travelling toward increasing density, would see the matter around him become infinitely dense in a finite amount of proper time. However, in the presently accepted viewpoint the points of infinite density (*singular points*) are not within the model but are treated as an additional structure, the *boundary* of the manifold proper.

At present there is no fully accepted method of defining the structure of the singular boundary points of a general manifold. What is accepted – and it is important to keep this concept in mind – is that each cosmological model is a well-defined manifold-with-metric, at each point of which the metric is non-singular, and that all singular points are on the boundary of this manifold. If no boundary points can be reached by any observer, the model itself is called non-singular, but if boundary points are not "at infinity," and the original model cannot be extended beyond these points to a non-singular model, then the model is called singular.

To define singularity it is first essential to have a well-formulated notion of non-singularity. The criteria for non-singularity are well-defined if the metric is positive definite (*Riemannian*) so we shall examine Riemannian theory first. In spite of the fact that a cosmological model is a *pseudo-Riemannian* manifold (it has a non-positive-definite metric), several

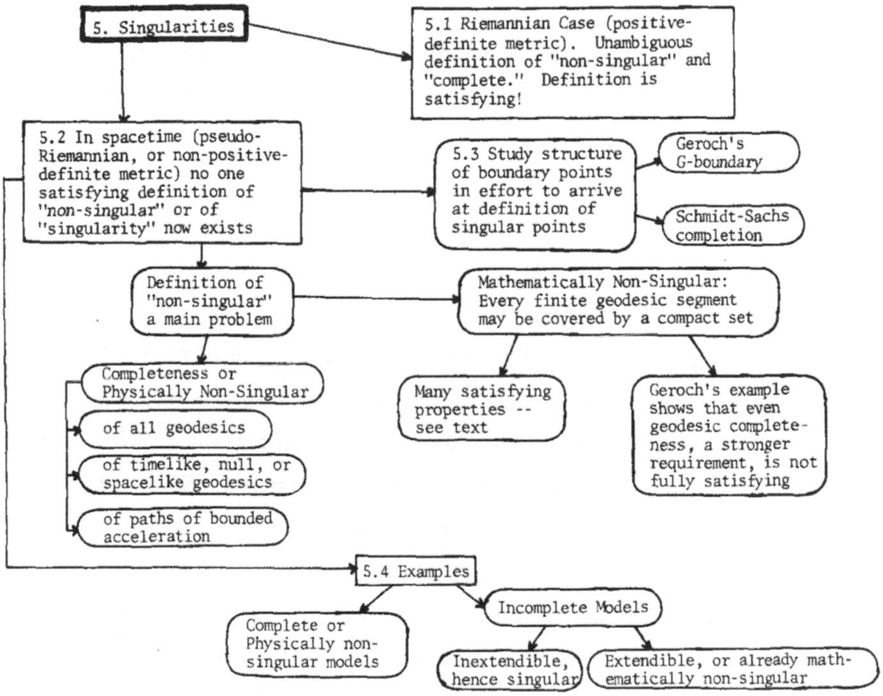

Fig. 5.1. Flow Chart of Chapter 5.

find common expression and usefulness in both cases. In addition, one method of investigating a pseudo-Riemannian manifold is to define from it a higher dimensional Riemannian manifold which is then treated by means of the methods described below. As usual, we present a chart of the ideas in this chapter (Figure 5.1). In addition we shall number various facts (F1, F2, $\cdots$), criteria for completeness (C1, $\cdots$), and other statements for comparison between the Riemannian and pseudo-Riemannian cases. In order to keep the discussion as intuitive as possible, we shall state theorems and facts without detailing many of the assumptions (such as smoothness and connectivity) necessary for rigor.

Metric and Topology in a Riemannian Manifold

The special feature of a Riemannian manifold M which makes M far simpler than the general pseudo-Riemannian manifold is the fact:

F(1) *A (connected) Riemannian manifold is a metric space*, a metric space being a set with a well-defined distance between any two points. Here the distance $d(P, Q)$ between two points P and Q is the minimum of the lengths of all lines from P to Q.

This distance function defines a topology, that is, an enumeration of the open sets of M. This *metric topology* is defined as follows: The sets {for fixed P, all Q such that $d(P, Q) < \varepsilon$} are open. All open sets are unions of these "basic" sets.

It is also true that: F(2) *The metric topology is the same as the manifold topology.* That is, the metric topology enumerates the same sets as being open as does the topology used to distinguish continuous from non-continuous functions (Section 2.1). The proof (Helgason, 1962) uses the fact that $d(P, Q)$ is continuous in both variables.

To see the consequences of F(1) and F(2) we must make a point about geodesics which we shall use later. This statement is true whether or not the metric is positive definite, and comes directly from the concepts of Section 3.2. A geodesic is a path whose tangent vector U obeys the geodesic equation, which if the path is parametrized by an affine parameter is:

$$\nabla_U U = 0 . \tag{5.1}$$

The equation is solvable, at least locally, to obtain U from initial data. The affine parameter λ (which in the Riemannian case measures path length) is itself determined by the vector U at our initial point P. In other words: F(3) *Given any vector* U *at a chosen point* P, *there exists a unique affinely parametrized geodesic passing through* P *whose tangent vector coincides with* U *at* P (Figure 3.2).

Completeness and Non-Singularity in a Riemannian Manifold

A Riemannian manifold M is non-singular if it satisfies either of two "completeness" criteria below. The equivalence of these criteria is a consequence of F(1), F(2), and F(3). That a complete manifold M is as extended as it can possibly be is also a consequence of these facts. This latter consequence, as well as others, show that a complete Riemannian manifold deserves the adjective non-singular.

If the completeness criteria are not satisfied, however, it may be possible to identify the given manifold M as a subset of a larger manifold M′. On M′ we shall have a metric tensor, non-singular at all points, which on the subset M coincides with the original metric on M. In this case M is *incomplete* but *not* necessarily *singular*. Consider the opposite instance where "not non-singular" *does* imply "singular." An incomplete ("not non-singular") manifold is said to be singular when it cannot be imbedded as an open subset of any larger manifold M′.

There are two distinct instances where a manifold M cannot be imbedded in a larger manifold M′ of the same dimension: The case where M is already infinite in extent, and therefore non-singular, and the case where M is singular. If M cannot be extended, whether or not M is singular, M is said to be *maximal*.

The criteria which a Riemannian manifold M must satisfy if it is to be *complete* or *non-singular* are (Figure 5.2):

(C1) *Complete metric topology*: Every Cauchy sequence converges. A Cauchy sequence is a sequence of points $P_i (i = 1, 2, \cdots)$ such that for any $\varepsilon > 0$, there exists a number N (depending on ε) such that $d(P_n, P_m) < \varepsilon$ for all n, m greater than N.

(C2) *Complete affine connection*: Every geodesic can be continued in both directions to infinite values of its affine parameter (the parameter measuring path length). A manifold satisfying (C2) may be called *infinite in extent* or *geodesically complete*.

In a Riemannian manifold criteria (C1) and (C2) are equivalent! The proof of this equivalence will not be given here (see Helgason, 1962).

The definition of a non-singular Riemannian manifold, as one which satisfies (C1) and (C2), is reasonable, as is shown by exhibiting several properties which hold in a complete Riemannian manifold M. These properties are (Helgason, 1962): (P1) M is inextendible: M cannot be

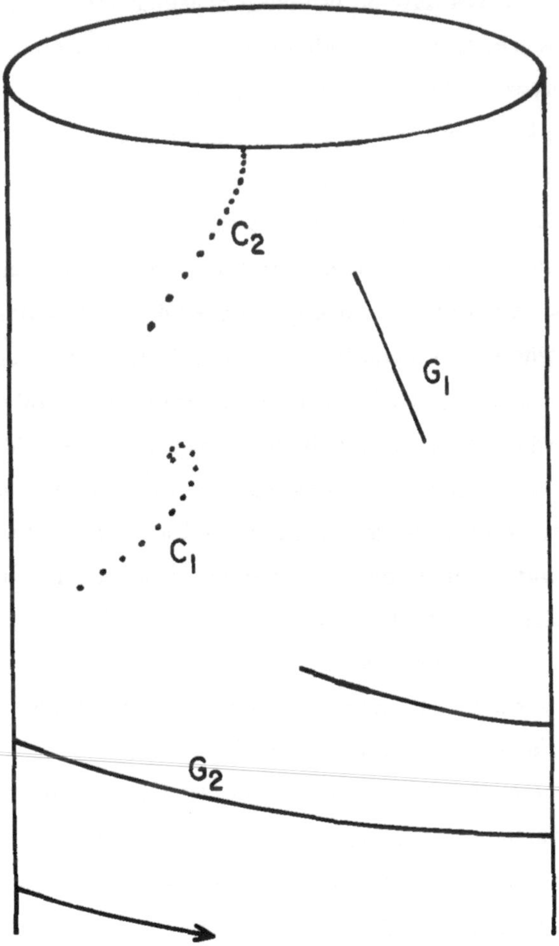

Fig. 5.2. A Half-Infinite Cylinder Illustrating Completeness Criteria in a Riemannian Manifold. C_1 and C_2 are Cauchy sequences. G_1 and G_2 are geodesic segments. C_2 cannot be completed since it converges to a point on the boundary. Although both G_1 and G_2 are either infinite or extendible or both, the equivalence of Cauchy completeness and geodesic completeness implies there are inextendible geodesic segments. This manifold can be imbedded in a larger connected manifold of the same dimension: a longer cylinder.

identified with a proper open submanifold of any connected Riemannian manifold. Thus M is *maximal*. (P2) M is *geodesically convex*: Any two points in M may be connected by a geodesic. (P3) If M is complete [satisfies (C1) or (C2)] then the manifold consisting of M with one point deleted is incomplete.

Furthermore, (P4) Any compact Riemannian manifold M is complete. A manifold M is compact if any given collection of open sets which covers M contains a finite subcollection of these open sets which is sufficient to cover M. A compact manifold is then a finite object in the topological sense, independent of any reference to the metric. This property of compact manifolds demonstrates that (C1) and (C2) are reasonable in a sense which is basically aesthetic.

One accepted procedure for examining a manifold M for singularities is: *On* M *certain coordinate patches are used to express the metric.* On certain subsets of the coordinate patches the metric will appear to be singular in all coordinate systems given. We must then study the manifold $\hat{M}$ obtained by eliminating from M those points of apparent singularity (if no point of apparent singularity exists, $\hat{M} = M$). If $\hat{M}$ satisfies (C1) and (C2), then $\hat{M}$ is non-singular. If, however, $\hat{M}$ does not satisfy (C1) or (C2), we extend it to obtain M′, the *largest non-singular* manifold extension of M (M′ may not, however, be unique). We must now focus on M′ as the manifold of greatest interest. If M′ does not satisfy (C1) or (C2), and thus is still not infinite in extent, and if M′ is inextendible, it is said to be *singular*.

This process is far from simple, a major difficulty being the extension of an incomplete manifold. The basic procedure can be performed on both Riemannian and pseudo-Riemannian manifolds, however, if criteria for non-singularity are given. Unfortunately, no fully accepted criterion for non-singularity exists in the pseudo-Riemannian case. However, a modification of (C2) is more widely used than alternative proposals.

5.2. Pseudo-Riemannian Manifolds – Completeness and Definitions of Non-Singular

In a pseudo-Riemannian manifold any two points may be joined by a (broken) line of zero length. Facts (F1) and (F2) are therefore not true, and criterion (C1) is meaningless. It is meaningful, however, to speak of geodesics, and Fact (F3), that a direction at a point defines a geodesic, is still true.

Criterion (C2), stating that a geodesic may be extended to indefinitely large values of its affine parameter therefore does make sense. For time-like or spacelike geodesics, the affine parameter is the proper time and length respectively. For lightlike geodesics, the element of length is always zero! An affine parameter may still be defined, however, as was pointed out in Section 3.3. In any of these three cases, if τ is an affine parameter, then $\hat{\tau} = a\tau + b\,(a \neq 0)$ is also an affine parameter, and infinite extension in τ is equivalent to infinite extension in $\hat{\tau}$. Criterion (C2) requires every geodesic segment, in some affine parameter, to be infinitely extendible. If (C2) holds for a manifold M, the manifold is called *complete*, and completeness is vulgarly used as the definition of non-singularity.

Other criteria for non-singularity have also been formulated. Some, like (C2), are conditions on all members of certain classes of geodesics (Kundt, 1963). These criteria make sense only in a pseudo-Riemannian manifold, where geodesics are classified as spacelike $(U \cdot U = 1)$, time-like $(U \cdot U = -1)$, or null $(U \cdot U = 0)$. An example is the criterion (C2t), *t-completeness*: Every timelike geodesic may be extended to infinite values of its proper time. Related to (C2t) is the stronger (C2tb): *completeness of paths of bounded acceleration*: Every timelike path of bounded acceleration may be extended indefinitely. Such a path need not be a geodesic. Its defining characteristic is that the quantity $(u^a{}_{;b}u^b)(u_{a;c}u^c)$ is a bounded function along the path if the path is parametrized by proper time $(u_a u^a = -1)$.

Kundt (1963) and Geroch (1967, 1968b) have shown by specific examples that there are pseudo-Riemannian manifolds in which these various criteria break down. An important example is due to Geroch (1968b): a geodesically complete model containing a finite but inextendible path of bounded acceleration. The path corresponds to a rocket which leaves the universe with a finite expenditure of fuel. It shows that (C2) is too weak a definition for non-singularity in a physical sense.

Nevertheless criterion (C2) is useful and simple. It has been applied in several general singularity theorems. We therefore give the status of the properties (P1) to (P4) of Section 5.1. (P1) is still a consequence of criterion (C2): If every geodesic in M may be indefinitely extended, then M may not be imbedded as an open submanifold in a larger M′. (P2), however, is not a consequence of (C2): The DeSitter (1917) universe is a counterexample (Calabi and Markus, 1962). On the other hand, (P3) is still true: If M is complete, M minus one point is incomplete. (P4) does not hold: A compact pseudo-Riemannian manifold is not necessarily complete (see Section 5.3 for an example).

Mathematical Non-Singularity

Completeness (as embodied in C2 or C2tb) is not a completely satisfactory definition of non-singularity because of the failure of (P4): (P4) was shown to fail when Misner (1963) exhibited a compact, incomplete manifold (Misner's example will be presented in the next section). This example, because it is compact, can be covered by a finite number of well-behaved coordinate patches; in this sense the model is uniformly well-behaved. Yet (C2) would deny this model the name "non-singular."

Misner (1963) proposed an alternate criterion for non-singularity. It is: (C3) Every finite segment of a given geodesic is contained in a compact subset of the manifold. A *finite segment* (which may be open or closed) of a geodesic is defined in terms of any affine parameter τ: It is a geodesic segment whose points have affine parameters τ taken from a finite segment of the real line. A closed segment is automatically compact and is itself a compact subset of M (Figure 5.3).

Criterion (C3) is applicable to both Riemannian and pseudo-Riemannian manifolds. In the Riemannian case (C3) is equivalent to (C2). Fact (F2) allows us in this case to extend the geodesic segment through the unique limit point of the open segment, and it is thus indefinitely extendible.

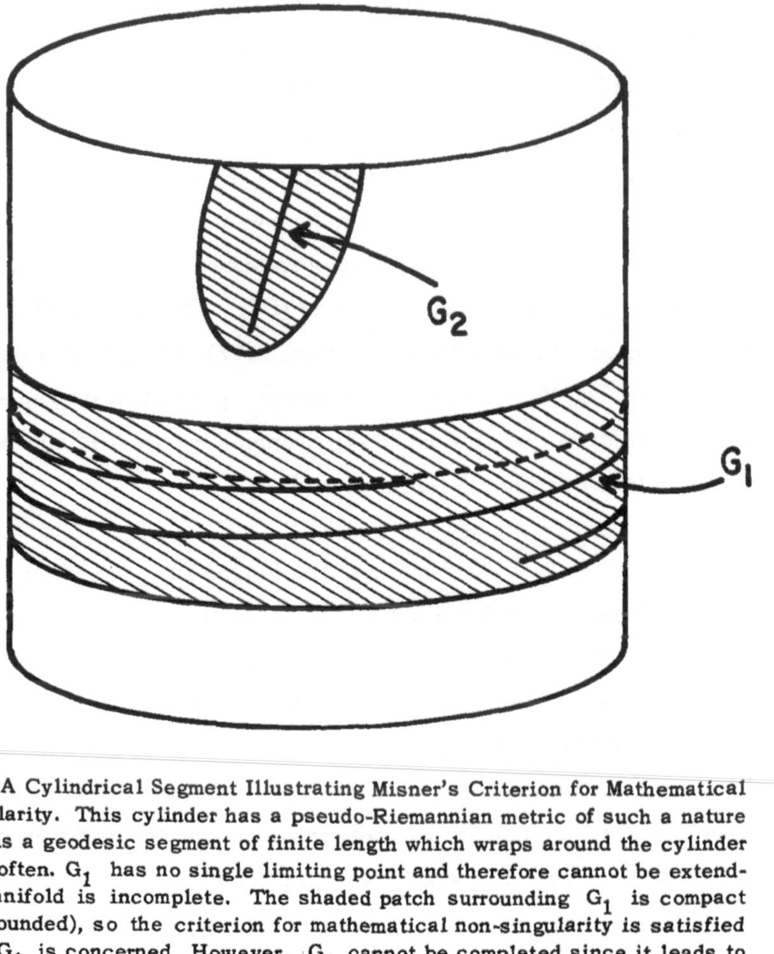

Fig. 5.3. A Cylindrical Segment Illustrating Misner's Criterion for Mathematical Non-Singularity. This cylinder has a pseudo-Riemannian metric of such a nature that G_1 is a geodesic segment of finite length which wraps around the cylinder infinitely often. G_1 has no single limiting point and therefore cannot be extended; the manifold is incomplete. The shaded patch surrounding G_1 is compact (closed, bounded), so the criterion for mathematical non-singularity is satisfied as far as G_1 is concerned. However, G_2 cannot be completed since it leads to the boundary. Any set surrounding G_2 cannot be closed, hence cannot be compact. Although this manifold is not non-singular, it can be extended to the mathematically non-singular example Misner's T^2 of Figure 5.7.

In an incomplete pseudo-Riemannian manifold M, an open geodesic segment may have many limit points. However, (C3) insures that any geodesic path which does go off to infinity (i.e., cannot be covered by a compact set) must involve an infinite affine parameter. Moreover, (C3) is a satisfactory definition in that (P1), (P3), and (P4) are satisfied (the failure of P2, geodesic convexity, is not an especially disturbing feature). We shall leave as an exercise the proof that (C3) implies (P1), (P3), and (P4).

There are two objections which we can raise to the use of (C3) as a definition of non-singular. First, (C3) deals with geodesic segments rather than segments of paths of bounded acceleration. If these paths do not obey the condition of criterion (C3) the manifold does not accurately describe all points which can be reached by a physical observer. It is not difficult to strengthen (C3) to include such paths, but in the models we shall deal with such added complexity is not justified. We shall retain (C3) as written.

A second objection to (C3) is more serious and throws doubt on applying any criterion *but* (C2) or (C2tb) as a definition of non-singular. Consider any incomplete manifold M which still satisfies (C3). There is in M a finite geodesic segment which is contained in a compact set, but which does not have a unique limit point and cannot be extended. The segment represents the world line of a test particle p. This particle, in a finite amount of proper time, finds itself approaching many different spacetime points at once. Clearly the particle p cannot be thought of as a material particle which responds only to the gravitational field.

Does this objection mean that incomplete manifolds are singular even if (C3) is satisfied? No, but it requires that incomplete manifolds be examined more closely than complete ones. As an example, the vacuum T-NUT-M space (Misner and Taub, 1968; see Chapter 8) is incomplete but satisfies (C3). It is also a vacuum model and when matter is added it becomes singular unambiguously (Section 10.2). The investigation of perturbations of this model (Misner and Taub, 1968) predicted that the model would be singular when matter was added. Criterion (C3) is thus more suited to the mathematician than to the physicist. We shall call (C3) *mathematical non-singularity*.

What is a Singular Model?

We have discussed two definitions of non-singularity, (C2) and (C3), each with strengths and weaknesses. A model which satisfies both and which also satisfies (C2tb) is unambiguously non-singular. A model which satisfies none of them and which is inextendible is unambiguously singular. Often the inextendibility is the result of the existence of a finite geodesic segment along which the density of matter is unbounded.

What, however, can we say of a model such as T-NUT-M? It satisfies (C3) but not (C2). We are calling such a model mathematically non-singular. At the same time we shall recognize that in T-NUT-M a potential observer may follow an incomplete geodesic path. We shall call the model *physically singular*. These two titles indicate that the model is of special complexity and must be examined more closely (for example, by examining perturbations in the model) to determine whether it is a viable cosmological model.

(C2) and (C3) are not as useful as one might hope when trying to prove a given manifold is singular. Using them, we can recognize whether a manifold is non-singular. However, even if a manifold is not non-singular, then the possibility of extending it to a non-singular manifold must be investigated. Only if M is *not non-singular* and *inextendible* can M be called *singular*. The proof that a true singularity exists is often extremely difficult.

In some cases, however, the proof of singularity is not difficult. If a manifold M is not non-singular (both (C2) and (C3) are not satisfied), then there is a geodesic segment p and a Cauchy (converging) sequence of points P_i on p which has no limit point in M. If there exists a scalar invariant $\bar{R}$ (one of the 14 given by Petrov, 1969) such that the values of $\bar{R}_i$ on the P_i do not approach a limit (for example the $\bar{R}_i$ may tend to infinity), it is clearly impossible to extend M, and M certainly is singular.

However, this method cannot be used without care. An arbitrary scalar invariant S cannot be employed. If, for example, $S = 1/\bar{R}$, and $\bar{R}$ tends to zero at a point P, then clearly S tends to infinity. The infinity in S does not indicate a singularity, while an infinity in $\bar{R}$ does (if $\bar{R}$ is one of the Petrov invariants). Moreover, there may be a singularity at which no scalar invariant becomes infinite, even in a Riemannian manifold. Consider an ordinary two-dimensional cone imbedded in three-dimensional

Euclidean space. This cone, without the central point, is singular in the terminology of this section, but it is flat (no scalar invariant is ever non-zero)! Thus our only hope of showing in general that a manifold is singular lies in showing it is non-singular and yet inextendible.

5.3. The Structure of Singular Points

It is sometimes not enough simply to say that a manifold is unambiguously singular. In a singular manifold a geodesic is prevented by a barrier from being extendible. This barrier is termed a *singular point*. The singular point is not part of the manifold but is an abstract point added to the manifold to give concrete realization to the notion of singularity.

The set of singular points – one for each inextendible path – may be given a topology, even a metric. Not all such points are distinct, and it is possible for many geodesics to be halted at the same singular point. It is the description of the structure of the singularity point set which forms a large part of the modern theory of singularities. We shall describe two types of structures, one due to Geroch, the other to Schmidt and Sachs. Other methods of determining structures of the singular points have been proposed. Since it is not known which methods lead to the same results, we can only say that much work along these lines remains.

Geroch's G-*Structure*

Geroch (1968b) gave structure to the singularity of a manifold M by structuring the set of incomplete geodesics. We have seen (Fact F2) that to each pair (P, U), P a point in M, U a non-zero vector at P (this set of pairs is the *reduced tangent bundle* of M), corresponds a geodesic ray with a unique affine parameter. Some of these geodesics may be incomplete and hence cannot be extended from P beyond a limiting value λ_0 of the affine parameter.

Suppose (P, U) gives rise to an incomplete geodesic, with λ_0 the limiting value of the affine parameter. Note that if α is a constant then (P, αU) is an incomplete geodesic whose limiting affine parameter is λ_0/α.

These two geodesics and many others define the same point in an abstract space S to be identified as the singularity points. Other geodesics emanating either from P or from other points in M may define other points in S. That is, if G_I is the set of incomplete geodesics, then S is a set of equivalence classes of these rays.

Roughly speaking, (P, U) and (Q, V) are equivalent if the distance between them becomes zero near the limiting values λ_0 and μ_0 of their respective affine parameters. To express the closeness of the endpoints of (P, U) and (Q, V) we cannot use distance as such, for Fact (F2) is not available in an incomplete manifold. Instead we must use special open sets of the manifold chosen to achieve the same equivalence relation that distance does in a Riemannian manifold.

To define the special open sets of M, we use the "natural topology" (for details see Geroch, 1968b) on the set of all pairs (P, U) given by the topology of M. We consider all geodesic rays (Q, V) near (P, U). If we try to travel along these rays to all affine parameters in some neighborhood of λ_0, we either obtain points in M or we are stopped because one or another of the rays is incomplete. To every open set of geodesics surrounding (P, U) and every open set of numbers containing λ_0 we associate that point set N which actually does lie in M and which is obtained by traveling along the rays as described above. Any open subset O of M containing such a set N is special. Any such special set O is said to be a *thickening of the end point* of (P, U).

Two incomplete geodesic rays are equivalent if *every* thickening of the end point of one is a thickening of the end point of the other (Figure 5.4). An equivalence class of geodesics defines a single point in the set of singularities of S. S may be given a topology, even a metric. If the procedure defining S is carried out in a Riemannian manifold, then S is found to have the same structure as that given by explicit use of the positive definite metric. There is reason to believe, therefore, that the structure assigned to S in a pseudo-Riemannian manifold is reasonable.

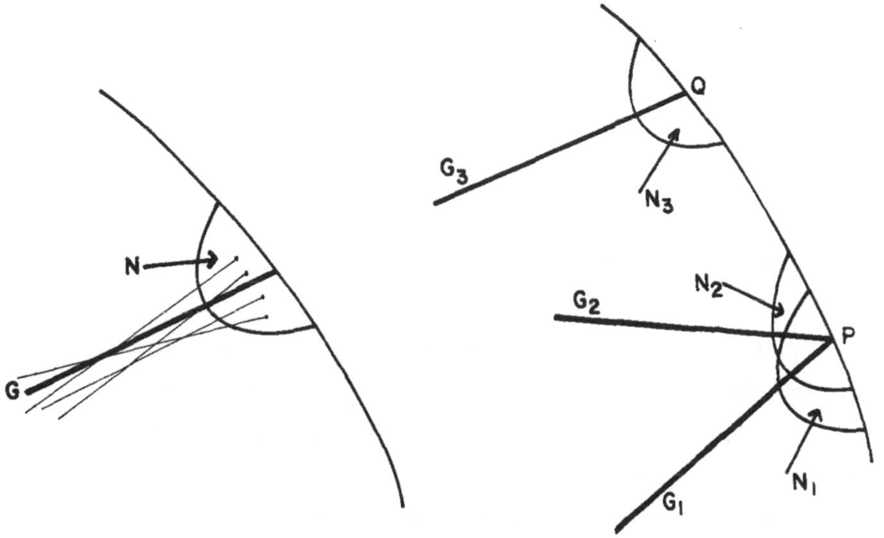

Fig. 5.4. Geroch's G-Boundary. The left figure illustrates a thickening N of the endpoint of the geodesic segment G. N consists of all points in the manifold which are endpoints of geodesics near G. The right figure shows two equivalent geodesics G_1 and G_2 — equivalent because every thickening N_1 of the end of G_1 overlaps every thickening N_2 of G_2. The boundary point P, previously undefined, is realized by the equivalence class of geodesic segments.

Schmidt-Sachs Completion Method

Geroch's G-boundary (the set S) is defined by the use of incomplete, geodesics and ignores non-geodesic paths. Schmidt (1971) and Sachs (see Eardley, Liang, and Sachs, 1972) have defined alternative methods of identifying singular points. The methods associate with a spacetime manifold M a higher dimensional manifold B which is Riemannian. Since B has a positive definite metric its singular points T_B are well-defined. T_B is a set of equivalence classes of Cauchy sequences which do not converge in B. From T_B is then defined a set T of equivalence classes of points in T_B. T is the set of singular points to be associated with M. Investigations of the relationship between T and S are currently underway.

With every timelike, unit contravariant vector field $U(U \cdot U = -1)$ can be associated a symmetric covariant positive-definite tensor field h_U defined by

$$h_U(X,Y) = g(X,Y) + 2g(U,X)G(U,Y). \tag{5.2}$$

However, the tensor field h_U while positive definite, depends on U for its definition and is therefore not canonically defined as a structure on M. The various h_U, however, are used in the Schmidt and Sachs constructions.

A natural metric may be put on a manifold whose dimension is sufficiently high that different points in it correspond to different vectors at the same point in M. The particular manifold of Schmidt has as points quintuples (P, X_0, X_1, X_2, X_3), where P is a point in M and the X_μ are a linearly independent quartet of vectors at P.

Sachs has modified Schmidt's method by taking as his manifold the *unit hyperboloid* in the *tangent bundle* of M (Figure 5.5). A point in this manifold is a pair (P, U), where P is a point of M and U is a unit timelike vector at P. A coordinate patch on this manifold B may be defined for every coordinate patch on M. If $\{x^\mu\}$ are coordinates in M in a neighborhood of P, then a vector V has components v^μ. We first define the 8-dimensional manifold $M \times R^4 \equiv A$ on which a coordinate patch is $\{x^\alpha, v^\mu\}$. The coordinate patch on B is in the hypersurface in A defined by $v^\mu = u^\mu$, where $g_{\alpha\beta} u^\alpha u^\beta = -1$.

A vector field in B is a linear differential operator W. In the coordinate system $\{x^\alpha, u^\mu\}$,

$$W = W_1{}^\alpha \frac{\partial}{\partial x^\alpha} + W_2{}^\mu \frac{\partial}{\partial u^\mu}, \tag{5.3}$$

W will be tangent to B if $dF(W) = 0$, where $F = g_{\mu\nu} u^\mu u^\nu + 1$. This condition implies

$$g_{\alpha\beta,\sigma} u^\alpha u^\beta W_1{}^\sigma + 2g_{\alpha\sigma} u^\alpha W_2{}^\sigma = 0. \tag{5.4}$$

Sachs' metric H on B is defined by means of the dot product of W with another vector V with components $(V_1{}^\alpha, V_2{}^\alpha)$:

$$H(W,V) = h_{\alpha\beta} W_1^{\alpha} V_1^{\beta} + h_{\mu\nu}(W_2^{\mu} + \Gamma^{\mu}_{\sigma\tau} u^{\sigma} W_1^{\tau})(V_2^{\nu} + \Gamma^{\nu}_{\pi\rho} u^{\pi} V_1^{\rho}) \,, \tag{5.5}$$

where $h_{\alpha\beta} = g_{\alpha\beta} + 2g_{\alpha\sigma} g_{\beta\tau} u^{\sigma} u^{\tau}$ (compare equation 5.2). The $\Gamma^{\alpha}_{\beta\nu}$ are the components of the affine connection derived from the metric on M in the coordinate system $\{x^{\mu}\}$. H is positive definite and uniquely determined by g on M. Its components transform like the components of a covariant tensor under changes of coordinates in B.

An important property of H is that if B is complete (according to C1, C2, C3), then M is complete. We shall not prove this statement here, but this theorem implies that completeness of B corresponds to geodesic completeness of M.

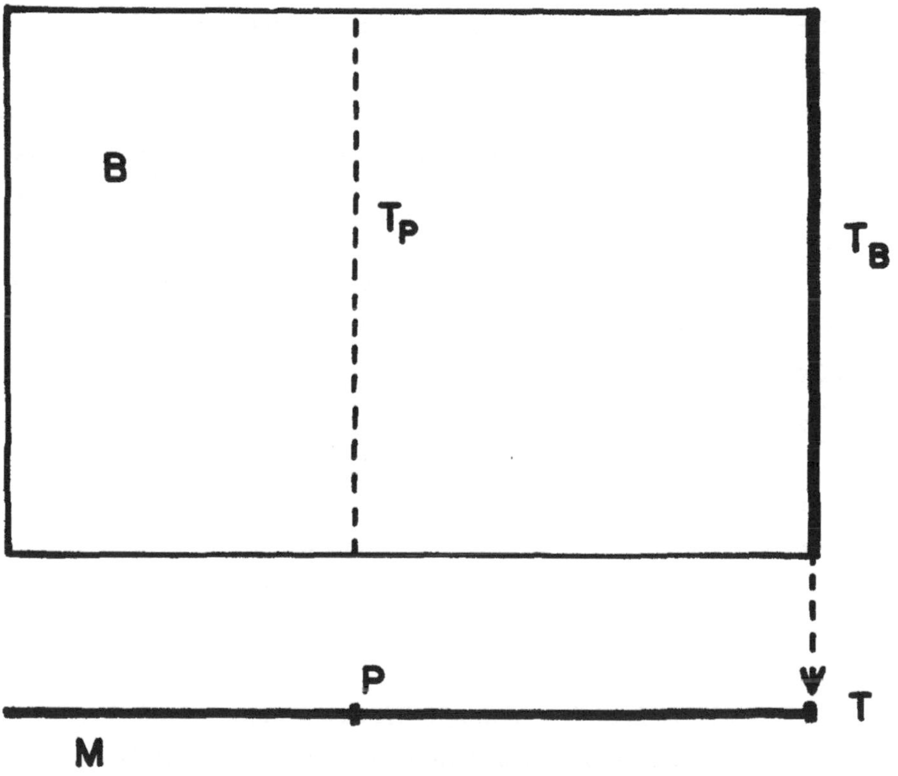

Fig. 5.5. The Unit Hyperboloid Tangent Bundle B of a Manifold M. T_P is the set of all unit-length vectors at a point P in M. B is the set of all unit-length vectors at all points of M. B is given a Riemannian (positive-definite) metric in the Sachs method of completion, and its boundary T_B is defined by a limiting procedure. A second limiting procedure projects T_B onto an abstract set of points, realizing the boundary T of the original manifold.

When B is incomplete, it is straightforward to give a structure to a set to be identified with the singular points of B, as we need not consider geodesics, but only Cauchy sequences in B. Some of these Cauchy sequences will have a limit point in B and some will not. Moreover, two distinct Cauchy sequences may approach the same point in B.

To distinguish endpoints of Cauchy sequences, we set up an equivalence relation among these sequences. If $\alpha = (P, U)$ and $\beta = (Q, V)$ are two points in B, let $D(\alpha, \beta)$ be the distance from α to β: The greatest lower bound of the lengths of all paths joining α to β. Two Cauchy sequences $L_1 = \{\alpha_1, \alpha_2, \cdots\}$ and $L_2 = \{\beta_1, \beta_2, \cdots\}$ are equivalent if for every positive real number ε there is an integer K such that $D(\alpha_i, \beta_i) < \varepsilon$ for every $i > K$. Each equivalence class of Cauchy sequences is taken to be a point in a set $\overline{B}$.

It is easy to show that B is contained in $\overline{B}$. The points in $\overline{B}$ which are not in B are called T_B, the *boundary* of B. Not all distinct points in T_B correspond to distinct points in T, the singularity set of M, however.

To define points in T, we must form equivalence classes in T_B. Two equivalence classes F and G are themselves equivalent if there are Cauchy sequences L_1 in F and L_2 in G of the form $L_1 = \{(P_1, U_1), (P_2\ U_2), \cdots\}$; $L_2 = \{(P_1\ V_1), (P_2, V_2), \cdots\}$. Thus, F and G are equivalent if one representative of F and one representative of G are formed over the same sequence of points P_i in M. If this equivalence relationship is applied to all points in $\overline{B}$, we obtain a set $\overline{M}$, some of whose points are equivalent to points in M. The ones which are not are the set of singular points T.

Other Methods of Associating a Riemannian Space with M

The method of Schmidt depends on the manifold whose points are (P, X_0, X_1, X_2, X_3). We can define a positive definite tensor field h_X on M by defining

$$h_X(U, V) = \sum_{\mu} u^{\mu} v^{\mu}, \quad \text{where } U = u^{\mu} X_{\mu}, \; V = v^{\mu} X_{\mu} .$$

Schmidt uses h_X in his definition of a metric analogous to that of Sachs. Schmidt's larger manifold is 20-dimensional where that of Sachs' is 7-dimensional. Schmidt's Riemannian manifold is the *bundle of frames*, and the group structure associated with linear transformations among basis vectors allows equivalence classes of incomplete paths to be formed an especially clean way.

A 10-dimensional manifold using only orthonormal frames for X_{μ} can also be defined. The equivalence of the methods of Schmidt and Sachs has yet to be demonstrated. It has been pointed out (by D. R. Brill) that we can place a pseudo-Riemannian metric on Sachs' manifold by replacing $h_{\alpha\beta}$ by $g_{\alpha\beta}$ in our definition of H. Sachs' process may then be applied to this manifold (resulting in a 13-dimensional Riemannian manifold). A whole series of pseudo-Riemannian manifolds with a final Riemannian manifold may be built up this way. The Riemannian one is used to define sets of singular points in M. At present it is unknown what the relationship between the sets of "singular" points defined by these various methods is, but we mention them all for completeness.

5.4. Examples of Singularities in Pseudo-Riemannian Manifolds

A) The simplest example is the Euclidean two-space R^2. Endowed with either a positive-definite or a Minkowskian flat metric, it is non-singular and complete. Minkowski two-space with a light-like line removed is an incomplete manifold which is of interest because it is completely homogeneous (has a simply transitive group of isometries; see Chapter 6, Section 7.1, and Wolf, 1967).

B) The next simplest example is the Riemannian Cone C. This cone is a two-space of all points with coordinates r, θ such that the metric is

$$ds^2 = dr^2 + a^2 r^2 d\theta^2 , \tag{5.6}$$

where r is restricted to be > 0, and $\theta = 0$ is identified with $\theta = 2\pi$. When $a = 1$, C is the same as R^2 with the usual metric, except that the point $r = 0$ is excluded (as a metric component is singular there). When $a = 1$, C is incomplete, but it is extendible. If $a \neq 1$, C is singular: incomplete and inextendible.

By rescaling θ we can always make ds^2 the metric of the plane in polar coordinates; thus C is flat. Thus geodesics are straight lines. Consider a straight line aimed at $r = 0$. Since the line must stop at $r=0$, there are geodesics in C with finite length. Hence C is incomplete and therefore not non-singular.

Can C be extended to a non-singular manifold? Consider a small geodesic triangle about $r = 0$ (Figure 5.6). As this triangle is made smaller, the lengths of its sides go to zero; hence if C is to be completed $r = 0$ must correspond to a single point. Because C is flat and because scalar curvature invariants are continuous functions, all these invariants must be zero at $r = 0$. However, the sum of the angles of any geodesic triangle must be π if the triangle has a flat interior (Eisenhart, 1926). That is not the case here: Every geodesic triangle with $r = 0$ inside it somewhere has $\pi(3-2a)$ as the sum of its angles. Thus it is impossible to complete C if $a \neq 1$. Hence C is a space which is both flat and singular.

C) The next example has already been discussed in Chapter 4. It is the closed FRW universe. Along the timelike geodesic which is the path of a dust particle, the scalar curvature R eventually becomes larger and larger. Finally after a finite proper time, R becomes infinite. The largest non-singular part of the solution is the space obtained by eliminating these singular points. It is clearly incomplete, and just as clearly not completable. Hence the Friedmann universe is singular.

D) The fourth example is the Schwarzschild metric (Tolman, 1934a). In coordinates t, r, θ, ϕ the metric is

$$ds^2 = \left(1-\frac{2m}{r}\right)^{-1} dr^2 + r^2(\sin^2\theta d\phi^2 + d\theta^2) - \left(1-\frac{2m}{r}\right) dt^2 . \qquad (5.7)$$

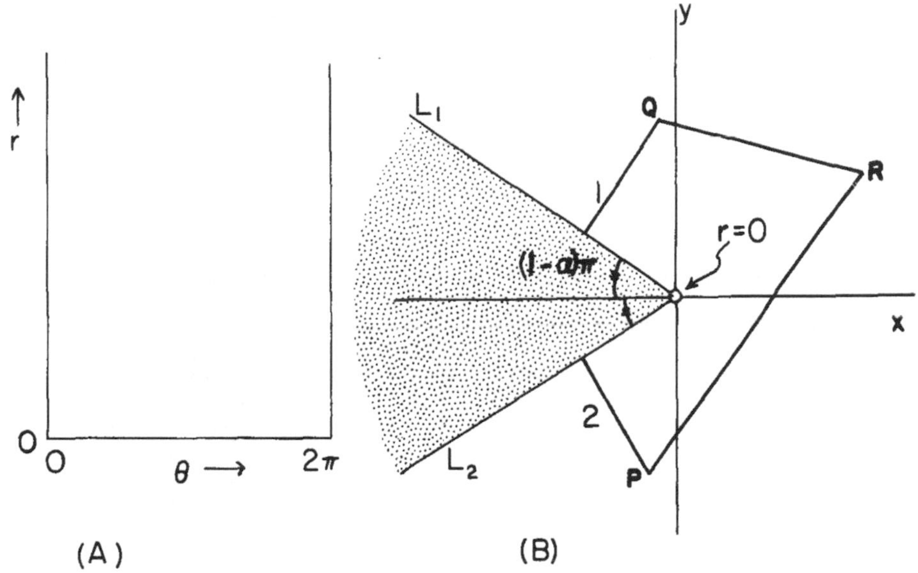

Fig. 5.6. The Cone. A) The cone C in polar coordinates, r, θ. The r coordinate is restricted to positive values: $r \geq 0$. The $\theta = 0$ line is identified pointwise with the $\theta = 2\pi$ line. The metric in these coordinates is $ds^2 = dr^2 + a^2 d\theta^2$ where $a \leq 1$. B) The cone in plane coordinates, x, y. The defining property of these coordinates is that the metric be of the form $ds^2 = dx^2 + dy^2$. Lines L_1 and L_2 are identified pointwise. The $r = 0$ point $(x = y = 0)$ is not part of the cone; neither is the dotted segment. A geodesic triangle PQR is drawn in. Note that lines 1 and 2 are two halves of the same geodesic segment PQ. The cone is singular: The $r = 0$ point may be reached by a geodesic of finite length. However, the $r = 0$ point is *not* part of C, and C cannot be extended to include $r = 0$.

The manifold is topologically the product of S^2, the two-dimensional sphere, and R^2, the two-plane. Equation (5.7) holds only for $r > 2m$! As $r \to 2m$ the metric components become singular. Moreover, the space of t, r, θ, ϕ with $r > 2m$ is incomplete.

This segment of a manifold can be extended. It has been known for a long time that as $r \to 2m$ no scalar invariants become infinite, and geodesics do not converge. Thus at $r = 2m$ there appears neither a conical singularity (as at $r = 0$ in C) nor a collapse singularity (as when $R \to \infty$ in the Friedmann models).

The definitive statement on the $r = 2m$ "singularity" was made independently by Szekeres (1960) and Kruskal (1960) who gave new coordinates (u, v) to extend (t, r) from the original segment of the manifold to the maximal solution which can be formed. This maximal extension, however, is still not non-singular. Curvature invariants become infinite along geodesic segments of finite length, at points corresponding to $r = 0$. Thus the Kruskal-Szekeres extension cannot be further enlarged, and the Schwarzschild solution is singular.

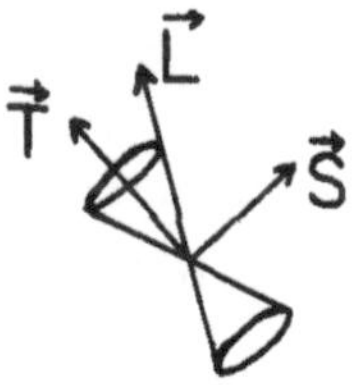

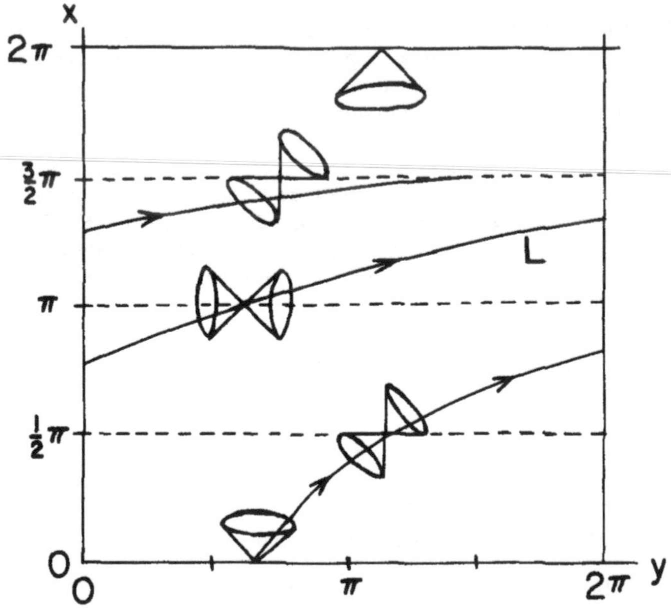

Fig. 5.7. The Misner Torus T^2. The lines $x = 0$ and $x = 2\pi$ are identified, as are the lines $y = 0$ and $y = 2\pi$. The metric we have placed on T^2 is $ds^2 = -\cos x\, dx^2 + 2 \sin x\, dxdy + \cos x\, dy^2$. L is a geodesic of finite total length which cannot be extended to infinite values of its affine path parameter. T^2 is mathematically *non-singular* but *incomplete*. The small light cone illustrated our convention — $\vec{T}$ is a timelike direction while $\vec{L}$ is lightlike, $\vec{S}$ spacelike.

E) The last example is one given by Misner (1963). It is an example of a pseudo-Riemannian manifold which is compact and therefore mathematically non-singular in the sense of (C3), but which is incomplete, and therefore physically singular in violation of (C2). It is a torus T^2 with coordinates (x, y); the lines $x = 0$ and $x = 2\pi$ are point wise identified as are the lines $y = 0$, $y = 2\pi$ (Figure 5.7). The metric is given by:

$$ds^2 = -\cos x\, dx^2 + 2 \sin x\, dx dy + \cos x\, dy^2 \,. \tag{5.8}$$

The metric has determinant $= -1$ everywhere, so is non-singular on T^2. However, T^2 with this metric is incomplete.

The geodesic equations for T^2 in terms of an affine parameter τ are $(\cdot \equiv d/d\tau)$:

$$\begin{aligned} &\ddot{x} + \tfrac{1}{2}\dot{x}^2 \sin x \cos x - \dot{x}\dot{y} \sin^2 x - \tfrac{1}{2}\dot{y}^2 \sin x \cos x = 0 \,, \\ &\ddot{y} + \dot{x}^2\left(1 - \tfrac{1}{2}\sin^2 x\right) - \dot{x}\dot{y} \sin x \cos x + \tfrac{1}{2}\dot{y}^2 \sin^2 x = 0 \,. \end{aligned} \tag{5.9}$$

Since τ is an affine parameter one first integral of (5.9) is

$$-\dot{x}^2 \cos x + 2\,\dot{x}\dot{y} \sin x + \dot{y}^2 \cos x = E = \pm 1, 0 \,. \tag{5.10}$$

Another first integral is

$$(\dot{x})^2 + E \cos x = P \,, \tag{5.11}$$

where P is a constant. Note that (5.10) and (5.11) imply $P = (\dot{x} \sin x + \dot{y} \cos x)^2$ so P is positive. Equations (5.10) and (5.11) are used in Figure 5.7 to give the qualitative behavior of the geodesics. The light cones at different points in T^2 are also shown in Figure 5.7. It is clear from the behavior of the geodesics and from the light cones that the only geodesics infinitely extendible in both directions are the lines $y = 0$ and $y = \pi$. And so T^2, although compact, is incomplete (see Miller and Kruskal, 1973).

6. ISOMETRIES OF SPACE AND SPACETIME

To seek the beauteous eye of heaven to garnish
Is wasteful and ridiculous excess
— WILLIAM SHAKESPEARE

6.1. The Lie Derivative

The field equations of general relativity are a complicated set of coupled, non-linear partial differential equations. In cosmology we simplify these equations by imposing symmetries on the solution. Moreover, a

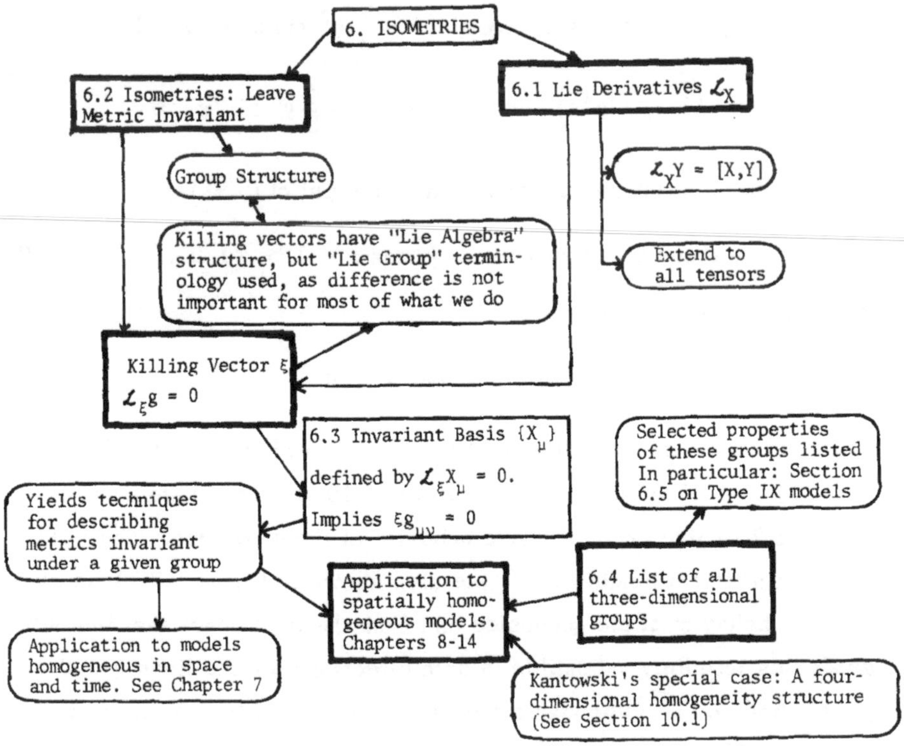

Fig. 6.1. Flow Chart of Chapter 6.

symmetrical, or *homogeneous*, model thus obtained will not merely be symmetrical in appearance (which might imply that a preferred coordinate frame is necessary). Rather the symmetry of the model will be expressed in a manner that is free of the encumbrance of special coordinates by the use of differential forms and vector fields. Figure 6.1 is an outline of this chapter.

A homogeneous cosmological model is a manifold M. The metric of M is invariant under a certain (specified) group of transformations. That is, each operation of the symmetry group corresponds to a map of M onto itself. This map carries a point P into another point Q at which the metric is the same when expressed in a coordinate-free manner.

Infinitesimal Transformations

The description of the invariance of a metric under a group (*Lie group*) of isometries is achieved by directing attention to the infinitesimal transformations (*Lie algebra*) in the group. Other members of the group can be obtained from the infinitesimal members by *exponentiation* (repeated application of the infinitesimal members, Helgason, 1962). Thus, a symmetric cosmological model is found by imposing the structure of a Lie *algebra*, although Lie *group* terminology is used.

To describe an infinitesimal transformation it is convenient temporarily to use a coordinate system (see Yano and Bochner, 1953; Misner, 1964). Consider a point P_0 in a neighborhood N in which coordinates $x^\mu (\mu = 1, \cdots, n)$ are used. A point P in N will have coordinates $x_P{}^\mu$. An *infinitesimal transformation* is of small effect and therefore carries points in N′, a small neighborhood of P_0 which lies within N, into other points of N. Our transformation may be described in N′ by n functions f^μ of the coordinates x^μ. The point P is carried to the point Q in N with the coordinates $\bar{x}_Q{}^\mu$:

$$f^\mu(x_P{}^\nu) = f^\mu(P) = \bar{x}_Q{}^\mu \; . \tag{6.1}$$

An infinitesimal transformation has the form

$$f^{\mu}(P) = x_P{}^{\mu} + \varepsilon a^{\mu}(P) . \tag{6.2}$$

The number ε is meant to be so small that points in N' are carried only to points in N. The vector field $X = a^{\mu}\partial_{\mu}$ describes the magnitude and the direction of the transformation.

A transformation acting on a space induces a transformation which carries a tensor at a point P into a tensor at the image point Q. This transformation of tensors defines a new tensor T_{new} whose value at Q is the "same" as the value of T at P. To find the description of T_{new}, consider that the map (6.1), which bodily carries the point P to the point Q, also applies as a coordinate change at P. (See Figure 6.2.) In that case the change in the tensor components of T is:

$$\bar{T}^{\alpha\beta\cdot\cdot}{}_{\gamma\delta\cdot\cdot} = J^{\alpha}{}_{\mu} J^{\beta}{}_{\nu}\cdots K^{\sigma}{}_{\gamma} K^{\tau}{}_{\delta}\cdots T^{\mu\nu\cdot\cdot}{}_{\sigma\tau\cdot\cdot} ,$$

where $J^{\alpha}{}_{\mu}$ and $K^{\sigma}{}_{\gamma}$ are the Jacobian matrix and inverse Jacobian matrix of the transformation. $\bar{T}^{\alpha\beta\cdot\cdot}{}_{\gamma\delta\cdot\cdot}$ is now identified as a component of T_{new} at the point Q:

$$T_{\text{new}}{}^{\alpha\beta\cdot\cdot}{}_{\gamma\delta\cdot\cdot}(Q) = T^{\mu\nu\cdot\cdot}{}_{\sigma\tau\cdot\cdot}(P) J^{\alpha}{}_{\mu} J^{\beta}{}_{\nu}\cdots K^{\sigma}{}_{\gamma} K^{\tau}{}_{\delta}\cdots \tag{6.3}$$

where

$$J^{\alpha}{}_{\mu} = \frac{\partial f^{\alpha}}{\partial x^{\mu}} ; \quad K^{\sigma}{}_{\gamma} = \frac{\partial x^{\sigma}}{\partial f^{\gamma}} . \tag{6.4}$$

For the transformation (6.2), the Jacobian and its inverse are:

$$J^{\alpha}{}_{\beta} = \delta^{\alpha}{}_{\beta} + \varepsilon a^{\alpha}{}_{,\beta} ; \quad K^{\mu}{}_{\nu} = \delta^{\mu}{}_{\nu} - \varepsilon a^{\mu}{}_{,\nu} + O(\varepsilon^2) . \tag{6.5}$$

Hence a vector $Y = b^{\mu}\partial_{\mu}$ will change by the formula

$$Y_{\text{new}} = b^{\mu}_{\text{new}}(Q)\partial_{\mu} = [b^{\mu}(P) + \varepsilon a^{\mu}{}_{,\nu} b^{\nu}(P)]\,\partial_{\mu} . \tag{6.6}$$

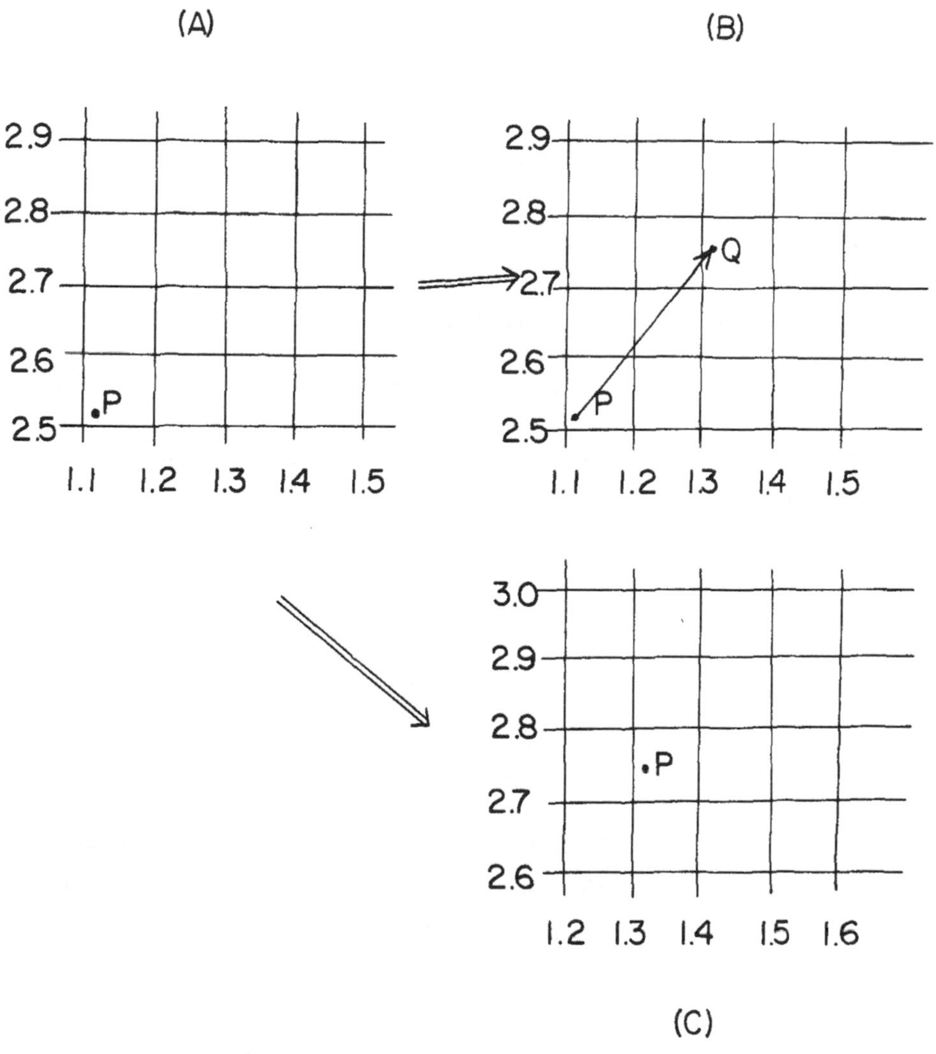

Fig. 6.2. Active and Passive Transformations. The same formulae describe the active transformation carrying P to Q and the passive transformation of renaming P by a change of coordinates. Q is the new point, with coordinates given by $x^i_Q = f^i(x^j_P)$. The "passive" transformation changes the coordinates of P to $x^i_{P,new} = f^i(x^j_{P,original})$.

The Lie Derivative

The value Y_{new} is what one would expect to see at Q if Y did *not* change in the direction given by X. $Y - Y_{new}$ is the observable change in the vector Y. The measure of this change is:

$$(Y-Y_{new})^{\mu} = b^{\mu}(Q) - b^{\mu}{}_{new}(Q) = b^{\mu}(Q) - b^{\mu}(P) - \varepsilon a^{\mu}{}_{,\nu}\, b^{\nu}(P)$$
$$= b^{\mu}(x^{a} + \varepsilon a^{a}) - b^{\mu}(x^{a}) - \varepsilon a^{\mu}{}_{,\nu}\, b^{\nu}(P)\ .$$

Dividing by ε and letting $\varepsilon \to 0$, we have the *Lie derivative* of Y with respect to X:

$$\mathcal{L}_X Y = (b^{\mu}{}_{,\sigma} a^{\sigma} - a^{\mu}{}_{,\sigma} b^{\sigma})\partial_{\mu}\ . \tag{6.7}$$

This expression is simply the commutator of X and Y, as defined in Chapter 2:

$$\mathcal{L}_X Y = [X, Y]\ . \tag{6.8}$$

We extend the Lie derivative to arbitrary tensors by requiring that $\mathcal{L}_X$ act as a differentiation with respect to the tensor product, and that $\mathcal{L}_X f =$ Xf, where f is a function. In the coordinated system used above, if T has components $T^{\alpha\beta}{}_{\gamma\delta}$, $\mathcal{L}_X T$ is given by:

$$(\mathcal{L}_X T)^{\alpha\beta}{}_{\gamma\delta} = T^{\alpha\beta}{}_{\gamma\delta,\sigma} a^{\sigma} - T^{\sigma\beta}{}_{\gamma\delta}\, a^{\alpha}{}_{,\sigma} - T^{\alpha\sigma}{}_{\gamma\delta}\, a^{\beta}{}_{,\sigma}$$
$$+ T^{\alpha\beta}{}_{\sigma\delta}\, a^{\sigma}{}_{,\gamma} + T^{\alpha\beta}{}_{\gamma\sigma}\, a^{\sigma}{}_{,\delta}\ . \tag{6.9}$$

All of the commas (partial derivatives) may be replaced by semi-colons (covariant derivatives) without affecting the truth of this relation. Thus the Lie derivative is independent of both metric and connection.

6.2. Killing Vectors of a Group of Isometries

A transformation which leaves the metric g invariant is called an *isometry*. An infinitesimal isometry is described by a vector ξ called a *Killing vector* (Killing, 1892), which is said to *generate isometries*. A Killing vector thus satisfies

$$\mathcal{L}_{\xi}\, g = 0\ . \tag{6.10}$$

In order to apply (6.10) to the components of g in a general basis, we make use of the fact that $\mathcal{L}_{\xi}$ operating on the function g(X, Y) can be written in terms of $\mathcal{L}_\xi g$, $\mathcal{L}_\xi X$, $\mathcal{L}_\xi Y$, where X and Y are arbitrary vector fields. Since $\mathcal{L}_\xi g = 0$, the derivative property of $\mathcal{L}_\xi$ implies

$$\begin{aligned} \xi[g(X,Y)] &= \mathcal{L}_\xi[g(X,Y)] = (\mathcal{L}_\xi g)(X,Y) + g(\mathcal{L}_\xi X,Y) + g(X,\mathcal{L}_\xi Y) \\ &= g([\xi,X],Y) + g(X,[\xi,Y]) \; . \end{aligned} \tag{6.11}$$

Suppose X_μ is a member of a general vector field basis which is invariant under $\mathcal{L}_\xi$; $\mathcal{L}_\xi X_\mu = 0$. In this basis:

$$\xi g_{\mu\nu} = \mathcal{L}_\xi[g(X_\mu, X_\nu)] = 0 \; . \tag{6.12}$$

In other words, the derivatives of the functions $g_{\mu\nu}$ in the direction ξ are zero.

Equation (6.10) also leads to the *Killing equation* for the components a_μ of the contravariant form of ξ in an arbitrary basis:

$$a_{\mu;\nu} + a_{\nu;\mu} = 0 \; . \tag{6.13}$$

We will not give the detailed calculations leading to (6.13) (see Yano and Bochner, 1953).

It is important to notice that if ξ_1 and ξ_2 are two Killing vectors, then the linear combination $a_1\xi_1 + a_2\xi_2$ is a Killing vector if a_1, a_2 are two constants. However, if a_1, a_2 are functions of position, $a_1\xi_1 + a_2\xi_2$ is a vector field, but not necessarily a Killing vector. $[\xi_1, \xi_2]$ is also a Killing vector.

Description of the Symmetry Group of a Manifold

We now turn to a manifold M whose metric is invariant under several isometries. The set of isometries of M has the structure of a group: An associative product is defined (the product of isometries A and B is A followed by B), an inverse exists for each element, and a unit transformation (the identity) exists. The group of isometries is the *symmetry group* of M.

Isometries are obtained from the Killing vectors by exponentiation in the same way that group elements are obtained from the infinitesimal generators which form the *Lie algebra* of the group. In the abstract group the commutators of the infinitesimal group elements define the structure constants of the group. Thus if η_i, $i = 1, \cdots, m$, are the basis elements of the Lie algebra, then the *structure constants* of G are defined by (Jacobson, 1962):

$$[\eta_i, \eta_j] = C^s{}_{ij}\eta_s \, . \tag{6.14}$$

The Jacobi equation (2.4) applied to η_i shows that the $C^s{}_{ij}$ must satisfy

$$C^s{}_{ij}C^a{}_{sk} + C^s{}_{jk}C^a{}_{si} + C^s{}_{ki}C^a{}_{sj} = 0 \, . \tag{6.15}$$

Further, any set of $C^s{}_{ij}$ (antisymmetric in i, j) which satisfy (6.15) are the structure constants for some Lie group.

The group of isometries of the manifold M is isomorphic to some abstract group G. The Lie algebra commutator of (6.14) is replaced by the commutator [,] of the Killing vectors ξ_i of the symmetry group. The m independent Killing vectors ξ_i (m may or may not be equal to n, the dimension of M) obey

$$[\xi_i, \xi_j] = C^s{}_{ij}\xi_s \, . \tag{6.16}$$

The structure coefficients in (6.16) are constants (functions independent of position) and are equal to the structure constants of the Lie algebra of the abstract group of isometries.

Suppose we are given a Lie group G, with structure constants $C^s{}_{ij}$. A manifold M is said to be invariant under the group G if there are m (the dimension of G) Killing vector fields ξ_i which obey the Lie algebra relation (6.16).

A group G of dimension m is called *simply transitive on subspaces* if the ξ_i's are linearly independent as vector fields ($\sum_i a_i\xi_i = 0 \Longrightarrow a_i = 0$, the a_i being functions). The *orbit* of a given point P_0 is the set of all points Q such that $A(P_0) = Q$ for some A in the isometry group G. The orbit is a subset H of M. These subspaces fill M, and two different subspaces have no points in common. H is called a *homogeneous* or *invariant* subspace. If G is simply transitive, the dimension of H is m, the dimension of G. If the dimension of H is less than m, G is called *multiply transitive* on H.

6.3. Generation of an Invariant Basis

The description of a manifold M with symmetry group G is simplest when an invariant basis is used. The members of such a basis are vector fields X_μ, each one of which is invariant under the group G. Therefore X_μ has zero Lie derivative with respect to any of the Killing vectors. If ξ_i is a member of a basis for the Killing vectors, we have

$$[\xi_i, X_\mu] = 0 \ (i = 1, \cdots, m = \dim G; \mu = 1, \cdots, n = \dim M) . \qquad (6.17)$$

An invariant basis is useful because: 1) Equation (6.12) shows that each metric component $g_{\mu\nu} = g(X_\mu, X_\nu)$ is group invariant. Thus $g_{\mu\nu}$ is constant on each homogeneous subspace generated by the group. 2) It can easily be shown that the structure coefficients of the X_μ are constant on each homogeneous hypersurface. The structure coefficients $D^\sigma{}_{\mu\nu}$ are defined by $[X_\mu, X_\nu] = D^\sigma{}_{\mu\nu} X_\sigma$.

Not every abstract group can be used as a symmetry group for an n-dimensional manifold. However, if G is the symmetry group of M, we must say when an invariant basis can be found. We must also exhibit the relation between the $D^\sigma{}_{\mu\nu}$ and the structure constants $C^i{}_{jk}$ of the group.

In this book we are primarily interested in three cases:

1) A manifold M of dimension 4 with the group G simply transitive on all of M. The dimension of G is therefore 4, also. M is called *homogeneous* or, for emphasis, *homogeneous in space and time* (ST-homogeneous).
2) The manifold M has dimension 4 but G has dimension 3 and is simply transitive. Thus G generates three-dimensional

invariant hypersurfaces H. M is called *spatially homogeneous*, or sometimes simply *homogeneous*. Some of the H's may not be spacelike, but they form a manifold-filling, one-parameter family. Therefore, the metric depends on only one variable, being independent of position on each H.

3) The third case is that of a manifold M on which G is multiply transitive, but traces out three-dimensional invariant hypersurfaces (and thus has dimension greater than 3). M is called *spatially-homogeneous* in this case, also (special techniques must be used; see Sections 6.4 and 11.4).

Existence of an Invariant Basis in a Homogeneous Manifold

In cases (1) and (2) an invariant basis $\{X_\mu\}$ may be found. Consider a homogeneous manifold M of dimension n, invariant under a (simply transitive) group of the same dimension. The Killing vectors $\xi_\mu (\mu = 1, \cdots, n)$ form a vector field basis of M. To construct an invariant basis $\{X_\mu\}$ we need only give the components of X_μ with respect to the ξ_μ. (The components will not, in general, be constants.)

An invariant basis $\{X_\mu\}$ is constructed from n independent vectors $X_{\mu 0}$ at a fixed point P_0 (vectors at P_0, not vector fields). We define vector fields on M by translating the $X_{\mu 0}$ using the Lie derivative, in other words, requiring that

$$[\xi_\mu, X_\nu] = 0 \quad \text{and} \quad X_\mu(P_0) = X_{\mu 0} \, .$$

The requirements are a set of first order differential equations. The integrability conditions for these equations are automatically satisfied if the $C^\mu{}_{\sigma\tau}$ are the structure constants of a group.

To determine the $D^\sigma{}_{\mu\nu}$ we must pick explicit values for the $X_{\mu 0}$. An especially natural choice is $X_{\mu 0} = \xi_\mu(P_0)$. The invariant vector fields X_μ are

$$X_\mu = a^\sigma{}_\mu \xi_\sigma , \quad a^\sigma{}_\mu(P_0) = \delta^\sigma{}_\mu \, .$$

The condition that X_μ be invariant implies that at P_0,

$$(\xi_\mu a^\sigma{}_\nu)(P_0) = -C^\sigma{}_{\mu\nu} .$$

We now use the equation

$$\begin{aligned}[X_\mu, X_\nu] &= [a^\sigma{}_\mu \xi_\sigma, a^\tau{}_\nu \xi_\tau] = D^\sigma{}_{\mu\nu} X_\sigma \\ &= a^\sigma{}_\mu(\xi_\sigma a^\tau{}_\nu)\xi_\tau + a^\sigma{}_\mu a^\tau{}_\nu [\xi_\sigma, \xi_\tau] - a^\tau{}_\nu(\xi_\tau a^\sigma{}_\mu)\xi_\sigma ,\end{aligned}$$

to find the values of the $D^\sigma{}_{\mu\nu}$ at P_0:

$$D^\sigma{}_{\mu\nu}(P_0) = -C^\sigma{}_{\mu\nu} .$$

Since $D^\sigma{}_{\mu\nu}$ is independent of position of M, we therefore have

$$[X_\mu, X_\nu] = -C^\sigma{}_{\mu\nu} X_\sigma . \tag{6.18}$$

We may now take duals ω^μ of the X_μ. The curl relations of the ω^μ are

$$d\omega^\mu = +\frac{1}{2} C^\mu{}_{\sigma\tau} \omega^\sigma \wedge \omega^\tau . \tag{6.19}$$

The $C^\mu{}_{\sigma\tau}$ are constants identical to the structure constants of the invariance group. Moreover, because the X_μ are invariant vectors, the metric is expressed by $ds^2 = g_{\mu\nu}\omega^\mu\omega^\nu$, the $g_{\mu\nu}$ being constants. We will use this basis later when discussing four-dimensional homogeneous manifolds which satisfy Einstein's equations for various stress-energy tensors.

An Invariant Basis in a Spatially Homogeneous Manifold

The second case we will discuss in this section is that of a four-dimensional manifold M invariant under a three-dimensional simply transitive Lie group. This group, whose structure constants are denoted $C^i{}_{jk}$ $(i, j, k = 1, 2, 3)$, generates three-dimensional homogeneous hypersurfaces. A one-parameter family of these hypersurfaces fills M, which is then considered the topological product $H \times R$, where H is a copy of the invariant three-dimensional subspaces, and R is a real line. For a given parameter $t \in R$, we will denote by H(t) the homogeneous hypersurface at t. The metric of M is independent of position in any invariant hypersurface H(t), but may depend on the value of the parameter t.

To generate an invariant basis (invariant under the three-dimensional group and consisting of four vector fields on M) choose one curve in M corresponding to the real line R in the topological product. The tangent to that curve is translated throughout each three-dimensional subspace H(t) by means of Lie differentiation. Call this vector field X_0. Three other vector fields X_1, X_2, X_3, tangent to H(t) itself, are chosen at the curve. They are translated to yield the three remaining vector fields needed for a basis.

By definition, we have

$$[X_0, X_i] = 0. \quad i = 1, 2, 3 . \tag{6.20}$$

Note that the three X_i have components independent of t, whereas X_0 is simply $\partial/\partial t$. Moreover, the X_1, X_2, X_3 may be chosen so that

$$[X_i, X_j] = -C^s{}_{ij} X_s \tag{6.21}$$

by using the procedures outlined in our discussion of homogeneous manifolds above.

Some Metrical Properties of the Invariant Basis

We can modify the invariant basis $\{X_\mu\}$ by finding a second invariant basis $\{Y_\mu\}$ with specified metrical properties. This additional specification is allowed by the freedom to line up the various homogeneous hypersurfaces H(t), in other words, the freedom of choice of t-axis.

The Y_μ are to be linear combinations of the X_μ:

$$Y_\mu = b^\sigma{}_\mu(t) X_\sigma .$$

Each $b^\sigma{}_\mu$ is a function of t alone because Y_μ must be invariant under G.

We wish to retain the properties that Y_i is tangent to $H(t)$ and commutes with Y_0 ("Y_i unaffected by translation in the direction Y_0"):

$$\begin{aligned}[Y_0, Y_i] &= [b^\sigma{}_0 X_\sigma, b^s{}_i X_s] = 0 \\ &= (b^0{}_0 \dot{b}^s{}_i - b^u{}_0 b^v{}_i C^s{}_{uv}) X_s \quad (\cdot = d/dt)\,.\end{aligned} \tag{6.22}$$

This first-order, ordinary differential equation has a solution $b^i{}_j(t)$ given any set of four functions $b^\mu{}_0$ (with $b^0{}_0 \neq 0$). Thus we can specify both the direction and the magnitude of Y_0.

Once Y_0 is chosen, the Y_i are found by solving (6.22). However, the Y_i must also satisfy

$$[Y_i, Y_j] = -C^s{}_{ij} Y_s\,. \tag{6.23}$$

This requirement is consistent with (6.22). The proof of consistency is that the t-derivative of (6.23) vanishes by virtue of (6.22) and the Jacobi identity applied to the X_i.

6.4. Allowed Isometry Groups; List of Three-Dimensional Groups

Not every group may serve as the isometry group of a four-dimensional manifold with metric signature $(-+++)$. The groups which are allowed were classified by Petrov (1969). We will first make a few general remarks and then will list all of the three dimensional groups (each of which may be used as the isometry group of a spatially-homogeneous cosmological model).

We will ignore such topological questions as whether a given four-dimensional manifold can be given a metric with signature $(-+++)$ at all. Moreover, the global structure of the group will not be treated and we will deal mainly with a Lie algebra structure, that is, with the Killing vectors.

Largest Groups – Isotropy

The *isotropy group* I_P of a point P is the set of all isometries which leave P fixed. It is a subgroup of the symmetry group G. If G is transitive (simply or multiply), all I_p are isomorphic. The isotropy group of a point, I_p, must be a subgroup of the (homogeneous) Lorentz group L. The proof is straightforward if at P coordinates are chosen such that $g_{\mu\nu}(P) = \eta_{\mu\nu} = \mathrm{diag}\,(-1,1,1,1)$. The isotropy group I_p then leaves $\eta_{\mu\nu}$ invariant and thus must be a subgroup of L. G is therefore restricted in that no isotropy group is allowed which is not a subgroup of L.

The dimension of I_p must be less than or equal to $6 = \dim L$. The dimension m of G itself for an n-dimensional manifold must be $m \leq n(n+1)/2$ (Eisenhart, 1926). If $n = 4$, this maximum is 10; if $n = 3$, this maximum is 6; and if $n = 2$, this maximum is 3.

Any 4-manifold with a 10 parameter group of isometries is a space of constant curvature, so that $R_{\mu\nu} = \lambda g_{\mu\nu}$, with $\lambda = \text{const.}$ If the right side of this equation is interpreted as a stress-energy tensor, T_{ij}, then T_{ij} corresponds to a fluid-filled universe with the pressure p equal to the negative of the energy density w. Thus a space of constant curvature is physically unrealistic unless λ vanishes, in which case the space is flat.

In a 4-manifold, for $10 \geq m > 6$, G must act transitively. These groups have been classified by Petrov (1969) and mostly do not interest us. It is only when G has a simply transitive subgroup that the methods of this section apply. An example of a space with a 10-dimensional group but no 4-dimensional simply transitive subgroup is the DeSitter universe (Calabi and Markus, 1962).

It is interesting to note that any four-dimensional group can serve as the isometry group for at least one manifold: itself. None of these manifolds are physically realistic, for none can exhibit the observed expansion of the universe (details in Chapter 7).

Spatially Homogeneous Models

If $m \leq 6$, the group may act transitively or else act on lower-dimensional subsets. If the group acts transitively on three-dimensional subsets, we call the manifold *spatially homogeneous*. An FRW model has a six-dimensional group of isometries containing a) a three-dimensional isotropy subgroup and b) a three-dimensional, simply transitive subgroup which acts on spacelike hypersurfaces.

The general spatially homogeneous cosmological model falls into one of the following two categories: (1) Those spaces in which G has a three-dimensional subgroup which acts simply transitively. (2) Spaces in which the homogeneous spacelike hypersurfaces H have a transitive group of isometries but not a simply transitive group.

The second category has been studies by Kantowski (1966). He found that category (2) consists only of spaces with an isometry group of order $m = 4$. All Lie algebras of order 4 have subalgebras of order 3 (see Kantowski, 1966). It is only when this subalgebra generates two-surfaces S of constant positive curvature (two-spheres) does a model in class (2) arise rather than a model from class (1). Because the spacelike hypersurfaces H do not have a simply transitive group of isometries the method outlined for obtaining an invariant basis is not directly applicable.

List of Three-Dimensional Groups

To conclude this section we list all of the three-dimensional Lie algebras. Each algebra uniquely determines the local properties of a three-dimensional group. Therefore the list is a compendium of all of the three-dimensional groups except for global topological considerations which do not concern us.

The list is given in Table 6.1. We use the classification and the notation given by Taub (1951). One coordinated representation is listed for each differential form. The numbering system is due to Bianchi (1897); for example, the first group is called "Bianchi Type I." If a space is spatially homogeneous and has the group of Bianchi Type N ($N = I, \cdots, IX$) as a simply transitive isometry group then the model will be said to be *Type* N*-homogeneous* or *Type* N. Thus the closed FRW universe is Type IX-homogeneous.

Table 6.1. List of Three-Dimensional Groups. This list is taken from Taub (1951) and is basically a list of the three-dimensional Lie algebras in canonical form — that is, global properties of the groups are not listed. The ξ_i are Killing vectors; $\{X_i\}$ is an invariant basis, so that $[\xi_i, X_j] = 0$. The basis dual to $\{X_i\}$ is $\{\omega^i\}$. The structure constants are defined by $[\xi_i,\xi_j]=C^s{}_{ij}\xi_s$, and we also have $[X_i, X_j] = -C^s{}_{ij}X_s$ and $d\omega^i = \frac{1}{2}C^i{}_{st}\omega^s \wedge \omega^t$. The coordinate system $\{x^i\}$ is used to express these vectors with the coordinated basis $\{\partial_i\}$ (where $\partial_i \equiv \partial/\partial x^i$) and dual basis $\{dx^i\}$ — of course, other coordinated bases are often used, too.

Bianchi Types VI and VII are each a family of groups parametrized by h within the limits listed.

Type I: $C^i{}_{jk} = 0.$

$\xi_1 = \partial_1.$ $\quad X_1 = \partial_1.$ $\quad \omega^1 = dx^1.$ $\quad d\omega^1 = 0$

$\xi_2 = \partial_2$ $\quad X_2 = \partial_2$ $\quad \omega^2 = dx^2$ $\quad d\omega^2 = 0$

$\xi_3 = \partial_3$ $\quad X_3 = \partial_3$ $\quad \omega^3 = dx^3$ $\quad d\omega^3 = 0$

Type II: $C^1{}_{23} = -C^1{}_{32} = 1.$

rest of $C^i{}_{jk} = 0$

$\xi_1 = \partial_2$

$\xi_2 = \partial_3$

$\xi_3 = \partial_1 + x^3\partial_2$

$X_1 = \partial_2$ $\quad \omega^1 = dx^2 - x^1 dx^3.$ $\quad d\omega^1 = \omega^2 \wedge \omega^3$

$X_2 = x^1\partial_2 + \partial_3$ $\quad \omega^2 = dx^3$ $\quad d\omega^2 = 0$

$X_3 = \partial_1$ $\quad \omega^3 = dx^1$ $\quad d\omega^3 = 0$

Type III: $C^1{}_{13} = -C^1{}_{31} = 1.$

rest of $C^i{}_{jk} = 0$

$\xi_1 = \partial_2$

$\xi_2 = \partial_3$

$\xi_3 = \partial_1 + x^2\partial_2$

$X_1 = e^{x^1}\partial_2.$ $\quad \omega^1 = e^{-x^1}dx^2.$ $\quad d\omega^1 = \omega^1 \wedge \omega^2$

$X_2 = \partial_3$ $\quad \omega^2 = dx^3$ $\quad d\omega^2 = 0$

$X_3 = \partial_1$ $\quad \omega^3 = dx^1$ $\quad d\omega^3 = 0$

Table 6.1. List of Three-Dimensional Groups (Continued)

Type IV: $C^1{}_{13} = -C^1{}_{31} = 1.$ $\qquad \xi_1 = \partial_2$.

$C^1{}_{23} = -C^1{}_{32} = 1$ $\qquad \xi_2 = \partial_3$

$C^2{}_{23} = -C^2{}_{32} = 1$ $\qquad \xi_3 = \partial_1 + (x^2 + x^3)\partial_2 + x^3\partial_3$

rest of $C^i{}_{jk} = 0$

$X_1 = e^{x^1}\partial_2$. $\quad \omega^1 = e^{-x^1}dx^2 - x^1e^{-x^1}dx^3.$ $\quad d\omega^1 = \omega^1 \wedge \omega^3 + \omega^2 \wedge \omega^3$

$X_2 = x^1e^{x^1}\partial_2 + e^{x^1}\partial_3$ $\quad \omega^2 = e^{-x^1}dx^3$ $\quad d\omega^2 = \omega^2 \wedge \omega^3$

$X_3 = \partial_1$ $\quad \omega^3 = dx^1$ $\quad d\omega^3 = 0$

Type V: $C^1{}_{13} = -C^1{}_{31} = 1.$ $\qquad \xi_1 = \partial_2$.

$C^2{}_{23} = -C^2{}_{32} = 1$ $\qquad \xi_2 = \partial_3$

rest of $C^i{}_{jk} = 0$ $\qquad \xi_3 = \partial_1 + x^2\partial_2 + x^3\partial_3$

$X_1 = e^{x^1}\partial_2.$ $\quad \omega^1 = e^{-x^1}dx^2.$ $\quad d\omega^1 = \omega^1 \wedge \omega^3$

$X_2 = e^{x^1}\partial_3$ $\quad \omega^2 = e^{-x^1}dx^3$ $\quad d\omega^2 = \omega^2 \wedge \omega^3$

$X_3 = \partial_1$ $\quad \omega^3 = dx^1$ $\quad d\omega^3 = 0$

Type VI: $C^1{}_{13} = -C^1{}_{31} = 1.$ $\qquad \xi_1 = \partial_2$.

$C^2{}_{23} = -C^2{}_{32} = h$ $\qquad \xi_2 = \partial_3$

$(h \neq 0, 1)$ $\qquad \xi_3 = \partial_1 + x^2\partial_2 + hx^3\partial_3$

rest of $C^i{}_{jk} = 0$

$X_1 = e^{x^1}\partial_2$. $\quad \omega^1 = e^{-x^1}dx^2$. $\quad d\omega^1 = \omega^1 \wedge \omega^3$

$X_2 = e^{hx^1}\partial_3$ $\quad \omega^2 = e^{-hx^1}dx^3$ $\quad d\omega^2 = h\omega^2 \wedge \omega^3$

$X_3 = \partial_1$ $\quad \omega^3 = dx^1$ $\quad d\omega^3 = 0$

Table 6.1. List of Three-Dimensional Groups
(Continued)

Type VII:

$$C^2_{13} = -C^2_{31} = 1 . \qquad \xi_1 = \partial_2$$
$$C^1_{23} = -C^1_{32} = -1 \qquad \xi_2 = \partial_3$$
$$C^2_{23} = -C^2_{32} = h \qquad \xi_3 = \partial_1 - x^3\partial_2 + (x^2 + hx^3)\partial_3$$
$$(h^2 < 4)$$

rest of $C^i_{jk} = 0$

$$X_1 = (A+kB)\partial_2 - B\partial_3 . \qquad \omega^1 = (C-kD)\,dx^2 - D\,dx^3 . \qquad d\omega^1 = -\omega^2 \wedge \omega^3$$
$$X_2 = B\partial_2 + (A-kB)\partial_3 \qquad \omega^2 = D\,dx^2 + (C+kD)\,dx^3 \qquad d\omega^2 = \omega^1 \wedge \omega^3 + h\omega^2 \wedge \omega^3$$
$$X_3 = \partial_1 \qquad \omega^3 = dx^1 \qquad d\omega^3 = 0$$

where: $A = e^{kx^1} \cos ax^1$; $B = -\frac{1}{a} e^{kx^1} \sin ax^1$;

$C = e^{-kx^1} \cos ax^1$; $D = -\frac{1}{a} e^{-kx^1} \sin ax^1$;

and where $k = \frac{h}{2}$ and $a = (1-k^2)^{\frac{1}{2}} = \frac{1}{2}(4-h^2)^{\frac{1}{2}}$.

Type VIII:

$$C^1_{23} = -C^1_{32} = -1 . \qquad \xi_1 = \frac{1}{2} e^{-x^3}\partial_1 + \frac{1}{2}[e^{x^3} - (x^2)^2 e^{-x^3}]\partial_2 - x^2 e^{-x^3}\partial_3$$
$$C^2_{31} = -C^2_{13} = 1 \qquad \xi_2 = \partial_3$$
$$C^3_{12} = -C^3_{21} = 1 \qquad \xi_3 = \frac{1}{2} e^{-x^3}\partial_1 - \frac{1}{2}[e^{x^3} + (x^2)^2 e^{-x^3}]\partial_2 - x^2 e^{-x^3}\partial_3$$

$$X_1 = \frac{1}{2}[1 + (x^1)^2]\,\partial_1 + \frac{1}{2}[1 - 2x^1x^2]\,\partial_2 - x^1\partial_3 .$$
$$X_2 = -x^1\partial_1 + x^2\partial_2 + \partial_3$$
$$X_3 = \frac{1}{2}[1 - (x^1)^2]\,\partial_1 + \frac{1}{2}[-1 + 2x^1x^2]\,\partial_2 + x^1\partial_3$$

$$\omega^1 = dx^1 + [1 + (x^1)^2]dx^2 + [x^1 - x^2 - (x^1)^2 x^2]dx^3 . \qquad d\omega^1 = -\omega^2 \wedge \omega^3$$
$$\omega^2 = 2x^1 dx^2 + (1 - 2x^1x^2)dx^3 \qquad d\omega^2 = \omega^3 \wedge \omega^1$$
$$\omega^3 = dx^1 + [-1 + (x^1)^2]dx^2 + [x^1 + x^2 - (x^1)^2 x^2]dx^3 \qquad d\omega^3 = \omega^1 \wedge \omega^2$$

Table 6.1. List of Three-Dimensional Groups
(Continued)

Type IX: $C^1{}_{23} = -C^1{}_{32} = 1.$ $\qquad \xi_1 = \partial_2$

$C^2{}_{31} = -C^2{}_{13} = 1 \qquad \xi_2 = \cos x^2 \partial_1 - \cot x^1 \sin x^2 \partial_2 + \dfrac{\sin x^2}{\sin x^1}\partial_3$

$C^3{}_{12} = -C^3{}_{21} = 1 \qquad \xi_3 = -\sin x^2 \partial_1 - \cot x^1 \cos x^2 \partial_2 + \dfrac{\cos x^2}{\sin x^1}\partial_3$

rest of $C^i{}_{jk} = 0$

$$X_1 = -\sin x^3 \partial_1 + \frac{\cos x^3}{\sin x^1}\partial_2 - \cot x^1 \cos x^3 \partial_3 .$$

$$X_2 = \cos x^3 \partial_1 + \frac{\sin x^3}{\sin x^1}\partial_2 - \sin x^3 \cot x^1 \partial_3$$

$$X_3 = \partial_3$$

$$\omega^1 = -\sin x^3 dx^1 + \sin x^1 \cos x^3 dx^2 . \qquad d\omega^1 = \omega^2 \wedge \omega^3$$

$$\omega^2 = \cos x^3 dx^1 + \sin x^1 \sin x^3 dx^2 \qquad d\omega^2 = \omega^3 \wedge \omega^1$$

$$\omega^3 = \cos x^1 dx^2 + dx^3 \qquad d\omega^3 = \omega^1 \wedge \omega^2$$

Three of the groups listed deserve further comment: A) The group of Type I is isomorphic to the three-dimensional translation group T_3 (the group of translations of Euclidean three-space). A "flat" FRW model is Type I-homogeneous. The notation "T_3-homogeneous" is also used, but should not be confused with a Russian usage for a different purpose.

B) The group of Type V is a simply transitive subgroup of an "open" FRW model. In an open FRW model, as well as in each closed and each flat FRW model, there is also a three-dimensional isotropy group (which is isomorphic to the three-dimensional rotation group).

C) The group of Type IX is isomorphic to $SO(3, R)$, the group of special (unit determinant), orthogonal, 3×3 matrices with real coefficients (isomorphic to the three-dimensional rotation group). A closed FRW model is "SO(3, R)-homogeneous" or Type IX-homogeneous. More general Type IX-homogeneous models are discussed in later chapters. The most general such model is anisotropic and has rotating matter.

The Subclassification Scheme of Ellis and MacCallum

It will be useful later to break up the groups we have discussed above into subclasses. Ellis and MacCallum (1969) have studied spatially homogeneous cosmologies based on the Bianchi groups and have classified them into subcategories for various uses. Actually these classifications are classifications of the underlying symmetry groups, so we list the various groups in Table 6.2 according to the Ellis-MacCallum scheme.

Table 6.2. The Classification Scheme of Ellis and MacCallum (1969). The structure constants are written in the form:

$$C^i_{jk} = \varepsilon_{jks} m^{si} + \delta^i_k a_j - \delta^i_j a_k$$

to define the matrix $m = (m^{ij})$ and the triplet (a_i).

Class $A(a_i = 0)$

Bianchi Type	m
I	$m = 0$
II	$m = \mathrm{diag}(1, 0, 0)$
VI_{-1}	$m = -\alpha$
VII_0	$m = \mathrm{diag}(-1, -1, 0)$
VIII	$m = \mathrm{diag}(-1, 1, 1)$
IX	$m = (\delta_{ij})$

Class $B(a_i \neq 0)$

Bianchi Type	m	a_i
III	$m = -\frac{1}{2}\alpha$	$a_i = -\frac{1}{2}\delta^i_3$
IV	$m = \mathrm{diag}(1, 0, 0)$	$a_i = -\delta^i_3$
V	$m = 0$	$a_i = -\delta^i_3$
$VI_{h\neq -1}$	$m = \frac{1}{2}(h-1)\alpha$	$a_i = -\frac{1}{2}(h+1)\delta^i_3$
$VII_{h\neq 0}$	$m = \mathrm{diag}(-1, -1, 0) + \frac{h}{2}\alpha$	$a_i = -\frac{h}{2}\delta^i_3$

where

$$\alpha = \begin{bmatrix} 0 & 1 & 0 \\ 1 & 0 & 0 \\ 0 & 0 & 0 \end{bmatrix}$$

6.5. The Three-Sphere in a Type IX Cosmology

Manifolds to which we will later devote much attention are the spaces invariant under SO(3, R) or the group of Bianchi Type IX. This group has as its structure constants

$$C^i_{\ jk} = \varepsilon_{ijk} \quad \text{or} \quad C^1_{\ 23} = 1 \quad \textit{et cyc} . \tag{6.24}$$

The underlying space of SO(3. R) is actually a three-sphere S^3 (Section 2.7) with antipodal points identified. However, for simplicity we may take as the topological prototype H of the invariant subspaces the "simply connected covering space" of SO(3, R), namely the three-sphere S^3 itself. Because of the importance of the FRW universe and other Type IX cosmologies, we here extend the discussion given in Section 2.7.

The three basis vectors Y_1, Y_2, Y_3 of (2.57) serve not as Killing vectors but as invariant vectors. Any metric placed on S^3 of the form $Y_i \cdot Y_j = g_{ij} = \text{const}$ is invariant under SO(3, R). We prove this invariance by finding three vectors ξ_1, ξ_2, ξ_3, on S^3 such that $[\xi_i, Y_j] = 0$. The structure constants of these vectors will be the $C^i_{\ jk}$ of (6.24). The three Killing vectors are

$$\begin{aligned} \xi_1 &= \tfrac{1}{2}\, x^2\partial_1 - \tfrac{1}{2}\, x^1\partial_2 - \tfrac{1}{2}\, x^4\partial_3 + \tfrac{1}{2}\, x^3\partial_4 \\ \xi_2 &= \tfrac{1}{2}\, x^3\partial_1 + \tfrac{1}{2}\, x^4\partial_2 - \tfrac{1}{2}\, x^1\partial_3 - \tfrac{1}{2}\, x^2\partial_4 \\ \xi_3 &= -\tfrac{1}{2}\, x^4\partial_1 + \tfrac{1}{2}\, x^3\partial_2 - \tfrac{1}{2}\, x^2\partial_3 + \tfrac{1}{2}\, x^1\partial_4 \end{aligned} \tag{6.25}$$

where, as in (2.53), the vectors ∂_μ of R^4 have been used as the basis. It may be noted that the Y_i's are obtained from the ξ_i's in the manner of Section 6.3. That is, $Y_i(P_0) = \xi_i(P_0)$ where P_0 is the point on S^3 with R^4 coordinates $x^4 = 1,\ x^1 = x^2 = x^3 = 0$.

Cosmologies

When a model universe M has the invariance group SO(3, R), the invariant hypersurfaces are taken to be three-spheres. If these spheres are spacelike, then any fourth invariant vector Y_0 will be timelike. As shown in Section 6.3, this vector may be chosen very freely. One convenient choice is to take this vector perpendicular to the spacelike S^3's and of unit length. In terms of the dual one-forms, the metric of M will then be of the form

$$ds^2 = -dt^2 + g_{ij}(t)\omega^i\omega^j . \tag{6.26}$$

Each g_{ij} is a function of the proper time t alone. The dt, ω^i are duals of Y_0, Y_i and obey

$$d(dt) = 0;\ d\omega^1 = \omega^2 \wedge \omega^3 \text{ et cyc or } d\omega^i = \frac{1}{2}\varepsilon_{ist}\omega^s \wedge \omega^t . \tag{6.27}$$

The equations immediately remind us of the closed FRW universe and the expression for its metric in terms of a basis dt, σ^i. In that example, the σ^i's were combinations of the ω^i's of (6.26), chosen to put the metric components in a simple form. Had we chosen to express the Friedmann metric in the ω^i basis, g_{ij} would have the form

$$g_{ij} = \frac{1}{16} G^2 \delta_{ij} \text{ with } G = G(t) . \tag{6.28}$$

The fact that g_{ij} is diagonal and has three equal entries shows that the metric of the FRW universe is *isotropic*. In other words, the FRW universe has symmetries in addition to the homogeneity of spacelike sections which is granted by invariance under SO(3, R) and expressed by (6.26) and (6.27). This additional symmetry of the FRW universe – its isotropy – may be expressed by the statement that its metric is invariant under rotations about any axis in the homogeneous three-space H(t).

A universe which is rotationally invariant about only one axis in each three space and which is invariant under SO(3, R) is the Taub universe (see Chapter 8). This (vacuum) model has the metric

$$ds^2 = -dt^2 + b_1^2(\omega^1)^2 + b_2^2[(\omega^2)^2 + (\omega^3)^2] . \tag{6.29}$$

The manifold again is $S^3 \times R$. As we see by the form of this metric, where the b's are functions of t only, the Taub universe is spatially homogeneous with invariance group SO(3, R).

The time evolution of the metric of the Taub universe is such that the basis of (6.29) is not globally valid. The basis breaks down beyond the Misner interface between the Taub geometry and the NUT geometry. At this interface the homogeneous subspaces change character. Previously spacelike, they become timelike (that is, orthogonal to a spacelike direction) in the region called "NUT space." In the NUT region it is impossible to take a timelike unit vector perpendicular to the invariant hypersurfaces. Another basis must be used. (Details in Chapter 8.)

A more general matter-filled SO(3, R)-homogeneous model may be imagined. In this model g_{ij} is not diagonal as a function of t nor may be made diagonal by changing the choice of ω^i. This model exhibits rotation and anisotropy. There is a seven parameter family of such models. Each one has a true singularity.

7. UNIVERSES HOMOGENEOUS IN SPACE AND TIME

А если что и остается
Чрез звуки лиры и трубы,
То вечности жерлом пожрется
И общей не уйдет судьбы.
— GAVRIIL ROMANOVICH DERZHAVIN

7.1. Exegesis and Exposition

We shall call a cosmological model in which the metric is the same at all points of space and time *homogeneous in space and time* (ST-homogeneous) (see Figure 7.1). Such a model is a manifold M on

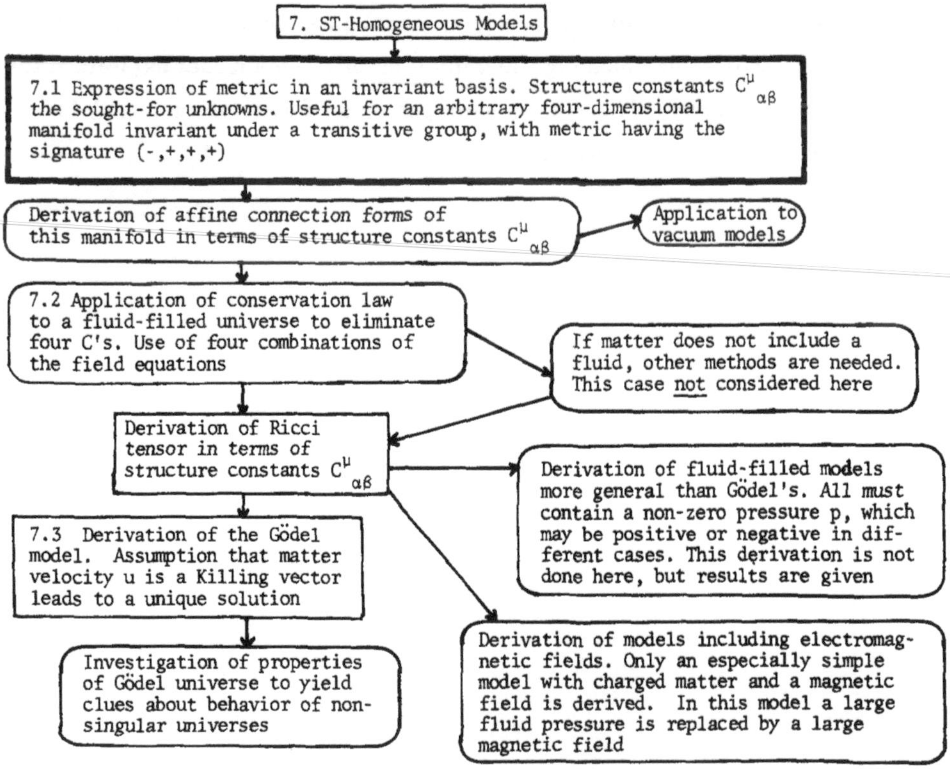

Fig. 7.1. Flow Chart for Chapter 7.

which a transitive (simply or multiply) group of isometries G acts. The metric of M is most easily expressed in an invariant basis where the $g_{\mu\nu}$ and the structure coefficients $C^{\alpha}{}_{\beta\gamma}$ are all constant. Einstein's equations become purely algebraic. We shall show that all the solutions are physically meaningless in some sense.

The archtypes of these universes are E, the Einstein universe (Einstein, 1917) and G, the Gödel universe (Gödel, 1949). Cahen, Debever, and Defrise (1967) have listed all possible vacuum models which are ST-homogeneous. Ozsvath (1965a, see also Farnsworth and Kerr, 1966) has found all models containing a perfect fluid for which G has a simply transitive subgroup.

An ST-*Homogeneous Manifold* M *May or May Not Be Complete;*
M *is Certainly Unrealistic*

An ST-homogeneous model may be incomplete. A homogeneous Riemannian manifold cannot be incomplete, but a homogeneous pseudo-Riemannian one may be (see manifold A of Section 5.4 and see Hermann, 1964).

The possibility of incompleteness is serious, but we are used to dealing with cosmologies represented by incomplete manifolds (the FRW universes). However, the existence of the nebular red shift (expansion) cannot be reconciled with a manifold which is ST-homogeneous. This incompatibility is to be expected, as a constant matter density cannot be "expanding." In fact, for all ST-homogeneous universes filled with a fluid, the expansion $\theta = u^{\sigma}{}_{;\sigma}$ is zero. The proof that $\theta = 0$ is instructive: The equation $T^{\mu\nu}{}_{;\nu} = 0$, in the case of a fluid, implies

$$\begin{aligned} w_{,\mu} u^{\mu} &= -(w+p)\theta \, , \\ p_{,\mu}(u^{\mu} u_{\alpha} + \delta^{\mu}{}_{\alpha}) &= -(w+p) u_{\alpha;\sigma} u^{\sigma} \, . \end{aligned} \tag{7.1}$$

If the universe is ST-homogeneous we must have $w_{,\mu} = p_{,\mu} = 0$. If we demand $(w+p) \neq 0$ then (7.1) reduces to

$$\theta = 0 ,$$
$$u_{a;\sigma} u^{\sigma} = 0 . \tag{7.2}$$

Equation (7.2) implies the fluid matter moves along geodesic lines.

The fact that $\theta = 0$ implies that there is no Hubble expansion in these universes (see Chapter 3). M is therefore unphysical in all cases. M can be very interesting in that it displays in simple form properties of more complex expanding universes. It is for this reason that we study these models.

Two Methods of Simplification: Canonical Metric and Canonical Structure Constants

We shall not consider spaces invariant under groups which are multiply transitive with no simply transitive subgroup, except as illustrative examples in later chapters. In case G is simply transitive there exist four linearly independent one forms ω^{μ} for which

$$d\omega^{\mu} = \frac{1}{2} C^{\mu}{}_{\sigma\tau} \omega^{\sigma} \wedge \omega^{\tau} , \quad C^{\mu}{}_{\sigma\tau} = \text{const.} \tag{7.3}$$

and for which the metric is

$$ds^2 = g_{\mu\nu} \omega^{\mu} \omega^{\nu} , \quad g_{\mu\nu} = \text{const.} \tag{7.4}$$

The $C^{\mu}{}_{\sigma\tau}$ and $g_{\mu\nu}$ are constants with respect to space and time.

Since the metric components are constant for the case above, we can easily make $g_{\mu\nu}$ the orthonormal metric $\eta_{\mu\nu} = \mathrm{diag}(-1, 1, 1, 1)$. We shall call this the *canonical metric*. Because of the invariance of $\eta_{\mu\nu}$ under Lorentz transformation we may take

$$\bar{\omega}^{\mu} = \Lambda^{\mu}{}_{\sigma} \omega^{\sigma}$$

for any arbitrary Lorentz transformation $\Lambda^{\mu}{}_{\sigma}$ and still retain the metric $\eta_{\mu\nu}$. We may use Lorentz transformations to expunge certain of the $C^{\mu}{}_{\sigma\tau}$.

The $C^{\mu}{}_{\sigma\tau}$, if treated as a set of unknowns must be chosen such that the vector basis dual to be basis $\{\omega^{\mu}\}$ satisfies

$$[[X,Y],Z] + [[Y,Z],X] + [[Z,X],Y] = 0 \ .$$

That is, one must have

$$C^{\sigma}{}_{\alpha\beta}C^{\mu}{}_{\sigma\gamma} + C^{\sigma}{}_{\beta\gamma}C^{\mu}{}_{\sigma\alpha} + C^{\sigma}{}_{\gamma\alpha}C^{\mu}{}_{\alpha\beta} = 0 \ . \tag{7.5}$$

Of course, the $C^{\mu}{}_{\sigma\tau}$ must satisfy the basic symmetry requirement:

$$C^{\mu}{}_{\sigma\tau} = -C^{\mu}{}_{\tau\sigma} \ . \tag{7.6}$$

It is well-known that any set of $C^{\mu}{}_{\sigma\tau}$ which satisfies (7.5) and (7.6) defines a Lie algebra of a Lie group (Helgason, 1962). Moreover, it is (7.5) which allows us to express a metric, given in an invariant basis, in a coordinated basis (Schouten, 1954).

We could, if we wish, choose the $C^{\mu}{}_{\sigma\tau}$ from a list of *canonical structure constants* for four-dimensional groups (see Petrov, 1969), and solve the field equations for $g_{\mu\nu}$. This method is algebraically complicated, so we shall restrict ourselves to choosing a canonical $g_{\mu\nu}$ and solving for various possible sets of $C^{\mu}{}_{\sigma\tau}$.

Once a set of $g_{\mu\nu}$ and $C^{\mu}{}_{\sigma\tau}$ is found which satisfies (7.5) and the Einstein equations we are through. A manifold invariant under a simply transitive Lie group is itself a Lie group (Helgason, 1962), so we may use standard group-theory techniques to construct a concrete example of the group. This example is equivalent to finding the metric of the spacetime in a coordinated basis. This final procedure is not always easy, but it is always possible.

Affine Connection Forms and Riemann and Ricci Tensors

Computing the affine connection forms, the curvature forms, $R^{\mu}{}_{\nu\alpha\beta}$ and $R_{\mu\nu}$ for universes homogeneous in space and time is an excellent exercise in the calculus of differential forms. We shall present the results of these calculations and leave details to the reader.

The first step is to solve

$$\omega^{\mu}{}_{\nu} = \Gamma^{\mu}{}_{\nu\sigma}\omega^{\sigma},$$

$$dg_{\mu\nu} = 0 = g_{\mu\sigma}\omega^{\sigma}{}_{\nu} + g_{\nu\sigma}\omega^{\sigma}{}_{\mu}, \tag{7.7}$$

$$d\omega^{\mu} = -\omega^{\mu}{}_{\sigma}\wedge\omega^{\sigma}$$

for the $\Gamma^{\mu}{}_{\nu\sigma}$. The result is

$$\Gamma^{\mu}{}_{\nu\sigma} = \frac{1}{2}\,(C^{\mu}{}_{\sigma\nu} - g^{\mu\tau}C^{\rho}{}_{\tau\nu}g_{\rho\sigma} - g^{\mu\tau}C^{\rho}{}_{\tau\sigma}g_{\rho\nu})\,. \tag{7.8}$$

Next we must compute the curvature forms $\theta^{\mu}{}_{\nu} = d\omega^{\mu}{}_{\nu} + \omega^{\mu}{}_{\alpha}\wedge\omega^{\alpha}{}_{\nu}$, and read off the Riemann tensor components from $\theta^{\mu}{}_{\nu} = \frac{1}{2}R^{\mu}{}_{\nu\alpha\beta}\omega^{\alpha}\wedge\omega^{\beta}$. The result is

$$R^{\mu}{}_{\nu\sigma\tau} = \Gamma^{\mu}{}_{\nu\rho}C^{\rho}{}_{\sigma\tau} + \Gamma^{\mu}{}_{\rho\sigma}\Gamma^{\sigma}{}_{\nu\tau} - \Gamma^{\mu}{}_{\rho\tau}\Gamma^{\rho}{}_{\nu\sigma}\,. \tag{7.9}$$

The final step is the calculation of $R_{\mu\nu} = R^{\sigma}{}_{\mu\sigma\nu}$ using all the symmetries of the $C^{\mu}{}_{\nu\alpha}$, including the Jacobi relation (7.5). We find

$$\begin{aligned} R_{\mu\nu} = & -\frac{1}{2}C^{\tau}{}_{\sigma\mu}C^{\sigma}{}_{\tau\nu} - \frac{1}{2}C^{\alpha}{}_{\sigma\mu}C^{\beta}{}_{\tau\nu}g^{\sigma\tau}g_{\alpha\beta} \\ & + \frac{1}{4}g^{\sigma\tau}g^{\pi\rho}C^{\alpha}{}_{\alpha\pi}C^{\beta}{}_{\tau\rho}g_{\alpha\mu}g_{\beta\nu} \\ & + \frac{1}{2}C^{\sigma}{}_{\sigma\rho}g^{\rho\tau}(C^{\pi}{}_{\tau\mu}g_{\pi\nu} + C^{\pi}{}_{\tau\nu}g_{\pi\mu})\,. \end{aligned} \tag{7.10}$$

7.2. Vacuum Models, Fluid Models, and a Charged Fluid Model

We can now write Einstein's equations for various kinds of matter with this form for $R_{\mu\nu}$. We shall present three types of solutions:

A) All ST-homogeneous vacuum models in which the group of isometries is simply or multiply transitive.

B) All ST-homogeneous fluid models in which the group is simply transitive, and

C) A one-parameter family of charged dust models.

ST-Homogeneous Vacuum Models

The vacuum field equations are

$$R_{\mu\nu} = 0 \, . \tag{7.11}$$

There are only two classes of solutions to these equations (Cahen, Debever, and Defrise, 1967; Cahen and McLenaghan, 1968):

(A) The first is given by the metric (in variables u, v, x, y)

$$ds^2 = \frac{1}{2} e^{-au} dxdy - \left[\frac{1}{2}(x^2 - y^2)\cos bu + xy \sin bu\right] du^2 + (dx)^2 + (dy)^2 \, . \tag{7.12}$$

This model has a multiply transitive group of isometries but no simply transitive group.

(B) The second class consists of the metric (in terms of x, y, x, t):

$$ds^2 = e^z[\cos(\sqrt{3}z)(dx^2 - dt^2) - 2\sin(\sqrt{3}z)\,dtdx] + e^{-2z}dy^2 + dz^2 \, . \tag{7.13}$$

In the orthonormal frame the structure coefficients of this metric are constant. This fact shows that this model is invariant under a simply transitive isometry group. Each of these metrics may be multiplied by a constant conformal factor to yield a new solution $k^2 ds^2$.

Still other "vacuum" solutions are possible if a cosmological constant Λ is allowed. The field equations for such a model read

$$R_{\mu\nu} = \frac{1}{4} \Lambda g_{\mu\nu} \, . \tag{7.14}$$

None of these models is of more than academic interest, however. They are most properly interpreted as fluid models, with unphysical densities and pressures.

Fluid Models

If we write the usual stress-energy tensor $T_{\mu\nu} = (w + p) u_\mu u_\nu + p g_{\mu\nu}$ and insert it into the Einstein equations we can solve for all possible fluid models. While there is no comprehensive list of all fluid ST-homogeneous models, Ozsvath (1966a; see Farnsworth and Kerr, 1966) has given a list of all models invariant under a simply transitive group. The simplest expression of these models uses an orthonormal basis and uses the Lorentz-transformation freedom to make

$$g_{\mu\nu} = \eta_{\mu\nu}, \quad u_\mu = (-1, 0, 0, 0) . \tag{7.15}$$

Equations (7.2) now place restrictions on the $C^\mu{}_{\nu a}$. They read

$$\theta = C^\tau{}_{\tau 0} = 0$$

$$u_{a;\sigma} u^\sigma = C^0{}_{a0} = 0 . \tag{7.16}$$

Table 7.1 is a list of these models classified by the vector $K_\mu = C^\sigma{}_{\sigma\mu}$.

Table 7.1. List of ST-Homogeneous Fluid Models. These models each have a simply transitive isometry group. The classification is due to Ozsvath (1965a), and results of Farnsworth and Kerr (1966) are used. The Einstein model E and the Gödel model G are singled out due to their historical importance. An over-all arbitrary multiplicative factor in the metric is omitted; moreover, the metrics are put into a form to emphasize similarities between models. The metric and structure constants are listed for one invariant basis, but a coordinated basis is not given. The matter variables w (energy density), p (pressure), and u^μ (fluid velocity) are given. The cosmological constant Λ used in the original papers is here set to zero; rather p is allowed here to be non-zero.

Category 1 ($K_\mu = C^\sigma{}_{\sigma\mu} = 0$)

Class G (The Gödel model)

metric: $ds^2 = -2(\omega^0)^2 + (\omega^1)^2 + (\omega^2)^2 + (\omega^3)^2$

structure: $d\omega^0 = -\omega^1 \wedge \omega^2$; $d\omega^1 = \omega^2 \wedge \omega^0$;
$d\omega^2 = \omega^0 \wedge \omega^1$; $d\omega^3 = 0$.

matter variables: $w = p = \frac{1}{2}$; $u^\mu = (\sqrt{2}, 0, 0, 0)$.

Class E (The Einstein model)

metric: $ds^2 = -(\omega^0)^2 + (\omega^1)^2 + (\omega^2)^2 + (\omega^3)^2$

structure: $d\omega^0 = 0$; $d\omega^1 = \omega^2 \wedge \omega^3$;
$d\omega^2 = \omega^3 \wedge \omega^1$; $d\omega^3 = \omega^1 \wedge \omega^2$.

matter variables: $w = -3p = \frac{3}{4}$; $u^\mu = (1, 0, 0, 0)$.

Table 7.1. List of ST-Homogeneous Fluid Models
(Continued)

Class I

metric: $ds^2 = -(\omega^0)^2 + (\omega^1)^2 + B^2(\omega^2)^2 + (1+B^2)(\omega^3)^2$

where $w + p > 0 \rightarrow \frac{1}{3} < B^2 < 3.$

structure: $d\omega^0 = D\omega^1 \wedge \omega^2;\ d\omega^1 = \omega^2 \wedge \omega^3;$
$d\omega^2 = \omega^3 \wedge \omega^1\ ;\ d\omega^3 = \omega^1 \wedge \omega^2.$

where $D^2 = [2(1+B^2)]^{-1}[8B^2 - (1+B^2)^2]$

matter variables: $w = [8B^2(1+B^2)]^{-1}[32B^2 - 5(1+B^2)^2];$

$p = -(8B^2)^{-1}(1+B^2);$ so that $p/w \leq -\frac{1}{3}.$

$$u_\mu = \left(\left[\frac{w-p}{w+p}\right]^{\frac{1}{2}}, 0, 0, \left[\frac{-2p(1+B^2)}{w+p}\right]\right).$$

note: when $B = 1$, the Einstein model results, but in a new basis

Class II

metric: $ds^2 = -(1+A^2)(\omega^0)^2 + (\omega^1)^2 + A^2(\omega^2)^2 + (\omega^3)^2$

structure: $d\omega^0 = -\omega^1 \wedge \omega^2;\ d\omega^1 = \omega^2 \wedge \omega^0;$
$d\omega^2 = \omega^0 \wedge \omega^1\ ;\ d\omega^3 = D\omega^2 \wedge \omega^1.$

where $D^2 = [2(1+A^2)^2]^{-1}[8A^2 - (1+A^2)^2]$

where $D^2 \geq 0$ and $w+p>0 \rightarrow 3<A^2 \leq 3+2\sqrt{2}$ while $w>0 \rightarrow A^2>\frac{1}{5}(11+4\sqrt{6})$

matter variables: $w = (8A^2)^{-1}[5(1+A^2)] - 4(1+A^2)^{-1};$

$p = (8A^2)^{-1}(1+A^2);$ so that $p \geq w.$

$$u_\mu = \left(\left[\frac{2p(1+A^2)}{w+p}\right], 0, 0, \frac{p-w}{w+p}\right).$$

note: when $A^2 = 3+2\sqrt{2}$, the Gödel universe results.

Class III

metric: $ds^2 = -(1+A^2)(\omega^0)^2 + (\omega^1)^2 + A^2(\omega^2)^2 + (\omega^3)^2$

structure: $d\omega^0 = -\omega^1 \wedge \omega^2;\ d\omega^1 = \omega^2 \wedge \omega^0;$
$d\omega^2 = \omega^0 \wedge \omega^1\ ;\ d\omega^3 = 0.$

matter variables: $w = p = (1+A^2)^{-1};$

$$u_\mu = ((1+A^2)^{-\frac{1}{2}}, 0, 0, 0).$$

note: when $A = 1$, the Gödel model results.

Table 7.1. List of ST-Homogeneous Fluid Models
(Continued)

Category 2 ($K_\mu = C^\sigma{}_{\sigma\mu} \neq 0$)

metric: $ds^2 = -(\omega^0)^2 + (\omega^1)^2 + (\omega^2)^2 + (\omega^3)^2$

structure: $d\omega^0 = [(T-A)\omega^1 - B\omega^2] \wedge \omega^3;$
$d\omega^1 = [(T+A)\omega^0 + (1+C)\omega^1 + (D+S)\omega^2] \wedge \omega^3;$
$d\omega^2 = [B\omega^0 + (D-S)\omega^1 + (1-C)\omega^2] \wedge \omega^3;$
$d\omega^3 = 0.$

where S and T are independent parameters, and where

$A = 2T(T^2 - 4S^2 - 4)/E;\quad B = 2ST(T^2 + 2S^2 + 2)/E;$

$C = T^2(4 - T^2 - 2S^2)/E;\quad D = -6ST/E;$

with $E = T^4 + 4T^2(S^2-2) + 4(1+S^2)(4+S^2).$

matter variables: $w = -\frac{5}{2}AT - 3;\ p = \frac{8}{5} + \frac{w}{5};$ so that $\frac{w}{p} > \frac{1}{5};\ u^\mu = (1, 0, 0, 0).$

note: when $S = 0,\ T = 2,$ the Gödel model results.

The Einstein universe (E) is of historical interest (Einstein, 1917). In this case the basis forms obey

$$d\omega^0 = 0, \qquad d\omega^i = \varepsilon_{ijk}\omega^j \wedge \omega^k, \quad i, j, k = 1, 2, 3. \tag{7.17}$$

These equations show that $\omega^0 = dt$, where t is a "cosmic time," and that the $t = \text{const.}$ surfaces are Bianchi Type IX spaces. We know that w and p are constant, and we find

$$w = -3p = \frac{3}{4}, \tag{7.18}$$

so the pressure is large and negative.

In fact in all of the models of Table 7.1 p is fairly large compared to w. In particular there is no pressureless (dust) model in the list.

Models with Electromagnetic Fields

Ozsvarth (1965b) has found all ST-homogeneous spacetimes filled with a null electromagnetic field plus a perfect fluid. A null electromagnetic field satisfies

$$F^{\mu\nu}F_{\mu\nu} = 0 \ .$$

The equations read for a general electromagnetic field-plus-fluid

$$R_{\mu\nu} = (w+p)u_\mu u_\nu + \frac{1}{2}(w-p)g_{\mu\nu} + F_{\mu\sigma}F^{\sigma}{}_{\nu} - \frac{1}{4}F_{\sigma\tau}F^{\sigma\tau}g_{\mu\nu}, \tag{7.19}$$

$$F^{\mu\sigma}{}_{;\sigma} = J^{\mu} ,$$

(see Weber, 1961), where J^{μ} are the components of the current density vector. A fluid may be expected to support a current density of the form (Lichnerowicz, 1967):

$$J^{\mu} = Qu^{\mu} + \sigma u_{\tau}F^{\tau\mu} , \tag{7.20}$$

where σ is the conductivity.

An especially simple model in the limit of zero conductivity $(\sigma = 0)$ is the following: In the orthonormal basis $(g_{\mu\nu} = \eta_{\mu\nu})$ this model has the structure constants

$$C^{0}{}_{12} = -\sqrt{2} = -C^{0}{}_{21},\ C^{1}{}_{20} = \frac{\sqrt{2}}{2}(1-B^{2}) = -C^{1}{}_{02} = C^{2}{}_{01} = -C^{2}{}_{10} \tag{7.21}$$

and all the rest are zero. The constant B is a free parameter, and we find for the electromagnetic field:

$$F_{12} = -F_{21} = B, \qquad \text{All others zero.} \tag{7.22}$$

The matter velocity U has the form $U = \omega^{0}$, and the current density is $J_{\mu} = Qu_{\mu}$, with $Q = \sqrt{2}\,B$. The energy density and pressure are

$$p = \frac{1}{2}(1-B^{2}), \quad w = \frac{1}{2}(1+B^{2}) . \tag{7.23}$$

Where $B = 0$ this model becomes the Gödel universe (see below). When $B \neq 0$, the form of $F_{\mu\nu}$ shows that B is the magnitude of a cosmic magnetic field. Thus this model replaces the large pressure in the Gödel universe with a magnetic field. In fact when $B = 1$, the pressure p vanishes.

7.3. The Gödel Universe

The Gödel universe G is an interesting example of an ST-homogeneous universe containing a perfect fluid. It contains closed time-like lines (it is *acausal*), and the matter velocity u is a Killing vector. Both w and p are positive.

We investigate G in the orthonormal basis and use the freedom of scaling and Lorentz transformation to write $u^\mu = (1,0,0,0)$ and $(w+p) = 1$. The field equations are

$$R_{\mu\nu} = \delta_{\mu 0}\delta_{\nu 0} + \left(\frac{1}{2} - p\right)\eta_{\mu\nu} \,. \tag{7.24}$$

Derivation of G

Let us define a matrix $m^\mu{}_\nu$ by $m^\mu{}_\nu = C^\mu{}_{0\nu}$. If we require u to be a Killing vector $(u_{\mu;\nu} + u_{\nu;\mu} = 0)$, we find $m^\sigma{}_\mu \eta_{\sigma\nu} + m^\sigma{}_\nu \eta_{\sigma\mu} = 0$, or $m^i{}_j = -m^j{}_i\,(i,j = 1,2,3)$ and $m^0{}_\mu = m^\mu{}_0 = 0$.

The Jacobi equation (7.5) implies that either $m^2{}_1$ or $C^2{}_{12}$ must vanish. It can be shown that these two possibilities describe the same model in different bases. These two bases are: (a) The one with $C^2{}_{12} \neq 0$ in which case the $C^\alpha{}_{\beta\gamma}$ are:

$$C^0{}_{12} = -C^0{}_{21} = \sqrt{2}\,, \quad C^2{}_{12} = -C^2{}_{21} = 1\,, \quad \text{rest zero}. \tag{7.25}$$

(b) The one with $m^2{}_1 \neq 0$; the $C^\alpha{}_{\beta\gamma}$ are:

$$-C^1{}_{02} = C^1{}_{20} = C^0{}_{12} = -C^0{}_{21} = \sqrt{2}; \; C^2{}_{01} = -C^2{}_{10} = -1/\sqrt{2}; \; \text{rest zero} \tag{7.26}$$

In both cases we find $p = w = \frac{1}{2}$.

We next define the twist three-vector (related to the rotation of matter) by

$$\bar{\omega}_1 = C^0{}_{23}\,, \quad \bar{\omega}_2 = C^0{}_{31}\,, \quad \bar{\omega}_3 = C^0{}_{12}\,. \tag{7.27}$$

In either (7.25) or (7.26) we find $\bar{\omega}_1 = \bar{\omega}_2 = 0$, whereupon the R_{00} field equation reads

$$R_{00} = \frac{1}{2}\,\overline{\omega}_3{}^2 = \frac{1}{2} + p = 1\ .$$

We have chosen p positive so that $\overline{\omega}_3 \neq 0$ (in the Einstein universe we choose $p < 0$ to get $\overline{\omega}_3 = 0$).

Originally G was presented (Gödel, 1949) in a coordinated basis in which the metric is

$$\begin{aligned} ds^2 = a^2\Big[-(dx^0)^2 + (dx^1)^2 - \frac{1}{2}\,e^{2x^1}(dx^2)^2 \\ + (dx^3)^2 - 2e^{x^1}dx^0dx^2\Big], \quad a = \text{const.} \end{aligned} \tag{7.28}$$

We can identify this metric with ours by letting $a = 1$ and

$$\omega^0 = dx^0 + e^{x^1}dx^2;\ \omega^1 = dx^1;\ \omega^2 = \frac{\sqrt{2}}{2}\,e^{x^1}dx^2;\ \omega^3 = dx^3\ . \tag{7.29}$$

This leads to the same structure constants as (7.25).

Because of the group symmetry, there are at least four Killing vectors on the manifold G, the generators of the group. In fact, there is a fifth Killing vector. In the coordinated basis of (7.28) the five Killing vectors are: $\partial_0, \partial_2, \partial_3, \partial_1 - x^2\partial_2$ and one other:

$$\xi = -2e^{-x^1}\partial_0 + x^2\partial_1 + \left[e^{-2x^1} - \frac{1}{2}\,(x^2)^2\right]\partial_2\,.$$

Acausality and Rotation

The Gödel universe contains closed timelike lines. This violation of causality is one of the most interesting features of the model. In order to consider this acausality let us first look at the hypersurface $x^3 = \text{const.}$ The structure of the forms ω^0, ω^1, ω^2, shows that this hypersurface is in many ways similar to the $t = \text{const.}$ hypersurface in a closed FRW model. In the FRW case, however, the $t = \text{const.}$ set is spacelike, and in the Gödel case the $x^3 = \text{const.}$ set is timelike. There is a second, more important difference: In the $k = +1$ FRW case the surface is S^3, and in the Gödel case it is L^3, a space of Bianchi Type VIII, whose group of isometries is the three-dimensional Lorentz group.

The space L^3 is the unit hyperboloid

$$(y^1)^2 + (y^2)^2 - (y^3)^2 - (y^4)^2 = 1 , \qquad (7.30)$$

conveniently visualized as a subset of the abstract space (y^1, y^2, y^3, y^4) or R^4. In the hyperboloid, the vectors X_0, X_1, X_2 dual to $\sigma^0, \sigma^1, \sigma^2$ are invariant vectors.

Unwrapping the Gödel Universe

What of closed timelike lines? The curve defined by the intersection of the plane $y^3 = y^4 = 0$ with the hyperboloid L^3 has X_0 as its tangent vector. Since $(y^1)^2 + (y^2)^2 = 1$, this curve is closed. This curve is a cheat however. Consider a cylinder imbedded in a Minkowski space. This manifold has closed timelike lines, but a cylinder locally has the metric of a plane. We may unwrap the cylinder and spread the resulting flat portion of a plane out until it fills the entire plane. Our closed curve (around the cylinder) becomes an open one, all without affecting the local metric a bit! We can do the same with the sample curve we proposed for a closed timelike line in G.

There are, however, curves in G which are closed and timelike and which cannot be unwrapped (though none are geodesic, see Chandrasekhar and Wright, 1961). One such curve (in the coordinates x^0, x^1, x^2, x^3 of 7.28) is

$$x^0 = A(\sin \tau - \frac{1}{2} \sin \tau \cos \tau); \; x^1 = -B \cos \tau; \; x^2 = -A \sin \tau; \; x^3 = 0 , \qquad (7.31)$$

where A and B are constants which need to be chosen properly and τ is unbounded. The existence of this curve shows that the Gödel universe does not obey a global causality principle.

The Role of Pressure in the Gödel Universe

Positive pressure in the Gödel universe accompanies rotation in G. We called the fact that $\overline{\omega}_3 \neq 0$ "the existence of rotation." What does this fact have to do with rotation in the sense of Chapter 3? There we defined a rotation vector Ω in terms of the curl of the fluid velocity u. In the basis (7.29) $u = \omega^0$ and

$$du = d\omega^0 = \frac{1}{2}\bar{\omega}_3\omega^1 \wedge \omega^2 = \sqrt{2}\omega^1 \wedge \omega^2 . \tag{7.32}$$

From our definition of Ω we find

$$\Omega^\mu = \frac{|g|^{-\frac{1}{2}}}{3!}\varepsilon^{\mu\alpha\beta\gamma}(du \wedge u)_{\alpha\beta\gamma} = \frac{1}{6}\varepsilon^{\mu\alpha\beta\gamma}(\sqrt{2}\omega^1 \wedge \omega^2 \wedge \omega^3)_{\alpha\beta\gamma} = (0,0,0,\sqrt{2}/6).$$

($g = -1$ in this basis). Thus Ω^μ is non-zero, the non-zero component being proportional to $\bar{\omega}_0$.

Rotation is necessary in the Gödel universe, as is shown by the Raychaudhuri equation (3.30). Since θ vanishes (3.30) requires that ω^2 be positive or else $R_{\sigma\tau}u^\sigma u^\tau$ be negative (or at least non-positive). In the Einstein universe $R_{\sigma\tau}u^\sigma u^\tau$ vanishes, but where $p > 0$, $R_{\sigma\tau}u^\sigma u^\tau$ is positive, and hence rotation cannot vanish in G.

One final remark: The magnitude of p, $p = w$, is large in G, but permissibly large. Zel'dovich (1961) showed that for a $p = w$ equation of state, sound waves propagate just at the speed of light c. For p any larger the sound velocity would be greater than c and the model unphysical.

The Gödel model is a gold mine of interesting properties which are useful in examining homogeneous universes. By itself, however, it cannot, any more than any other ST-homogeneous universe, be a model for the real universe. The spatially homogeneous cosmologies discussed in the remainder of the book have proved to be of immense importance and relevance to the study of the real universe.

8. T-NUT-M SPACE – OPEN TO CLOSED TO OPEN

Hic coquus scite ac munditer condit cibos
– PLAUTUS

8.1. Taub Space and NUT Space

The models of the preceding chapters have no expansion, and therefore other models must be used to describe the actual universe. We now consider *spatially homogeneous* universes, those homogeneous in space but not in time.

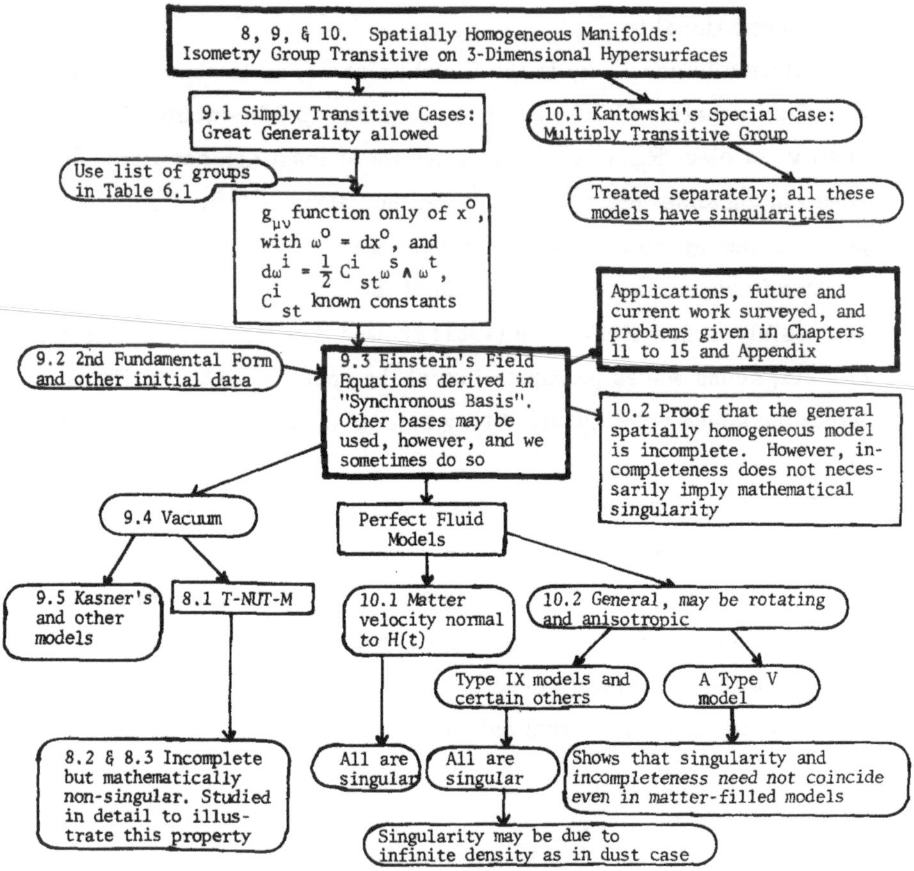

Fig. 8.1. Flow Chart for Chapters 8, 9, 10.

We might hope that a vacuum metric mocks the behavior of the actual universe, but this is not the case. Any amount of matter added to a vacuum model changes its character drastically. Nevertheless, we shall consider some vacuum models because the field equations are simpler and their features can be studied in detail.

The most important vacuum spatially-homogeneous model is T-NUT-M (Taub, 1951; a similar model of Newman, Tamburino and Unti, 1963; and their connection by Misner and Taub, 1968). In this model the spacelike sections evolve in a finite proper time from open to closed to open. Unlike the Friedmann models T-NUT-M has no singularity but expands and contracts in a non-singular framework. Unfortunately, T-NUT-M space is incomplete: Some timelike geodesics cannot be extended indefinitely. Moreover, if any amount of matter is added to T-NUT-M it becomes singular (see Chapter 10). Figure 8.1 is a logical outline of Chapters 8, 9, and 10.

Taub's Vacuum Model

In 1951 Taub published an example of a vacuum solution which is spatially homogeneous. That is, through each point passes a spacelike section on which the metric is independent of position. The group structure of these space sections is "Bianchi-Type IX." The structure constants of this group are

$$C^i_{jk} = \varepsilon_{ijk} \quad i,j,k = 1,2,3 \,. \tag{8.1}$$

The group is also called SO(3, R), the group of special (unit determinant) orthogonal 3×3 matrices with real coefficients (SO(3, R) is isomorphic to the group of rotations in three-dimensional Euclidean space).

On the spacelike sections the metric is

$$^3d\sigma^2 = g_{ij}\omega^i\omega^j \,, \tag{8.2}$$

where g_{ij} is a matrix of constants in space. This matrix varies in time (from one hypersurface to the next). If the time coordinate is chosen to be proper time t measured perpendicular to these hypersurfaces, the full space-time metric is

$$ds^2 = -dt^2 + g_{ij}(t)\omega^i\omega^j \ . \tag{8.3}$$

Without loss of generality we may assume g_{ij} diagonal for all time. (The diagonal character is also true for such a model filled with a non-rotating perfect fluid; see Chapter 9.) The metric, which will be extremely useful later, can therefore be written:

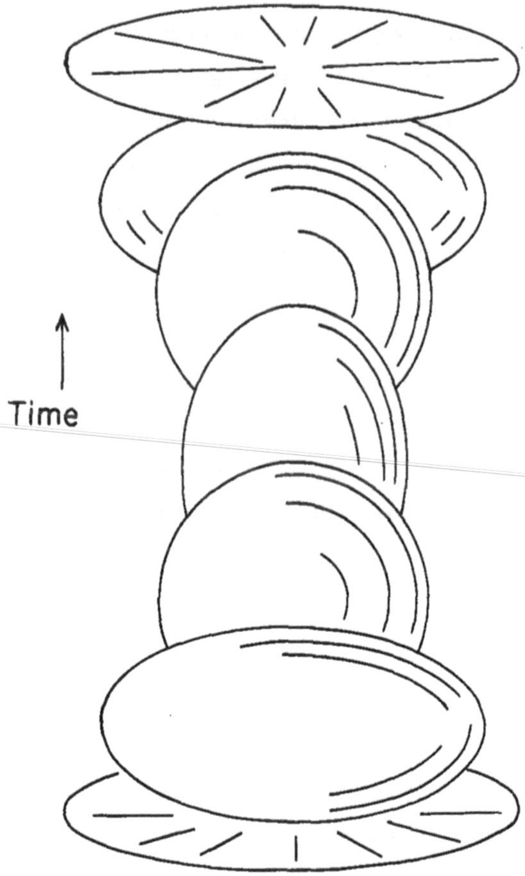

Fig. 8.2. Taub-NUT-Misner Space. T-NUT-M is representable as a disc, which becomes a flattened ellipsoid of revolution, thickens to a cigar-shaped ellipsoid as its volume reaches a maximum, and collapses again to a disc. This description is for the synchronous system, which is no longer a valid descriptive device after the collapse to zero volume has taken place. T-NUT-M space does *not* have a geometric singularity at the time of collapse.

$$ds^2 = -dt^2 + e^{-2\Omega}(e^{2\beta_1}(\omega^1)^2 + e^{2\beta_2}(\omega^2)^2 + e^{2\beta_3}(\omega^3)^2). \tag{8.4}$$

Taub's model requires $\beta_1 = \beta_2$, or

$$ds^2 = -dt^2 + g(\omega^1)^2 + \gamma^2[(\omega^2)^2 + (\omega^3)^2]. \tag{8.5}$$

Figure 8.2 illustrates Taub space in much the same way as we earlier portrayed the Friedmann universe. The function $g = g(t)$ becomes zero after a finite proper time, but γ does not. The zero of g was originally thought to indicate a singular point.

The NUT Space of Newman, Tamburino, and Unti

In 1963 Newman, Tamburino, and Unti described a model which they called a "generalization" of the Schwarzschild metric. Misner (1963), however, showed that the topology of "NUT space" is different from the topology associated with the Schwarzschild model. He discovered that the topology of NUT space is $S^3 \times R$, the Cartesian product of a three-sphere with a line, the metric being independent of position on the three-sphere section.

If we let ω^i be the same as before and use t to label the S^3 sections we find

$$ds^2 = +dt^2 - g(\omega^1)^2 + \gamma^2[(\omega^2)^2 + (\omega^3)^2], \tag{8.6}$$

where g is a positive function of the coordinate t. This metric differs from Taub space in that one of the directions in S^3 is timelike and the direction of the t coordinate is spacelike. Both metrics share the property that g becomes zero at some bounded value of t while γ remains finite.

The Misner Bridge

The similarity of (8.5) and (8.6) hints that these are actually the same metric. Misner and Taub (1968) have shown that we can label the homogeneous hypersurfaces in such a way that Taub space and NUT space are both portions of one non-singular manifold, *T-NUT-M space*. In T-NUT-M there are two regions with the NUT metric and one with the Taub metric. The two boundaries between NUT and Taub are lightlike hypersurfaces with the topology of S^3. These homogeneous (invariant under $SO(3, R)$) hypersurfaces are called *Misner bridges*.

In the evolution coordinate to be used in the next section, T-NUT-M consists of a one-parameter family of copies of S^3. We will call the evolution parameter t and the S^3 copies $H(t)$. As t varies $H(t)$ "evolves" in that the metric of $H(t)$ varies. At large negative values of t, $H(t)$ is timelike (a NUT region). Near $t = 0$, $H(t)$ is spacelike (the Taub region). At large positive values of t, $H(t)$ is again timelike (the second NUT region).

Through each point in the Taub region passes a compact (or *closed*) spacelike hypersurface, namely $H(t)$. In contrast, in each NUT region no spacelike hypersurface is closed (Misner, 1963). (In these regions, $H(t)$, which is closed, is timelike.) Consequently, T-NUT-M represents a universe which evolves "from open to closed to open again."

8.2. The Metric of T-NUT-M

The point of view we will follow to obtain the metric of the T-NUT-M model is: Postulate the existence of a Type IX vacuum model which is mathematically non-singular. Some metric properties of this general model suggest certain simplifications. Once the simplifications (which cause the metric to be expressed in a form similar both to (8.5) and to (8.6)) are imposed, the T-NUT-M metric may be straightforwardly and explicitly derived. We will then show (in the next section) that T-NUT-M is indeed mathematically non-singular and yet incomplete. We will defer many technical details to later chapters, where they will be discussed in terms of matter-filled models in order to present a concise discussion of T-NUT-M itself.

We have pointed out that in the general vacuum Type-IX model the metric of the hypersurface $H(t)$ can be written as

$$d\sigma^2 = \sum_{s=1}^{3} a_s(t)(\omega^s)^2 .$$

No off-diagonal element ever appears as t varies if the g_{0i} metric components are all zero. We must allow the a_i to be positive, negative, or zero, as one a_i is less than zero in NUT space and greater than zero in Taub space.

We shall show in another chapter that if H(t) is spacelike at some time, it must become lightlike and then timelike in a finite proper time. If the four geometry is to be mathematically non-singular at the time t_0 of transition then the determinant of $g_{\mu\nu}$ must remain non-zero. At t_0 therefore, no more than one of the a_i can be zero. However, at t_0, since H(t) is lightlike, one of the a_i does vanish.

We have shown that there is great freedom in choosing the zeroth basis one-form. This freedom appears as a freedom in the forms of $g_{0\mu}$. Ryan (1970) discusses this point in the vacuum case; we will exploit it in the matter-filled case later. In the T-NUT-M model, the most convenient basis is one in which $g_{00} = g_{02} = g_{03} = 0$. Further, by calculation of R_{11} at $t = t_0$, it can be shown that $a_2 = a_3$ then. In both Taub space and NUT space, $a_2 = a_3$ at all values of t. This property implies the existence of a fourth Killing vector in addition to the three of SO(3, R) symmetry.

Can $a_2 = a_3$ only at $t = t_0$? The work of Misner (1969a) shows that this possibility does not occur. We therefore write the metric as

$$ds^2 = 2g_{01}\omega^1 dt + g_{11}(\omega^1)^2 + e^{2\beta}(\omega^2)^2 + e^{2\beta}(\omega^3)^2 .$$

Taub space corresponds to $g_{11} > 0$, and NUT space to $g_{11} < 0$. Further, we may paramotrize the t coordinate so that $g_{01} = 1$. In the basis

$$\sigma^1 = \omega^1, \ \sigma^2 = e^{\beta}\omega^2, \ \sigma^3 = e^{\beta}\omega^3 , \tag{8.7}$$

with $g \equiv g_{11}(t)$, the metric components are

$$g_{\mu\nu} = \begin{bmatrix} 0 & 1 & 0 & 0 \\ 1 & g & 0 & 0 \\ 0 & 0 & 1 & 0 \\ 0 & 0 & 0 & 1 \end{bmatrix} \quad \text{and} \quad g^{\mu\nu} = (g_{\mu\nu})^{-1} = \begin{bmatrix} -g & 1 & 0 & 0 \\ 1 & 0 & 0 & 0 \\ 0 & 0 & 1 & 0 \\ 0 & 0 & 0 & 1 \end{bmatrix}. \tag{8.8}$$

Ricci Tensor and Field Equations

We can now use the Cartan equations to compute the Riemann tensor in the usual fashion. First, notice that of the connection forms $\sigma^{\mu}{}_{\nu}$ only $\sigma^1{}_1$, $\sigma^0{}_A$, $\sigma^1{}_A$, $\sigma^2{}_3$ are independent $(A = 2, 3)$; all the others are given by the equations relating the $\sigma_{\mu\nu}$ $(dg_{\mu\nu} = \sigma_{\mu\nu} + \sigma_{\nu\mu})$. The First Cartan Equation $d\sigma^{\mu} = -\sigma^{\mu}{}_{\sigma} \wedge \sigma^{\sigma}$ may be solved by inspection to yield

$$\begin{aligned}
\sigma^1{}_1 &= -\tfrac{1}{2}\,\dot{g}\sigma^1 \quad (\text{where } \cdot = \tfrac{d}{dt}) \\
\sigma^0{}_2 &= g\dot{\beta}\sigma^2 \\
\sigma^0{}_3 &= g\dot{\beta}\sigma^3 \\
\sigma^1{}_2 &= -\dot{\beta}\sigma^2 + \tfrac{1}{2}\, e^{-2\beta}\sigma^3 \\
\sigma^1{}_3 &= -\tfrac{1}{2}\, e^{-2\beta}\sigma^2 - \dot{\beta}\sigma^3 \\
\sigma^2{}_3 &= -\tfrac{1}{2}\, e^{-2\beta}dt + \left[1 - \tfrac{1}{2}\, g e^{-2\beta}\right]\sigma^1 .
\end{aligned} \tag{8.9}$$

From the Second Cartan Equation $\theta^{\mu}{}_{\nu} = \frac{1}{2} R^{\mu}{}_{\nu\alpha\beta}\,\sigma^{\alpha} \wedge \sigma^{\beta}$ and from $R_{\mu\nu} = R^{\sigma}{}_{\mu\sigma\nu}$ the next step is to compute $R_{\mu\nu}$. The vacuum field equations read:

$$\begin{aligned}
R_{00} &= -2\ddot{\beta} - 2(\dot{\beta})^2 + \tfrac{1}{2}\, e^{-4\beta} = 0\,, \\
R_{01} &= \tfrac{1}{2}\,\ddot{g} + \dot{g}\dot{\beta} + \tfrac{1}{2}\, g e^{-4\beta} = 0\,, \\
R_{22} = R_{33} &= \dot{g}\dot{\beta} + g\ddot{\beta} + 2g(\dot{\beta})^2 + e^{-2\beta} - \tfrac{1}{2}\, g e^{-4\beta} = 0\,, \\
R_{11} = gR_{01} &= R_{02} = R_{03} = R_{12} = R_{13} = R_{23} = 0\,.
\end{aligned} \tag{8.10}$$

All the field equations are identically satisfied except for the R_{00}, R_{01} and R_{22} equations. Moreover, the R_{01} equation is easily shown to be a consequence of the R_{00} and R_{22} equations, these being the two independent equations which give g and β.

The Solution

The general solution of $R_{00} = 0$ is

$$e^{\beta} = \left(Bt^2 + \frac{1}{4B}\right)^{\frac{1}{2}}, \tag{8.11}$$

where B is a positive constant chosen so $t = 0$ is a minimum of β. This result, that e^{β} never becomes zero, shows the non-singularity of the metric. At $g = 0$, e^{β} is non-zero, so the structure coefficients and the determinant of $g_{\mu\nu}$ do not vanish. Hence $g_{\mu\nu}$ has no singularity at $g = 0$.

Inserting (8.11) into the R_{22} equation we find

$$g = \frac{At + 1 - 4B^2t^2}{B(4B^2t^2 + 1)} \quad (A, B \text{ const.}) . \tag{8.12}$$

As a reminder, the full metric is

$$ds^2 = -2dt\omega^1 + g(\omega^1)^2 + e^{2\beta}[(\omega^2)^2 + (\omega^3)^2] \text{ with } d\omega^i = \frac{1}{2}\varepsilon_{ijk}\omega^j \wedge \omega^k. \tag{8.13}$$

8.3. Complete and Incomplete Geodesics of T-NUT-M

T-NUT-M space is a vacuum space-time whose topology is $S^3 \times R$. It is a mathematically non-singular but incomplete model, and as such very interesting and important. It serves as a simple example of such a cosmological model.

Since T-NUT-M is non-singular, every geodesic segment can be covered by a compact (closed) subset of the manifold. We shall show that this property holds. Since T-NUT-M is incomplete, some geodesic segments, although coverable by a compact set, cannot be extended. We shall demonstrate this property also.

Group Symmetry Applied to Geodesics

A geodesic has a tangent vector X for which

$$\nabla_X X = 0 . \tag{8.14}$$

The points along the geodesic are labelled by an affine parameter u, and the components of X in a given basis are functions of u. Given these functions we can reconstruct the geodesic. In T-NUT-M let us use the basis dual to (dt, σ^i) and let the components of X in this basis be a^μ $(\mu = 0, 1, 2, 3)$. Equation (8.14) is a set of four equations for the four unknowns a^μ (it contains function of t, but t is to be considered a function of u given by $a^0 = dt/du$).

Symmetries in the metric simplify (8.14), each symmetry implying a constant of motion. In T-NUT-M these constants are all that are needed to describe completely the behavior of geodesics. We choose u such that

$$\begin{aligned} a^\sigma a_\sigma &= E = \pm 1, 0 \\ &= 2a^0a^1 + g(a^1)^2 + (a^2)^2 + (a^3)^2 . \end{aligned} \tag{8.15}$$

This is one constant of motion for (8.14).

If ξ is a Killing vector of T-NUT-M space write

$$X(\xi \cdot X) = \xi_{\mu;\nu} a^\mu a^\nu + a^\mu{}_{;\nu} a^\nu \xi_\mu .$$

Since ξ is a Killing vector $\xi_{\mu;\nu} + \xi_{\nu;\mu} = 0$, and since X is geodesic $a^\mu{}_{;\nu} a^\nu = 0$. Hence

$$d(\xi \cdot X)/du \equiv X(\xi \cdot X) = 0 ; \tag{8.16}$$

that is, the dot product $\xi \cdot X$ is constant along the geodesic.

There are four Killing vectors of T-NUT-M, three expressing spatial homogeneity and one due to the fact that $\alpha_3 = \alpha_2$. This one is the vector dual to the one-form σ^1 (notice that its contravariant components are $\xi^\mu = \delta_1{}^\mu$ so $\xi_\mu = g_{1\mu}$). Since $\xi^\mu = \delta_1{}^\mu$, $a_1 = \text{const}$. This implies

$$a_1 = a^0 + ga^1 = \text{const.} \equiv p_{\parallel} . \tag{8.17}$$

There are also three constants of motion

$$p_i = \xi_i \cdot X , \tag{8.18}$$

where the ξ_i are the three generators of SO(3, R). We know that $\xi_i = O_i{}^j Y_j$, where Y_i is dual to ω^i, and $O_i{}^j$ is an orthogonal transformation (here a rotation matrix). Because $O_i{}^j$ is a rotation,

$$p^2 = \sum_i (p_i)^2 = \text{const.}$$

Since $\sigma^1 = \omega^1$, $\sigma^2 = e^{\beta}\omega^2$, $\sigma^3 = e^{\beta}\omega^3$, we have the formulas $Y_1 \cdot X = a_1$, $Y_2 \cdot X = e^{\beta} a_2$, $Y_3 \cdot X = e^{\beta} a_3$. Therefore,

$$p^2 = a_1{}^2 + e^{2\beta} a_2{}^2 + e^{2\beta} a_3{}^2 = \text{const.}$$

and since a_1 is constant

$$p_{\perp}{}^2 \equiv p^2 - p_{\parallel}{}^2 = e^{2\beta}((a^2)^2 + (a^3)^2) = \text{const.} \tag{8.19}$$

(Remember $a_2 = a^2$, $a_3 = a^3$ in the $\{dt, \sigma^i\}$ basis.)

The Geodesic "Potential"

From our definitions of $p_{\perp}$, $p_{\parallel}$, and E we find

$$p_{\parallel}{}^2 - (a^0)^2 + gp_{\perp}{}^2 e^{-2\beta} = gE .$$

Since the a^{μ} are functions of u and $a^0 = dt/du \equiv \dot{t}$ we find

$$p_{\parallel}{}^2 = (\dot{t})^2 + g(E - p_{\perp} e^{-2\beta}) . \tag{8.20}$$

Notice that this equation looks exactly like the equation of a particle moving in a potential, where u takes the place of time, t takes the place

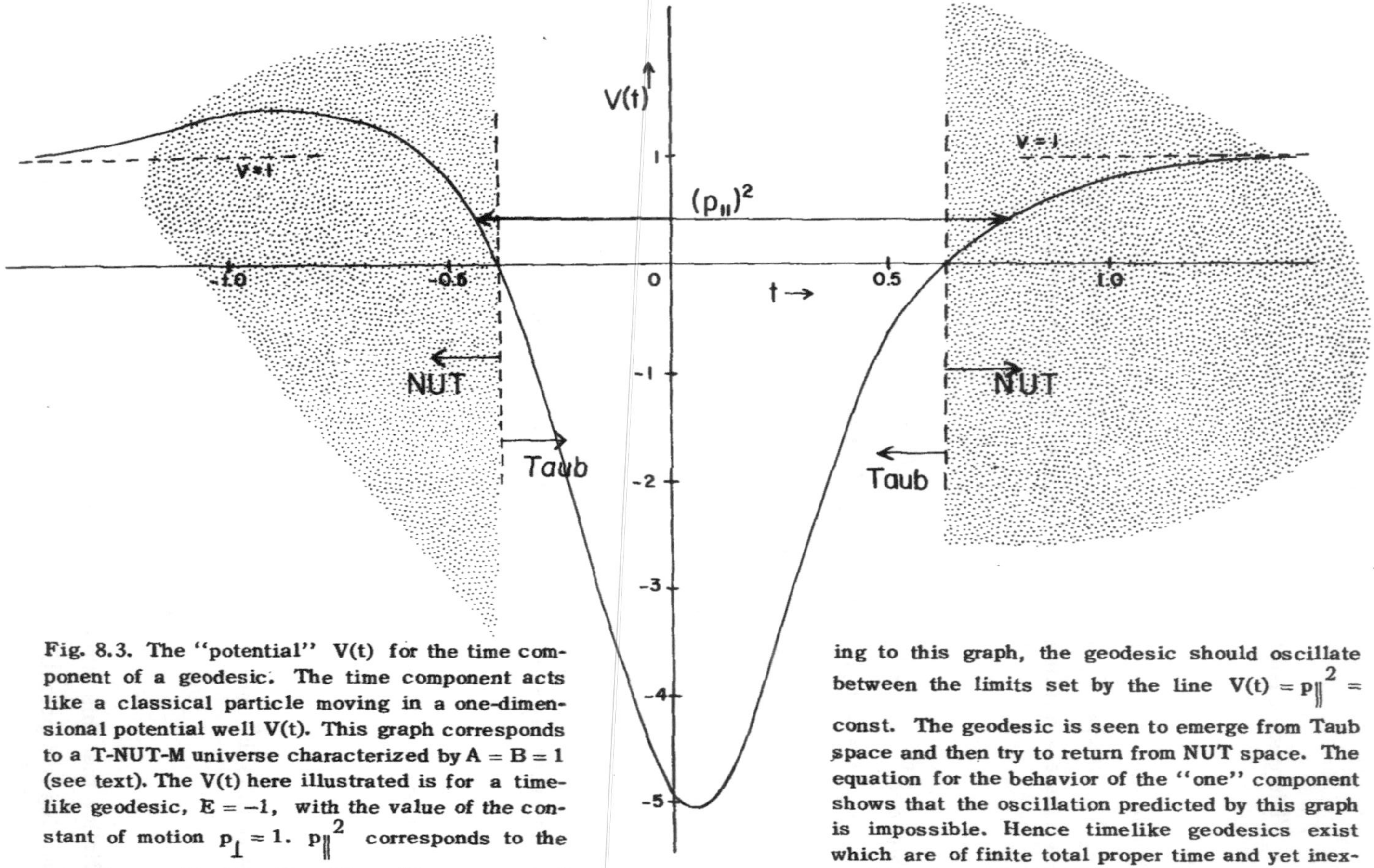

Fig. 8.3. The "potential" V(t) for the time component of a geodesic. The time component acts like a classical particle moving in a one-dimensional potential well V(t). This graph corresponds to a T-NUT-M universe characterized by $A = B = 1$ (see text). The V(t) here illustrated is for a timelike geodesic, $E = -1$, with the value of the constant of motion $p_{\perp} = 1$. ${p_{\parallel}}^2$ corresponds to the total "energy" of the time ("zero") component of this path's tangent, and is always ≥ 0. According to this graph, the geodesic should oscillate between the limits set by the line $V(t) = {p_{\parallel}}^2 =$ const. The geodesic is seen to emerge from Taub space and then try to return from NUT space. The equation for the behavior of the "one" component shows that the oscillation predicted by this graph is impossible. Hence timelike geodesics exist which are of finite total proper time and yet inextendible, i.e., T-NUT-M is *incomplete*.

of position, $p_{\parallel}^2$ takes the place of "total energy," and $g(E - p_{\perp}^2 e^{-2\beta})$ is a "potential" function $V(t)$. Inserting our values for g and β from (8.11) and (8.12) we find

$$V(t) = \left(\frac{At + 1 - 4B^2t^2}{4B^3t^2 + B}\right)\left(E - \frac{p_{\perp}^2}{Bt^2 + \frac{1}{4B}}\right). \tag{8.21}$$

(Figure 8.3 shows $V(t)$ for one set of constants.)

In (8.21) the constants play different roles. A and B are fixed for all discussion of any given manifold. $E = \pm 1, 0$ determines the character of the geodesic (spacelike, timelike, null). The constant $p_{\perp}^2$ is a continuously adjustable constant roughly equivalent to a measure of the components of the geodesic in the 2 and 3 directions.

Incompleteness Without Singularity

T-NUT-M is *mathematically non-singular* in the sense of Chapter 5. Any finite segment of a geodesic is coverable by a compact subset of the manifold. This is true because $V(t)$ becomes constant for $t \to \infty$, so only geodesics of infinite parameter length lead to $t = +\infty$. (See Figure 8.3 for the timelike case.) Any finite geodesic segment must remain within finite bounds of the t coordinate and so is coverable by a compact subset of the manifold.

While T-NUT-M is non-singular, it is still incomplete, (in Chapter 5 we indicated why incompleteness is thought of as "physically singular"). In the T-NUT-M case incompleteness unambiguously leads to singularity once any amount of matter is added. What is the easiest way to see that T-NUT-M is incomplete? In Figure 8.3 we see that a timelike geodesic with $p_{\parallel}^2 < 1$ does not have enough energy to surmount the "potential barrier." From the form of the potential, t should oscillate as a function of u, the affine parameter, but we shall show that in fact the geodesic terminates at a finite value of u.

Consider equation (8.17) for $p_{\parallel}$:

$$a^1 = (p_{\parallel} - \dot{t})/g\,.$$

From equation (8.20) for $\dot{t}$ we have

$$a^1 = \frac{1}{g}\left[p_{\|} \mp \left(p_{\|}^{\,2} - gE + gp^{\,2}/b^2\right)^{\frac{1}{2}}\right].$$

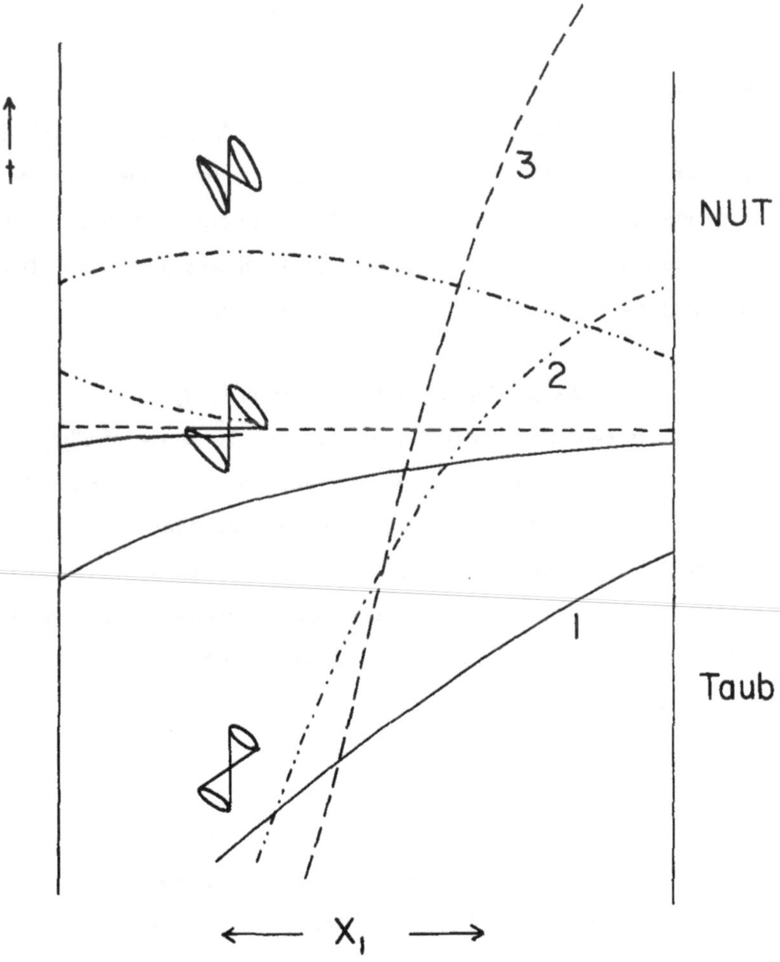

Figure 8.4. T-NUT-M Space with Three Timelike Geodesics. Vertical directions are the t coordinate (always lightlike, see text). Horizontal directions correspond to the "one" direction in S^3 (a closed direction). The "one" direction is periodic – the solid vertical lines are identified pointwise. Geodesics 1 and 2 terminate with total finite length. No. 1 cannot cross the Misner boundary: It winds up too fast. No. 2 crosses the boundary but bounces back. The light cone behavior shows it cannot return and must be terminated. Geodesic 3 has infinite total length because it has enough "energy" to travel to $t = +\infty$.

If we take the + sign, (and $p_{\|} > 0$) then as $t \to t_0$, $g(t_0) \to 0$ and $a^1 \to \infty$. This infinite value of a^1 corresponds to geodesic No. 1 in Figure 8.4. This geodesic "wraps" around the "one" direction an infinite number of times and does not cross the boundary between Taub and NUT space. However, potential-well considerations show that the approach to the Misner boundary takes place in a finite proper time. Hence this geodesic is of finite extent and inextendible. T-NUT-M space is incomplete!

Not all geodesics are incomplete. Consider geodesics 2 and 3 in Figure 8.4. Both originate in Taub space, cross into NUT space. Geodesic No. 2 bounces off the potential and tries to return to Taub space. It encounters the same type of problem as No. 1 and is incomplete. Geodesic No. 3, however, surmounts the barrier and travels to $t = +\infty$.

The Collapse of Taub Space

In Taub's (1951) model the volume of the spacelike homogeneous hypersurface $H(t)$ becomes zero at finite values of t. Rather than indicating a singularity, however, this zero volume simply means that the Taub model is that portion of T-NUT-M space in which $g(t) > 0$. T-NUT-M space itself is a mathematically non-singular model, as we have seen.

T-NUT-M at first glance appears to be a useful model of the real universe. The behavior of T-NUT-M, with regard to collapse – the opening of closed spacelike section – is exactly the behavior we might expect in a non-singular, matter-filled universe. In a universe filled with matter, it will be shown that the volume of spacelike hypersurfaces must become zero in some observer's finite proper time. In T-NUT-M this zero volume is not indicative of a mathematical singularity; the closed spacelike sections become open with only the difficulty that some (not all!) of the timelike geodesics are incomplete. As shown by work of Hawking (1967), Geroch (1967), and others, non-singular cosmological models in which all geodesics (in addition to the ones along which matter flows) are complete do not exist in general.

However, T-NUT-M is not adequate to describe the real universe near a time of maximum contraction. Since T-NUT-M space is a vacuum model, it can only be thought of as an approximation to the cosmos in which matter is treated as test particles. Because some of the test particles may travel on incomplete geodesics in T-NUT-M, the model itself, as it stands, is not adequate to portray the cosmos. If some way could be found to introduce matter into T-NUT-M in an explicit way so that none of the matter travels on incomplete geodesics, then a non-singular cosmological model would be the result. That it is *impossible* to introduce matter into T-NUT-M without causing a singularity is a fact that will be shown later.

One final remark should be made: Misner and Taub (1968) have shown that there are two inequivalent ways of joining Taub space onto NUT space through a lightlike boundary. In both ways the join is analytic. As we saw, half of the timelike geodesics with origin in Taub space cross the boundary to NUT space and half do not. In the second method of gluing T and NUT together, these two sets of geodesics are (roughly) interchanged. The half that made it before are brought to a halt, and those geodesics that originally ran out of steam suddenly find themselves given free passage. Non-analytic extensions may be considered also. A recent mathematical discussion of this ambiguity in joining solutions across a lightlike barrier has been carried through by Miller (1973) and Miller and Kruskal (1973). A physical interpretation of this recent work has yet to be given.

9. THE GENERAL SPATIALLY HOMOGENEOUS MODEL IN THE SYNCHRONOUS SYSTEM

They take a serpentine course, their arms flash in
the sun – hark to the musical clank –
– WALT WHITMAN

9.1. The Metric and Connection Forms

The T-NUT-M model shows that a spatially homogeneous model can have exciting characteristics. To extend our discussion to the general spatially homogeneous model we must first develop some useful equations. We shall compute the Ricci tensor of homogeneous models in a particularly simple basis – the synchronous system. These equations will be applicable to both matter-filled and vacuum models. In succeeding chapters we shall use the synchronous system to discuss the existence of singularities in these models. Figure 8.1 included a logical outline of Chapters 9 and 10.

The Proper Time Orthogonal to Homogeneous Hypersurfaces

Through every point in a spatially homogeneous model M passes an invariant or homogeneous, three-dimensional hypersurface H. This hypersurface is generated by the three-dimensional isometry group G of the model. A one-parameter family of these hypersurfaces fill M. The direction of the axis of this parameter t may be chosen quite freely. Once this choice has been made, the one-forms ω^i, $i = 1,2,3$, are found as in Chapter 6. The ω^i satisfy the curl relations appropriate to the group G.

One very useful choice for the direction of t is the timelike direction perpendicular to each hypersurface H. This choice defines the *synchronous system*, so called because clocks are synchronized throughout the spacelike surfaces (see Lifshitz and Khalatnikov, 1963). The existence of such a timelike normal vector assumes that the H(t) are spacelike.

We parametrize by proper time, so that the four-dimensional metric is

$$ds^2 = -dt^2 + g_{ij}\omega^i\omega^j .$$
$$g_{ij} = g_{ij}(t) . \tag{9.1}$$

The one-forms ω^i obey

$$d\omega^i = \frac{1}{2} C^i{}_{st}\omega^s \wedge \omega^t , \tag{9.2}$$

where the $C^i{}_{jk}$ are the structure constants of G. The group structure of the manifold implies the existence of a vector field basis dual to $\{-dt, \omega^i\}$, such that

$$[Y_0, Y_i] = 0, \ [Y_i, Y_j] = -C^s{}_{ij} Y_s ,$$

$$Y_0 \cdot Y_0 = -1, \ Y_0 \cdot Y_i = 0, \ Y_i \cdot Y_j = g_{ij}(t) .$$

We could have begun with the existence of this vector field basis and proceeded to (9.1) and (9.2) as in Chapter 6.

The synchronous basis is essentially unique as long as the homogeneous hypersurfaces remain spacelike. That is, the hypersurfaces are picked out by the group action unambiguously. The vector Y_0 is the unique normal to these surfaces.

If the H(t) change from spacelike to timelike (as in T-NUT-M), the synchronous system breaks down and another basis must be used. If a singularity prevents this change of signature (as in any matter-filled model with the same symmetry as T-NUT-M, for example) the synchronous system is useful to describe the entire evolutionary history of the model.

The Orthonormal Basis

We shall now compute the Ricci tensor. Since this form of the Ricci tensor will be applicable to the FRW and T-NUT-M solutions as well as the generally spatially homogeneous model, we shall derive it in some detail.

We change bases to write (9.1) in the *orthonormal synchronous basis*, defined by

$$\sigma^0 = dt, \quad \sigma^1 = b_{is}(t)\omega^s, \tag{9.3}$$

where $b_{is}b_{sj} = g_{ij}$ and $b_{ij} = b_{ji}$. The matrix $B = (b_{ij})$ is the symmetric square root of $G = (g_{ij})$. This choice of B puts the metric in diagonal or Minkowski form. In the σ^μ basis

$$ds^2 = \eta_{\mu\nu}\sigma^\mu\sigma^\nu = -(\sigma^0)^2 + (\sigma^1)^2 + (\sigma^2)^2 + (\sigma^3)^2. \tag{9.4}$$

Following Misner we shall write the scalar $(\det B)^{\frac{1}{3}}$ as $(\det B)^{\frac{1}{3}} \equiv e^{-\Omega(t)}$, Ω a scalar. Also, we write $e^\Omega B \equiv (e^{\beta(t)}{}_{ij})$, where $(\beta_{ij}(t))$ is a 3×3 traceless matrix (e^β signifies matrix exponentiation; note that $\det e^\beta = e^{\mathrm{tr}\beta} = 1$). Therefore

$$B = (e^{-\Omega}e^{\beta}{}_{ij}). \tag{9.5}$$

The curls of the σ^μ will be needed to compute affine connection forms. These curls are

$$\begin{aligned} d\sigma^i &= (-\dot\Omega e^{-\Omega}e^{\beta}{}_{is} + e^{-\Omega}(e^{\beta}{}_{is})^{\cdot})dt\wedge\omega^s + e^{-\Omega}e^{\beta}{}_{is}\,d\omega^s \\ &= (-\dot\Omega\delta_{iu} + (e^{\beta}{}_{it})^{\cdot}e^{-\beta}{}_{tu})e^{-\Omega}e^{\beta}{}_{us}\,dt\wedge\omega^s \\ &+ \tfrac{1}{2}e^{-\Omega}e^{\beta}{}_{is}C^s{}_{tu}\omega^t\wedge\omega^u. \end{aligned} \tag{9.6}$$

Using the expression of ω^i in terms of σ^i,

$$\omega^i = e^{\Omega}e^{-\beta}{}_{is}\sigma^s, \tag{9.7}$$

we have

$$d\sigma^i = k_{is}\,dt\wedge\sigma^s + \tfrac{1}{2}d^i{}_{st}\sigma^s\wedge\sigma^t, \tag{9.8}$$

where

$$\begin{aligned} k_{ij} &= -\dot\Omega\delta_{ij} + (e^{\beta}{}_{it})^{\cdot}(e^{-\beta}{}_{tj}), \\ d^i{}_{jk} &= e^{\Omega}e^{\beta}{}_{is}C^s{}_{tu}e^{-\beta}{}_{tj}e^{-\beta}{}_{uk}. \end{aligned} \tag{9.9}$$

The expression for k_{ij} simplifies if $\dot{\beta}$ commutes with β, for then $(e^{\beta}{}_{it})^{\cdot} e^{-\beta}{}_{tj} = \dot{\beta}_{ij}$. Notice also that $d^i{}_{jk}$ has the same symmetry as $C^i{}_{jk}$ and satisfies the Jacobi equation:

$$d^i{}_{jk} = -d^i{}_{kj}; \quad d^a{}_{si} d^s{}_{jk} + d^a{}_{sj} d^s{}_{ki} + d^a{}_{sk} d^s{}_{ij} = 0 . \tag{9.10}$$

Connection Forms in the Orthonormal Basis

We shall now compute the connection forms $\sigma^{\mu}{}_{\nu} = \Gamma^{\mu}{}_{\nu\sigma}\sigma^{\sigma}$ using the first Cartan equation and the vanishing of the covariant derivatives of the metric. The latter equation reads, in our case,

$$\sigma_{\mu\nu} + \sigma_{\nu\mu} = 0 \tag{9.11}$$

(to lower or raise an index use $\eta_{\mu\nu}$). This equation implies

$$\sigma^0{}_0 = 0, \quad \sigma^0{}_i = \sigma^i{}_0, \quad \sigma^i{}_j = -\sigma^j{}_i .$$

The first Cartan equation reads

$$\Gamma^0{}_{\mu\nu}\sigma^{\mu} \wedge \sigma^{\nu} = 0; \quad k_{is}\sigma^0 \wedge \sigma^s + \frac{1}{2} d^i{}_{st}\sigma^s \wedge \sigma^t = \Gamma^i{}_{\mu\nu}\sigma^{\mu} \wedge \sigma^{\nu} .$$

Enforcing the symmetry (9.11), we have

$$k_{ij} = \Gamma^i{}_{0j} - \Gamma^i{}_{j0}; \quad d^i{}_{st} = \Gamma^i{}_{st} - \Gamma^i{}_{ts}; \quad \Gamma^0{}_{\mu\nu} - \Gamma^0{}_{\nu\mu} = 0 .$$

The solution of these equations is

$$\begin{aligned}
\Gamma^0{}_{i0} &= 0, \\
\Gamma^0{}_{ij} &= \Gamma^0{}_{ji} = \frac{1}{2}(k_{ij} + k_{ji}) \equiv \ell_{ij}, \\
\Gamma^i{}_{j0} &= \frac{1}{2}(k_{ji} - k_{ij}) \equiv m_{ji} = -m_{ij}, \\
\Gamma^i{}_{st} &= \frac{1}{2}(d^i{}_{st} - d^t{}_{is} - d^s{}_{it}),
\end{aligned} \tag{9.12}$$

whence

$$\begin{aligned} \sigma^0{}_i &= \ell_{is}\sigma^s \\ \sigma^i{}_j &= -m_{ij}\sigma^0 + \frac{1}{2}(d^i{}_{js} - d^j{}_{is} - d^s{}_{ij})\sigma^s . \end{aligned} \tag{9.13}$$

9.2. Aside: The Second Fundamental Form of an Invariant Hypersurface

We have defined the symmetric and antisymmetric parts of k_{ij} as ℓ_{ij} and m_{ij} respectively, that is, in matrix language:

$$L = \frac{1}{2}(K+K^T); \quad M = \frac{1}{2}(K-K^T) . \tag{9.14}$$

From $K = -\dot{\Omega}1 + (e^{\beta})^{\cdot}e^{-\beta}$ one can show that

$$e^{\beta}L - Le^{\beta} = e^{\beta}M + Me^{\beta} . \tag{9.15}$$

Initial Data

Equation (9.15) is a matrix equation relating M, L and e^{β}; e^{β} being symmetric and positive-definite, L being symmetric and M being antisymmetric. Because the matrices are all 3×3, we may solve (9.15) uniquely for M. Because of this fact it is necessary only to give L, β and Ω at some time t_0 to get a complete set of initial data. We shall find this enumeration useful later when we wish to give initial conditions for numerical solutions.

It is not surprising that Ω, β, and L form a set of initial data for homogeneous cosmologies. The equations of relativity, being hyperbolic, require two items of initial data (Choquet, 1962): First, a "function" (the metric of an initial three-surface); and second, the "first derivative" of this function (the second fundamental form of the initial surface, its extrinsic curvature; see Wheeler, 1962b). The matrix $e^{-\Omega}e^{\beta}$ is directly related to the three-dimensional metric of the homogeneous hypersurface. From our definition of K we see that L is related to the time derivative of $e^{-\Omega}e^{\beta}$ (both L and $e^{-\Omega}e^{\beta}$ have six independent components).

We can show directly that L is the second fundamental form of the invariant hypersurface H(t) imbedded in the four-dimensional spacetime.

The *second fundamental form* of a submanifold M' in a manifold M is an operation $N(X,Y)$ acting on vectors X, Y of M'. If we covariantly differentiate Y of M' in the direction X using the connection forms of M the result is a vector Z in M. We define $N(X,Y)$ as the component of Z perpendicular to M' (Eisenhart, 1926). We write

$$N(X,Y) = (\nabla_X(Y))^{\perp}$$

where ∇ is covariant differentiation in M and the superscript $\perp$ (superperp) extracts the component of the result which is perpendicular to M'.

We can use the torsionless nature of $\nabla_X Y$ to show that

$$N(X,Y) = N(Y,X) .$$

If M' is a hypersurface, the perpendicular vectors form a one-dimensional space, so N maps two vectors of M' into a real number (the one component of $Z^{\perp}$). Thus N is a *tensor* on M' (depending on the detailed manner in which M' is imbedded in M).

For a spatially homogeneous manifold in the synchronous system, perpendicular means along the σ^0 direction (here M' is $H(t)$). The components of N in the basis X_i dual to the σ^i are

$$N_{ij} = N(X_i, X_j) = \Gamma^0{}_{ij} = \ell_{ij} . \tag{9.16}$$

This equation shows that L is indeed the second fundamental form of $H(t)$.

9.3. The Ricci Tensor

We now calculate the components of the Ricci tensor in the $\{\sigma^\mu\}$ basis.

The (00) component is (indices raised and lowered with $\eta_{\mu\nu}$; use the symmetries of $R_{\mu\nu\alpha\beta}$):

$$R_{00} = R^\sigma{}_{0\sigma0} = -R^0{}_{101} - R^0{}_{202} - R^0{}_{303} .$$

Since $\theta^0{}_i = R^0{}_{i\mu\nu}\sigma^\mu \wedge \sigma^\nu$, we compute $\theta^0{}_i = d\sigma^0{}_i + \sigma^0{}_\alpha \wedge \sigma^\alpha{}_i$ and take its $0i$ component:

$$R^0{}_{i0i} = \dot{\ell}_{ii} + \ell_{is}k_{si} - m_{si}\ell_{is} \text{ (no sum on } i) . \tag{9.17}$$

Therefore

$$R_{00} = -\dot{\ell}_{ss} - \ell_{st}\ell_{ts} . \tag{9.18}$$

Similarly we find

$$\begin{aligned} R_{0i} = R^\sigma{}_{0\sigma i} &= \ell_{st}d^t{}_{si} + \ell_{st}\Gamma^t{}_{si} - \ell_{ti}\Gamma^t{}_{ss} \\ &= \ell_{st}d^t{}_{si} + \ell_{ti}d^s{}_{ts} . \end{aligned} \tag{9.19}$$

Note that

$$d^s{}_{ts} = e^\Omega e^\beta{}_{sa}C^a{}_{bc}e^{-\beta}{}_{bt}e^{-\beta}{}_{cs} = e^\Omega e^{-\beta}{}_{bt}C^s{}_{bs} . \tag{9.20}$$

This equation implies that the second term on the right in (9.19) vanishes for some groups. For example, it is zero for the groups T_3, L_3, and $SO(3, R)$ (Bianchi Types I, VIII, and IX).

Example: R_{0i} *and Rotation in Bianchi Type I Models*

As an example, consider a space invariant under $T_3(C^i{}_{jk} = 0)$. Since $d^i{}_{jk} = 0$, (9.19) implies $R_{0i} = 0$. For a fluid-filled universe $T_{0i} = (w+p)u_0 u_i$ in our basis. Hence $R_{0i} = 0$ implies $u_i = 0$ (or, since $u^\mu u^\nu \eta_{\mu\nu} = -1$, $u_\mu = (-1, 0, 0, 0)$). Consequently, $du = d(-dt) = 0$, so that rotation is impossible in Type I universes.

The Spatial Components of the Ricci Tensor

A tedious computation shows

$$\begin{aligned} R_{ij} &= \dot{\ell}_{ij} + \ell_{ij}\ell_{ss} + \ell_{is}m_{sj} + \ell_{js}m_{si} \\ &\quad + \tfrac{1}{2}d^t{}_{ts}(d^i{}_{sj} + d^j{}_{si}) - \tfrac{1}{2}d^s{}_{it}(d^s{}_{jt} + d^t{}_{js}) + \tfrac{1}{4}d^i{}_{st}d^j{}_{st} . \end{aligned} \tag{9.21}$$

We have used the fact that the three-dimensional Jacobi identity implies $d^t{}_{ts} d^s{}_{ij} = 0$.

This completes the calculation of the Ricci tensor in the orthonormal synchronous basis. For many purposes the synchronous system is very useful. For some purposes we shall use other bases, as we did with the T-NUT-M model.

9.4. Vacuum Models – Existence and Examples

In vacuum the Einstein field equations read

$$R_{\mu\nu} = 0 .$$

Taub (1951) showed that solutions exist for all nine Bianchi types. These models are generally of little physical interest, and investigations of their global properties, especially the existence of singularities have not been carried through.

Taub's existence proof is outlined as follows: From $R_{ij} = 0$ we have

$$\begin{aligned}\dot{\ell}_{ij} = -\ell_{ij}\ell_{ss} - \ell_{is}m_{sj} - \ell_{js}m_{si} - \tfrac{1}{2}d^t{}_{ts}(d^i{}_{sj} + d^j{}_{si}) \\ + \tfrac{1}{2}d^s{}_{it}(d^s{}_{jt} + d^t{}_{js}) - \tfrac{1}{4}d^i{}_{st}d^j{}_{st} .\end{aligned} \tag{9.22}$$

Our definition of $e^{-\Omega}e^{\beta}$ gives

$$-\dot{\Omega}e^{-\Omega}e^{\beta}{}_{ij} + e^{-\Omega}(e^{\beta}{}_{ij})^{\cdot} = (\ell_{is} + m_{is})e^{-\Omega}e^{\beta}{}_{sj} . \tag{9.23}$$

From (9.9) we have

$$d^i{}_{jk} = e^{\Omega}e^{\beta}{}_{is}C^s{}_{tu}e^{-\beta}{}_{tj}e^{-\beta}{}_{uk} ,$$

and (9.15) gives

$$m_{ij} = m_{ij}(e^{\beta}, L) . \tag{9.24}$$

Thus the (ij) field equations may be put into the form

$$
\begin{aligned}
\dot{L} &= f(\Omega, \beta, L)\,, \\
\dot{\Omega} &= g(\Omega, \beta, L)\,, \\
\dot{\beta}_{ij} &= h(\Omega, \beta, L)\,.
\end{aligned}
\tag{9.25}
$$

Given Ω, β_{ij}, and L_{ij} (one, five and six quantities respectively) at time t_0 we can integrate this set of equations to find Ω, β, and L at any time t. Note that Ω, β_{ij}, $\dot{\Omega}$, and $\dot{\beta}_{ij}$ are sufficient initial data for these differential equations since L_{ij} may be determined from these quantities.

Restrictions on the Initial Data

In the discussion above we used only the (ij) Einstein equations, ignoring the four (0μ) equations. It is a general property of the Einstein equations that the $R_{0\mu}$ equations are consistent with the solution of the R_{ij} equations, and only act as restriction on the initial data (see Wheeler, 1962b). In our case they restrict the values of Ω, β, and L we may specify at t_0.

In the spatially homogeneous case Taub shows the effect of the restriction by examining

$$R_{0i} = \ell_{st} d^t{}_{si} + \ell_{ti} d^s{}_{ts} = 0 \tag{9.26}$$

and

$$
\begin{aligned}
2S &\equiv R_{00} + R_{11} + R_{22} + R_{33} \\
&= (\ell_{ss})^2 - \ell_{st}\ell_{ts} + d^t{}_{ts} d^u{}_{su} - \tfrac{1}{4} d^s{}_{ut} d^s{}_{ut} - \tfrac{1}{2} d^s{}_{ut} d^t{}_{us} = 0\,.
\end{aligned}
\tag{9.27}
$$

(S is defined with a factor of two to conform to a convention used in fluid-filled models.) These two equations must be satisfied if a set of Ω, β, L that we pick as initial data is to be allowable.

We can readily show that $R_{0i} = 0$ and $R_{00} = 0$ are satisfied for all time if they are satisfied at $t = t_0$ and if $R_{ij} = 0$ for all time. The proof uses the twice-contracted Bianchi identity:

$$g^{\mu\nu}(R_{\alpha\mu} - \frac{1}{2} R g_{\alpha\mu})_{;\nu} = 0$$

which implies:

$$R^0{}_{\mu,0} = -R^i{}_{\mu;i} - R^\nu{}_\mu \Gamma^0{}_{\nu 0} + R^0{}_\nu \Gamma^\nu{}_{\mu 0} \, . \tag{9.28}$$

It is easy to see that $R_{\mu\nu} = 0$ at $t = t_0$ implies $R^0{}_{\mu,0} = 0$ so $R^0{}_\mu = 0$ remains true as long as $R_{ij} = 0$. The establishment of this fact completes the demonstration that the (ij) vacuum field equations are consistent with the conditions imposed by the (0μ) equations, and that solutions therefore exist for vacuum models for any of the Bianchi types.

A Bianchi Type III Homogeneous Model

Exact solutions of the vacuum field for various homogeneous cosmologies do not exist in any great number. The T-NUT-M and Kasner solutions are important examples. Taub (1951) has given a solution for Type II. As an example we shall give a Type III solution in which the metric is given except for one component function which is defined by an integration. It represents a simple illustration of the above formal treatment.

Let us investigate the special case where $e^\beta{}_{ij}$ is diagonal, with only one independent component:

$$e^\beta = \mathrm{diag}(e^{2\beta}, e^{-\beta}, e^{-\beta}); \;\; \beta = \beta(t) \, . \tag{9.29}$$

L then has the form

$$L = \mathrm{diag}(\ell_1, \ell, \ell) = \mathrm{diag}(2\dot\beta - \dot\Omega, -(\dot\Omega + \dot\beta), -(\dot\Omega + \dot\beta)) \, . \tag{9.30}$$

The structure constants of the Type III group are

$$C^2{}_{23} = -C^2{}_{32} = 1 \, , \quad \text{all others zero.} \tag{9.31}$$

The field equations reduce to three equations, the S equation and the R_{ij} equations:

$$\begin{aligned} 2\ell\ell_1 + \ell^2 - e^{2(\beta+\Omega)} &= 0 \\ \dot{\ell}_1 + \ell_1(\ell_1 + 2\ell) = \dot{\ell} + \ell(\ell_1 + 2\ell) + \tfrac{1}{2}e^{2(\beta+\Omega)} &= 0\,. \end{aligned} \tag{9.32}$$

If we solve the S equation and one of the two lower equations, the second is automatically satisfied because of the consistency proof above.

The solution we find yields $\beta + \Omega$ as an implicit function of t:

$$t - t_0 = e^{-(\beta+\Omega)/2} + A \log\left[e^{-(\beta+\Omega)/2} + (e^{-(\beta+\Omega)} + A)^{\frac{1}{2}}\right], \tag{9.33}$$

where t_0 and A are constants. We can find $2\beta - \Omega$ from the S equation by

$$\ell_1 = 2\dot{\beta} - \dot{\Omega} = -\tfrac{1}{2}[Ae^{3(\beta+\Omega)} + e^{2(\beta+\Omega)}]^{\frac{1}{2}} + \tfrac{1}{2}[Ae^{-(\beta+\Omega)} + e^{2(\beta+\Omega)}]^{-\frac{1}{2}}. \tag{9.34}$$

We have $\beta+\Omega$ as a function of t so we can integrate this equation to find $2\beta-\Omega$ (this is the integration referred to above). Adding and subtracting these solutions, we find β and Ω.

This model is of interest primarily as an example. Its derivation is typical of the method by which exact solutions for spatially homogeneous vacuum models may be found. Although we have used the orthonormal basis, a coordinated basis is readily found using Table 6.1.

Diagonalization of the General Bianchi Type IX Vacuum Model

The most important of the vacuum models is the Bianchi Type IX (SO(3, R)-homogeneous) model, a special case of which is T-NUT-M. We can show that the most general *vacuum* type IX model has a diagonal metric in the synchronous basis; that is

$$ds^2 = -dt^2 + e^{-2\Omega}[e^{2\beta_1}(\omega^1)^2 + e^{2\beta_2}(\omega^2)^2 + e^{2\beta_3}(\omega^3)^2]\,. \tag{9.35}$$

If e^β is a diagonal matrix, then so is β, and this diagonal traceless matrix may be parametrized by

$$\beta = \mathrm{diag}(\beta_+ + \sqrt{3}\beta_-, \beta_+ - \sqrt{3}\beta_-, -2\beta_+) . \tag{9.36}$$

It is easy to show that β may be chosen diagonal at any one time by transforming to a basis $\bar{\omega}^i = 0^i{}_j\omega^j$, where $0^i{}_j$ is an orthogonal matrix of unit determinant. The $\bar{\omega}^i$ satisfy the Type IX structure equations. Therefore, by a suitable choice of $0^i{}_j$ we may diagonalize e^β and hence β at the instant t_0.

The fact that $R_{0i} = 0$ proves that β remains diagonal. These equations read

$$\ell_{12} \sinh(\beta_{11} - \beta_{22}) = 0, \textit{ et cyc.} \quad \text{at } t = t_0 . \tag{9.37}$$

Thus either $\ell_{12} = 0$ or $\beta_{11} = \beta_{22}$. In general either by the choice of t_0 we may be sure that $\beta_{11} \neq \beta_{22}$, so $\ell_{12} = 0$; or accomplish $\ell_{12} = 0$ by a further rotation of basis forms, so that L is diagonal at t_0.

When e^β and L are diagonal at $t = t_0$, M vanishes by (9.15), and (9.22) shows that L is diagonal. Therefore, L remains diagonal for all time, and hence e^β does also. Thus in the general vacuum Type IX model we may assume β is diagonal.

In Chapter 8 we found no singularity if we took $\beta_1 = \beta_2$. Let us look at this requirement from the standpoint of the general vacuum Type IX equations. The field equations reduce to

$$R_{00} = 0 = 3\ddot{\Omega} - 3\dot{\Omega}^2 - 6\dot{\beta}_+^{\,2} - 6\dot{\beta}_-^{\,2}$$

$$R_{11} = 0 = -\ddot{\Omega} + \ddot{\beta}_+ + \sqrt{3}\ddot{\beta}_- + 3\dot{\Omega}(\dot{\Omega} - \dot{\beta}_+ - \sqrt{3}\dot{\beta}_-)$$
$$+ \frac{1}{2} e^{2\Omega}\left[e^{4(\beta_+ + \sqrt{3}\beta_-)} - e^{4(\beta_+ - \sqrt{3}\beta_-)} - e^{-8\beta_+} + 2e^{-2(\beta_+ + \sqrt{3}\beta_-)}\right]$$

$$R_{22} = 0 = -\ddot{\Omega} + \ddot{\beta}_+ - \sqrt{3}\ddot{\beta}_- + 3\dot{\Omega}(\dot{\Omega} - \dot{\beta}_+ + \sqrt{3}\dot{\beta}_-) \tag{9.38}$$
$$+ \frac{1}{2} e^{2\Omega}\left[e^{4(\beta_+ - \sqrt{3}\beta_-)} - e^{4(\beta_+ + \sqrt{3}\beta_-)} - e^{-8\beta_+} + 2e^{-2(\beta_+ - \sqrt{3}\beta_-)}\right]$$

$$R_{33} = 0 = -\ddot{\Omega} - 2\ddot{\beta}_+ + 3\dot{\Omega}(\dot{\Omega} + 2\dot{\beta}_+)$$
$$+ \frac{1}{2} e^{2\Omega}\left[e^{-8\beta_+} - e^{4(\beta_+ + \sqrt{3}\beta_-)} - e^{4(\beta_+ - \sqrt{3}\beta_-)} + 2e^{4\beta_+}\right] .$$

From these equations it can be shown that $e^{-\Omega}$ becomes zero at a finite value of t, say t′. If no singularity is to exist then, the hypersurface H(t′) must have a metric of signature (+,+,0), so that only one of the coefficients in (9.35) vanishes at t′. If that coefficient is the term multiplying $(\dot{\omega}^3)^2$, we can say that at t′ we have $\Omega + 2\beta_+ \to \infty$, but $\Omega - \beta_+$ and β_- must each be finite. We can write $R_{11} + R_{22} - R_{33} - R_{00}$ as:

$$\begin{aligned} 4(\ddot{\beta}_+ - \ddot{\Omega}) + 6(\dot{\beta}_+ - \dot{\Omega})^2 + 6\dot{\beta}_-^{\,2} - \tfrac{3}{2}\, e^{4(\Omega-\beta_+)}\, e^{-2(\Omega+2\beta_+)} & \\ + e^{2(\Omega-\beta_+)}\left[e^{-2\sqrt{3}\beta_-} + e^{2\sqrt{3}\beta_-}\right] & \\ = -\frac{1}{2}\, e^{2(\Omega+\beta_+)}\left[e^{2\sqrt{3}\beta_-} - e^{-2\sqrt{3}\beta_-}\right] . & \end{aligned} \tag{9.39}$$

At $t = t'$ the condition that no singularity appears requires the left side of (9.39) to be finite. The right side will then be finite only if $\beta_- = 0$ at $t = t'$. This result is only a rough indication of why the T-NUT-M model requires $\beta_- = 0$ for all t, for near t′ the basis used in (9.39) strictly speaking becomes degenerate.

9.5. The T_3-Homogeneous Model of Kasner

The general Type I vacuum model is especially simple and important. It is also known as the T_3-homogeneous model because the Bianchi Type I group is isomorphic to the three-dimensional translation group T_3. This model has been treated by Kasner (1921); Taub (1951); Misner (1967d); and Lifshitz and Khalatnikov (1963). This metric is important in the discussion of matter-filled models near the moment of maximum compactification (see Lifshitz and Khalatnikov, 1963, and Chapter 13). Each vacuum T_3-homogeneous model turns out to be either singular or flat: In the absence of a singularity the vanishing of the Ricci tensor implies the vanishing of the full Riemann tensor (Taub, 1951).

The Vacuum Field Equations for Type I Models

The structure constants of the group T_3 are zero: $C^i{}_{jk} = 0 = d^i{}_{jk}$. Notice that this group structure implies $R_{0i} = 0$. If we examine the previous section we see that the proof of the diagonal form of the general vacuum metric may be carried out in the Type I case also. Thus we need only consider the metric

$$ds^2 = -dt^2 + e^{-2\Omega} e^{2\beta}{}_{ij}\omega^i\omega^j \text{ here } \omega^i = dx^i ; \tag{9.40}$$

with $\beta = \mathrm{diag}(\beta_+ + \sqrt{3}\beta_-, \beta_+ - \sqrt{3}\beta_-, -2\beta_+)$. For this metric the vacuum field equations read

$$\begin{aligned} \dot\Omega^2 - \dot\beta_+{}^2 - \dot\beta_-{}^2 &= 0 \\ \ddot\beta_+ - 3\dot\Omega\dot\beta_+ = \ddot\beta_- - 3\dot\Omega\dot\beta_- &= 0 \\ 2\ddot\Omega - 3\dot\Omega^2 - 3\dot\beta_+{}^2 - 3\dot\beta_-{}^2 &= 0 . \end{aligned} \tag{9.41}$$

Equations (9.41) combine to imply $\ddot\Omega - 3\dot\Omega^2 = 0$, or

$$\Omega = -\ell n \, [H(t-t_0)^{\frac{1}{3}}], \quad H = \text{const}, \quad \text{or } \Omega = 0 . \tag{9.42}$$

If we again look at (9.41) we find

$$\beta_\pm = \ell n(B_\pm(t-t_0)^{\sigma\pm}), \quad B_\pm, \sigma_\pm \text{ const.} \tag{9.43}$$

The solution for the full spatial metric is

$$g_{ij} = e^{-2\Omega} e^{2\beta}{}_{ij} = (t-t_0)^{2\sigma_i} A_i \delta_{ij} \text{ (no sum)} ,$$

where the A_i and the σ_i are each constant. The σ_i obey (from 9.41)

$$\sigma_1 + \sigma_2 + \sigma_3 = 1 , \tag{9.44}$$

and

$$\sigma_1^2 + \sigma_2^2 + \sigma_3^2 = 1 . \tag{9.45}$$

We can now see the final details of the general T_3-homogeneous vacuum model: First, if Ω is zero the result is a flat model, $g_{ij} = \text{const.}$ Second, if Ω is non-zero the models are given by (9.43) where the σ_i satisfy (9.44) and (9.45). Note that if only one of the σ_1 is non-zero we would have (since $d\omega^i = 0$ for Type I, the ω^i are a coordinated basis dx, dy, dz)

$$ds^2 = -dt^2 + t^2dx^2 + dy^2 + dz^2 , \tag{9.46}$$

but a coordinate transformation shows that this is just the metric of a flat manifold.

Equations (9.44) and (9.45) do not allow exactly one of the σ_i to vanish; therefore, all other models have all three σ_i non-zero. In this case the Riemann tensor is non-zero. Note that at $t = t_0$ there is a singularity: The determinant of g_{ij} vanishes ($\Omega \to +\infty$). Moreover, at $t = t_0$ two of the components of g_{ij} vanish (but one becomes infinite: one σ_i must be negative). It is therefore impossible for the spacelike hypersurface H(t) to become a nonsingular null hypersurface at $t = t_0$, and then to become a timelike hypersurface. The "metric" in a nonsingular null hypersurface would be reducible to the form

$$(g_{ij}) = \text{diag}(0, 1, 1) \tag{9.47}$$

in a suitable coordinate system. The conclusion is that $t = t_0$ is a *true singularity*. (The conclusion that $t = t_0$ is a singular time in this model may be obtained in another way: Calculate the scalar $R^{\alpha\beta}{}_{\gamma\delta} R^{\gamma\delta}{}_{\alpha\beta}$ and show that it becomes infinite at $t = t_0$!)

In summary, the T_3-homogeneous vacuum models either have a singularity or are flat. It is interesting and important to contrast the behavior of a T_3-homogeneous vacuum solution with the behavior of T-NUT-M space. In T-NUT-M space the matrix $b_{ij}(t)$ does have vanishing determinant at a value of t (actually the determinant vanishes at two different values of t). However, when the determinant vanishes, g_{ij} is of the form of (9.47). T-NUT-M actually has no metric singularity – the homogeneous hypersurface H(t) merely becomes lightlike at those values of t.

The T_3-homogeneous models are interesting in their own right, and not merely as models to contrast with T-NUT-M. An equation just like (9.43) describes the behavior of anisotropy in a fluid-filled model near a point of maximum contraction. Kasner's model thus proves to be of importance in any discussion of the beginning of the real universe.

10. SINGULARITIES IN SPATIALLY HOMOGENEOUS MODELS

A l'horizon, par les brouillards
Les tintamarres des hasards,
Vagues, nous armons nos démons
Dans l'entre-deux sournois des monts

– ALFRED JARRY

10.1. Singularities in Selected Models

A cosmological model containing a perfect fluid is a manifold on which the metric obeys the field equations

$$R_{\mu\nu} = (w+p)u_\mu u_\nu + \frac{1}{2}(w-p)g_{\mu\nu} , \qquad (10.1)$$

where w is the energy density and p the pressure of the fluid; the u_μ are the components of the fluid velocity field. It is accepted by most cosmologists that a perfect-fluid model can represent the real universe very well (however, see deVaucouleurs, 1970, and Ellis, 1973).

Often the additional requirement of spatial homogeneity is imposed on a fluid model. If isotropy is also required the Friedmann-Robertson-Walker (FRW) models result (see Chapter 4). If homogeneity, but not isotropy, is required, the resulting model will still be fairly tractible from a computational point of view. Such a model may well provide the best description of large scale features of the real universe.

Any model which is used to describe the early stages of the universe must be investigated for singularities. In this chapter we shall treat the existence of singularities in spatially homogeneous, perfect-fluid models. Figure 8.1 includes an outline of this discussion.

We shall take care to distinguish among: 1) coordinate singularities, 2) incompleteness, and 3) mathematical singularities which prevent the extension of the manifold. Let us recall our definition of a mathematically singular manifold M (Chapter 5). It contains a geodesic segment G of finite affine parameter length, and there is no manifold M′ in which M can be imbedded and in which G can be covered by a compact set.

Gravitational Collapse and R_{00}

We shall write the Einstein equations in the orthonormal synchronous system of the previous chapter. Of the Ricci tensor components, the R_{00} is of special importance, and we repeat it here:

$$R_{00} = \dot{\ell}_{ss} - \ell_{st}\ell_{st} , \tag{10.2}$$

where the dot means d/dt, and ℓ_{ij} is given by (9.12). The fluid velocity U is a timelike unit vector (the four components u_μ are functions of time only, as are w and p):

$$u_\mu u^\mu = -1 = -(u_0)^2 + (u_1)^2 + (u_2)^2 + (u_3)^2 . \tag{10.3}$$

The R_{00} field equation is therefore

$$-\dot{\ell}_{ss} - \ell_{st}\ell_{ts} = \frac{1}{2}(w+3p) + (w+p)(u_1^{\ 2}+u_2^{\ 2}+u_3^{\ 2}). \tag{10.4}$$

The energy density w must be a positive definite function, and while p could be slightly negative it is unlikely that p will be less than $-\frac{1}{3}w$. We make the assumption that $p > -\frac{1}{3}w$. This assumption implies that the right side of (10.4) is strictly positive unless the model is a vacuum model:

$$-\dot{\ell}_{ss} - \ell_{st}\ell_{ts} > 0 \ \ (\text{or} = 0 \ \text{ if } \ w = 0) . \tag{10.5}$$

Collapse of the Invariant Hypersurface H(t)

The inequality (10.5) may be written

$$-\dot{\ell}_{ss} - \frac{1}{3}(\ell_{ss})^2 - \frac{1}{3}[(\ell_{11}-\ell_{22})^2 + (\ell_{11}-\ell_{33})^2 + (\ell_{22}-\ell_{33})^2]$$
$$-2[(\ell_{12})^2 + (\ell_{13})^2 + (\ell_{23})^2] > 0 \ \ (\text{or} = 0 \ \text{ if } \ w = 0) ,$$

which implies

$$-\dot{\ell}_{ss} - \frac{1}{3}(\ell_{ss})^2 > 0 \quad (\text{or} \geq 0 \text{ if } w = 0) .$$

If $\ell_{ss} \neq 0$ we have

$$\frac{d}{dt}\left(\frac{1}{\ell_{ss}}\right) > \frac{1}{3} , \tag{10.6}$$

Equation (10.5) shows that ℓ_{ss} cannot vanish for all t if $w \neq 0$. Even if $w = 0$ (vacuum case), ℓ_{ss} cannot vanish for all t, presuming the model is evolutionary.

From (10.6) we see that $|\ell_{ss}| \to \infty$ in a finite interval of (proper) time t, either toward the past or toward the future. If we look at (9.12) we see that

$$\ell_{ss} = -3\dot{\Omega} ; \tag{10.7}$$

thus ℓ_{ss} becomes infinite when $\dot{\Omega}$ becomes infinite. It is important to examine the sign of ℓ_{ss}. We know that ℓ_{ss} is non-zero at some value of t, say t_1. If ℓ_{ss} is positive (expansion) then $\ell_{ss} \to \infty$ at $t_2 < t_1$. If ℓ_{ss} is negative (contraction) then $\ell_{ss} \to -\infty$ at $t_3 > t_1$. Furthermore, suppose there existed a time t_4 at which ℓ_{ss} is zero. From the R_{00} equation for $w \neq 0$ we find $\ell_{ss} > 0$ for $t < t_4$ and $\ell_{ss} < 0$ for $t > t_4$; in this case $|\ell_{ss}| \to \infty$ at a finite time in the past and at a finite time in the future. It is not necessary, however, that ℓ_{ss} ever be zero: In the flat and open FRW models one sign of ℓ_{ss} prevails throughout and $|\ell_{ss}|$ becomes infinite at one time only. These models are infinitely expandible to times greater than this time of infinite $|\ell_{ss}|$.

Equation (10.7) requires $\dot{\Omega} \to \infty$; the possibility that $\dot{\Omega} \to -\infty$ does not occur, as can be seen from the sign of ℓ_{ss}. To show that $\Omega \to \infty$, that $|\ell_{ss}| \to \infty$ does not simply lead to a cusp (Ω finite but $\dot{\Omega}$ infinite) requires other field equations. The R_{ij} equations in fact are inconsistent with $|\ell_{ss}| \to \infty$ unless $e^{-\Omega}$ becomes zero. This point is treated in greater detail for specific models later. Thus collapse may be either toward the past or toward the future, but it is clear that some sort of breakdown occurs in a finite time.

What is Collapse – Singularity or Mere Coordinate Effect?

"Collapse" ($\Omega \to \infty$) indicates that the matrix $e^{-\Omega} e^{\beta}$ becomes singular. It may be, however, that there is no physical singularity, but that we have come to a point at the edge of a coordinate patch where the basis forms are no longer linearly independent. This latter occurrence happens in T-NUT-M space. If the synchronous system is used to describe the Taub universe, then this same "collapse" is found, but it only means spacelike $H(t)$ has become lightlike. In this case the limit of the unique unit timelike vector normal to the spacelike surfaces is a lightlike vector. This vector lies within the invariant surface at the "Misner" boundary. As we have seen in Chapter 8, the use of a lightlike evolution parameter eliminates the apparent singularity of the metric of T-NUT-M space.

Let us call t_0 the time when $\Omega \to \infty$. We will sometimes use the common terminology that the "volume" of $H(t)$ becomes zero when $e^{-\Omega} \to 0$. The terminology stems from the connection of $e^{-3\Omega}$ with a three-dimensional volume element. The zero "volume" of $H(t_0)$ may indicate a singularity or it may be that one direction in $H(t_0)$ is lightlike.

In fact, if $e^{-\Omega} \to 0$ does not correspond to a true singularity, then $H(t_0)$ must be lightlike. If $H(t_0)$ remained spacelike the unique timelike normal could be used to construct a new synchronous system in which Ω would be finite. $H(t_0)$ is lightlike and non-singular in T-NUT-M space, but in a fluid-filled Type IX model, t_0 is a time of singularity. In a fluid-filled Bianchi Type V model $e^{-\Omega} = 0$ does not necessarily correspond to a singularity. (It should be pointed out that the example mentioned later does have both a mathematical and physical singularity at another time when $e^{-\Omega} \neq 0$.)

The global question of whether a general cosmological model need be singular in the mathematical sense or whether there are non-singular theoretical models remains unanswered. Hawking and Ellis (1968) have shown that any model which closely approximates the real universe now (if our observations are correct) must have incomplete geodesics. The time when incompleteness occurs is in our past and is analogous to the time when $e^{-\Omega} \to 0$ in a spatially homogeneous model.

The general spatially homogeneous model is incomplete. We shall prove this fact below.

Models with Velocity Orthogonal to H(t)

In the case where the fluid velocity of a perfect-fluid spatially homogeneous model is orthogonal to each invariant hypersurface, collapse is especially simple to describe. Assume that the fluid velocity u is orthogonal to the H(t), that is, in the synchronous system

$$u_\mu = (-1, 0, 0, 0) . \tag{10.8}$$

This hypothesis implies that $u = -dt$, and $du = 0$ (rotation vanishes). In Type I models u is necessarily of the above form ($R_{0i} = (w+p)u_0u_i = 0$) but in other spatially homogeneous models this need not be so. Note, however, that it is possible to have $u_i \neq 0$ and $du = 0$, so $u = -dt$ is a sufficient but not a necessary condition for non-rotation.

In Chapter 3 we showed that $T^{\mu\nu}{}_{;\nu} = 0$ implies, for fluids,

$$(w+p)u^\sigma{}_{;\sigma} = -w_{,\sigma}u^\sigma ; \tag{10.9}$$

$$(w+p)u_{\mu;\sigma}u^\sigma = -p_{,\sigma}(u^\sigma u_\mu + \delta^\sigma{}_\mu) . \tag{10.10}$$

Equation (10.10) says nothing new in our case: Inserting $u^\mu = \delta^\mu{}_0$ we find $u_{\mu;\sigma}u^\sigma = 0$, but this equation comes directly from the fact that $u = -dt$.

Hypersurface Orthogonal Velocity – Collapse Means Singularity

If we write (10.9) out in full we find

$$-\dot{w}u^0 = (w+p)u^\sigma{}_{;\sigma} = (w+p)(\dot{u}^0 + u^0\ell_{ss} + u^i d^s{}_{is}) . \tag{10.11}$$

With (10.8) this equation becomes

$$\frac{\dot{w}}{w+p} + \ell_{ss} = \frac{\dot{w}}{w+p} - 3\dot{\Omega} = 0 \,. \tag{10.12}$$

We now assume an equation of state $p = p(w)$. (Since p and w are both functions of t only, the existence of such a relation is certainly reasonable.) We can define the "baryon number density" n such that

$$\frac{dn}{n} = \frac{dw}{w+p} \,. \tag{10.13}$$

This definition allows us to solve (10.12):

$$ne^{-3\Omega} = M = \text{const.} \tag{10.14}$$

Thus when $\Omega \to \infty$ at t_0, $n \to \infty$.

This infinite value of n does indeed correspond to a singularity as n is a matter variable defined solely in terms of the fluid's internal properties. We have assumed that $p > -\frac{1}{3}w$, and for sound velocity in the fluid to be less than the speed of light we must have $|p| < w$. By our definition of n, these two conditions imply that if n is infinite then w, the rest energy density, is infinite.

We now return to our definition of mathematical singularity. The models of this section *do* possess a finite geodesic segment, namely the path generated by u from some arbitrary time to t_0. The limit point t_0 of the segment does not lie in the manifold. The manifold cannot be extended to cover a point at which n would be infinite and consequently the manifold has a mathematical singularity.

The Kantowski-Sachs Model

Kantowski (1966) and Kantowski and Sachs (1966) have discussed the one class of spatially-homogeneous models which cannot be described by the formalism we have developed heretofore. The above formalism is dependent on the existence of a three-dimensional group which is transitive on each $H(t)$. The Kantowski-Sachs model has hypersurfaces $H(t)$ which are invariant under a four-dimensional group, but not a three-dimensional one.

From Chapter 6, if H(t) is to be a homogeneous three-space, it must be invariant under a group of dimension greater than or equal to three. A five dimensional group implies the existence of a six-dimensional one (Eisenhart, 1926). A six-dimensional symmetry group is the group of highest dimension possible and must contain a transitive, three-dimensional subgroup. Consequently, only if the invariance group is of four-dimensions does there exist the possibility that the formalism we have been using could be inappropriate.

Because each four-dimensional Lie algebra contains a three-dimensional subalgebra (Kantowski, 1966), there exists a three-dimensional isometry group if a four-dimensional one exists. If this three-dimensional group acts on three-dimensional surfaces, our usual formalism may be used. Only if the three-dimensional subgroup acts on two-dimensional surfaces need we adopt new methods. The Kantowski-Sachs model is the only spatially-homogeneous cosmology in which this situation occurs.

Ricci Tensor Components in a Synchronous Basis

Since the three-dimensional subgroup acts on two-spaces, the two spaces must be surfaces of constant curvature. Kantowski showed that two-surfaces of zero and negative curvature give four-dimensional invariance groups which have transitive, three-dimensional subgroups.

The only spaces that are of interest are those where the two-surfaces are two-spheres. The commutation relations of the three Killing vectors on each two-sphere are

$$[\xi_1, \xi_2] = \xi_3 \quad et\ cyc., \tag{10.15}$$

and therefore the three-dimensional group is of Bianchi Type IX, but it is not transitive on H(t). As Kantowski shows, the fourth Killing vector must commute with these three. If we call this fourth vector η, we have

$$[\eta, \xi_i] = 0, \quad i = 1, 2, 3 . \tag{10.16}$$

In order to describe these models we shall use a synchronous basis with the homogeneous hypersurfaces labelled by a parameter t, as before. We shall label the two surfaces by a parameter r, so $\eta = \partial/\partial r$. Because η commutes with ξ_i, g_{12} and g_{13} are zero, and g_{11} is a function of t only. The metric can therefore be written

$$\begin{aligned} ds^2 &= -dt^2 + e^{-2\Omega}(e^{2\beta}dr^2 + e^{-\beta}(d\theta^2 + \sin^2\theta d\phi^2)), \\ \Omega &= \Omega(t), \ \beta = \beta(t) . \end{aligned} \tag{10.17}$$

The Ricci tensor components in the basis $\{dt, dr, d\theta, \sin\theta\, d\phi\}$ are readily computed to be

$$\begin{aligned} R_{00} &= 3\ddot{\Omega} - \tfrac{3}{2}(\dot{\beta})^2 + \dot{\beta}\dot{\Omega} - 3(\dot{\Omega})^2 , \\ R_{11} &= \ddot{\beta} - \ddot{\Omega} - 3\dot{\beta}\dot{\Omega} + 2(\dot{\Omega})^2 , \\ R_{22} = R_{33} &= -\ddot{\Omega} - \tfrac{1}{2}\ddot{\beta} + 3(\dot{\Omega})^2 + \tfrac{5}{2}\dot{\beta}\dot{\Omega} , \\ R_{01} = R_{02} = R_{03} &= R_{12} = R_{13} = R_{23} = 0 . \end{aligned} \tag{10.18}$$

Consequently, when use is made of the field equations for a fluid, (10.1), we see that u has only a t-component, and it is for this reason we have included the model here:

$$u^\mu = \delta^\mu{}_0 .$$

Singularities

The Kantowski-Sachs model has a metric very similar to that of the other models we have treated in this chapter. As might be suspected the proof that the model has a true singularity is essentially the same as in the previous case.

First, use R_{00} to show that $e^{-6\Omega}$ must vanish within a finite proper time. We shall not give the details here. Next write (10.9) for this metric and find

$$-\frac{\dot{n}}{n} = -3\dot{\Omega}, \quad \text{or} \quad ne^{-3\Omega} = M = \text{const.}$$

as in (10.14). The fact that $n \to \infty$ is a finite proper time leads to the immediate conclusion that a mathematical singularity exists.

10.2. Is the General Spatially Homogeneous Model Mathematically Singular or Simply Incomplete, and Thus Physically Singular, Without Being Mathematically Singular?

Hawking and Ellis (1965) showed that the general fluid-filled spatially homogeneous model is incomplete. Incompleteness may or may not be an indication of a mathematical singularity, as was shown in Chapter 5. We shall therefore proceed to examine the possibility that a non-singular, fluid-filled, spatially homogeneous model exists.

Although no such model is found we shall prove that all models invariant under certain groups (including Bianchi Types I and IX) do have a true singularity (this singularity may or may not involve an infinity in some matter variable). Incompleteness is not always accompanied by a singularity.

Collapse of the H(t)

We shall prove later that a model has a singularity if the matter contained in some three-dimensional volume is compressed into a two-dimensional set, and we shall discuss the general method of proving incompleteness.

We saw above that for the spatially homogeneous metric there is a time t_0 when $|\ell_{ss}| = \infty$. At other times this metric is represented by the non-singular matrix $e^{-2\Omega}e^{2\beta}(t)$. Since the volume element is proportional to $g^{\frac{1}{2}}$, g being the determinant of the metric, we see that at t_0, when $\Omega \to \infty$, the "volume" of H(t) vanishes. At t_0 either a singularity appears or it is discovered that certain geodesic segments cannot be completed (Hawking and Ellis, 1965).

There are several possible explanations of the zero volume of H(t). First, H(t) may be compressed from a three-dimensional to a two-dimensional hypersurface. Later in this section we shall show that this change in topology cannot occur without a singularity. Second, there may be a singularity at t_0, whether it can be directly proved that $H(t_0)$ has undergone a change in topology or not. Third, $H(t_0)$ may simply have changed from spacelike to lightlike without singularity. This last possibility is very interesting since it occurs in the mathematically non-singular but incomplete T-NUT-M model.

Homeomorphism Between $H(t)$ *and* $H(t_0)$

The compression of $H(t)$ from a three-dimensional to a two-dimensional set (that is, a true zero volume) cannot occur in a non-singular model. That is, the topology of $H(t)$ cannot change with t. This *conservation of topology* implies that the only alternative to a singularity at t_0 is for $H(t_0)$ to be lightlike. $H(t_0)$ has the "same" topology as $H(t_1)$ if $H(t_0)$ and $H(t_1)$ can be put into a one-to-one correspondence in a continuous manner (open sets map onto open sets). $H(t_0)$ is then *homeomorphic* to $H(t_1)$ and the correspondence or mapping is a *homeomorphism*.

If we assume $H(t_0)$ is non-singular we can construct a homeomorphism between $H(t_0)$ and $H(t_1)$ by means of the fluid velocity u. First, consider the point p_1 in $H(t_1)$. At p_1 we construct L, the path of a particle in the fluid. This path maps the real line R into a one-dimensional set whose tangent at each point is the vector u. We continue L until it hits $H(t_0)$ and let p_0 be the point at which L hits. Since u is timelike and unit, and since t_0 is the first value of t for which $H(t)$ may be non-spacelike, L will certainly hit $H(t_0)$. In fact, $H(t_0)$ cannot itself contain a timelike line, hence L must pass through $H(t_0)$ (conservation of mass implies that L cannot spontaneously stop short of $H(t_0)$). Thus $h(p_1) = p_0$ defines a map h which maps $H(t_1)$ to $H(t_0)$.

It is essentially a necessary and sufficient condition for non-singularity that the world lines of fluid particles never cross. This criterion makes h continuous (a homeomorphism). Intuitively one can understand this relationship by considering a volume of fluid material in $H(t_1)$. This volume may neither become infinite or zero, nor develop holes if the fluid is to remain continuous. In a spatially homogeneous model all fluid variables are independent of position in H(t), and thus the fluid does remain continuous. $H(t_0)$ is topologically equivalent to $H(t_1)$ so long as no singularity develops. In particular, the equivalence holds even if the character of the metric of $H(t_0)$ is different from that of $H(t_1)$.

The homeomorphic invariance of the hypersurfaces H is closely connected with the physically reasonable motion of a continuous fluid. We explicitly do not allow fluid discontinuities, such as shock waves. In a non-homogeneous model, such an effect should properly be considered (see Grishchuk, 1966; Taub, 1957). That the topology of H(t) cannot change is important in a second sense. The invariance of topology represents in part the fact that we are using a classical theory. One of the features of a quantized theory would be that changes in the topology of spacelike hypersurfaces could perhaps be possible (Wheeler, 1962b).

The Necessity of Incompleteness

If a singularity appears at time t_0 the model is incomplete. Even if no singularity in the mathematical sense appears, we shall show that it is impossible for the model to be complete. The method (Hawking and Ellis, 1965) assumes that the model is complete and draws a contradiction with the stipulation that all homogeneous hypersurfaces are three-dimensional. The assumptions of completeness and non-singularity mean that $H(t_0)$ exists and is lightlike. As we showed $H(t_0)$ must have the topology of each spacelike H(t), that is, be three-dimensional. $H(t_0)$, therefore, contains a basis in which the metric is

$$(g_{ij}) = \mathrm{diag}(0,1,1)\ . \tag{10.19}$$

Thus there is a vector field T in $H(t_0)$ which is null and perpendicular to all vector fields in the hypersurface; that is, $T \cdot X = 0$ if X is tangent to $H(t_0)$. The contravariant vector field T is in the unique normal direction to $H(t_0)$. Because T is perpendicular to $H(t_0)$ as well as tangent to it, the one form corresponding to T (also written T) is proportional to the curl of a function τ:

$$T = \alpha d\tau .$$

Without loss of generality we may take $\alpha = 1$. T is group invariant, curl-free, null, and lies in $H(t_0)$, so

$$\nabla_T T = 0 ,$$

(T is geodesic). T is parametrized by an affine parameter s, so

$$T^\mu = dx^\mu/ds; \quad T^\mu{}_{;\sigma} T^\sigma = 0 = T^\sigma T_\sigma .$$

We know that

$$T_{\mu;\sigma\tau} - T_{\mu;\tau\sigma} = T_\nu R^\nu{}_{\mu\sigma\tau} ,$$

and contracting on μ and σ and contracting the result with T^τ we find

$$(T^\sigma{}_{;\sigma})_{;\tau} T^\tau + T^\sigma{}_{;\tau} T^\tau{}_{;\sigma} = -T^\sigma T^\tau R_{\sigma\tau} . \tag{10.20}$$

It is from this equation that we will draw the contradiction that proves incompleteness.

The general vector in $H(t_0)$ is a linear combination of T and two other independent, spacelike vectors X and Y which may be assumed orthonormal and orthogonal to T. Because $T_{\mu;\nu}$ is symmetric (T curl-free) and orthogonal to T we may expand it as

$$T_{\mu;\nu} = \frac{1}{2}\overline{\theta}(X_\mu X_\nu + Y_\mu Y_\nu) + S_{\mu\nu} ,$$

where $S^\mu{}_\mu = 0$, $S_{\mu\nu} = S_{\nu\mu}$, $S_{\mu\nu} T^\nu = 0$. The function $\overline{\theta}$ is the *expansion* of T.

Equation (10.20) becomes

$$\overline{\theta}_{,\tau} T^\tau + \frac{1}{2}\overline{\theta}^2 + S^\sigma{}_\tau S^\tau{}_\sigma = -T^\sigma T^\tau R_{\sigma\tau} . \tag{10.21}$$

Since $S_{\sigma\tau}$ is orthogonal to T, and since only spacelike and null vectors are orthogonal to T, $S_{\sigma\tau}$ is positive definite: $S^\sigma{}_\tau S^\tau{}_\sigma > 0$. Since $R_{\mu\nu}$

is the Ricci tensor of a fluid-filled universe, $R_{\sigma\tau} T^{\sigma} T^{\tau} = T^{\sigma} T^{\tau}(T_{\sigma\tau} + \frac{1}{2} g_{\sigma\tau} T)$ and $T_{\sigma\tau} = (w+p) u_{\sigma} u_{\tau} + p g_{\sigma\tau}$ implies

$$T^{\sigma} T^{\tau} R_{\sigma\tau} = (w+p)(T^{\sigma} u_{\sigma})^2 > 0 .$$

This inequality implies that

$$\overline{\theta}_{,\tau} T^{\tau} < \frac{1}{2} \overline{\theta}^2, \quad \text{or} \quad (\overline{\theta}^{-1})_{,\tau} T^{\tau} > \frac{1}{2} .$$

Thus $\overline{\theta}$ must become infinite for some value of s, the affine path parameter.

The Caustic Surface M_s *and the Contradiction Implied by its Existence*

By our hypothesis of completeness, a set of points M_s (the caustic surface) must exist on which $\overline{\theta}$ is infinite. For each individual path defined by T there is one value of s (one point) where $\overline{\theta}$ is infinite, so M_s is of lower dimension than $H(t_0)$. Because T is a group invariant vector and $\overline{\theta}$ is a scalar, the set M_s must be transformed into itself by the action of the group G.

However, the existence of M_s of lower dimension than $H(t_0)$ is a contradiction, because all homogeneous hypersurfaces are topologically the same. As we saw above, the existence of M_s of lower dimension than three would imply that all of the matter in a three-dimensional volume would be compressed infinitely. But the existence of M_s is implied by the hypothesis that each geodesic segment in $H(t_0)$ has unique endpoints. Thus this hypothesis is impossible in a non-singular cosmological model.

Consequently all fluid-filled, spatially-homogeneous models are incomplete. They may or may not be singular, *vide* T-NUT-M where T is incomplete and does not necessarily have a unique endpoint. In fact, in T-NUT-M the points where $\overline{\theta} = \infty$ are the entire hypersurface $H(t_0)$. Hawking (1967) and others have shown by a similar technique which avoids symmetry assumptions that a wide variety of cosmological models are incomplete.

The Bianchi Type IX Model: Mathematically Singular as Well as Incomplete

We shall show that the general Bianchi Type IX model is not only incomplete, but is singular in the sense of Chapter 5, that is, it cannot be extended beyond $t = t_0$. We shall show for a Bianchi Type IX model, or indeed for any Bianchi Type model in which $C^s{}_{is} = 0$, that the fluid field equations cannot be satisfied at $t = t_0$.

We construct a general basis which would be valid if $H(t_0)$ were nonsingular and light-like. Let $H(t)$ be invariant under a group G, and suppose $H(t_0)$ has a lightlike geometry, i.e.: one lightlike and two spacelike eigendirections. Then in a neighborhood of a point in $H(t_0)$ there exists a basis $\{\sigma^\mu\}$ of one-forms such that: 1) σ^0 is a gradient, $\sigma^0 = dt$, where t is the parameter labeling the H's and 2) $\sigma^i = e^{-\Omega} e^{\beta}{}_{ij}\,\omega^j$ where the ω^j are the three one-forms invariant under the group ($d\omega^i = \frac{1}{2} C^i{}_{st}\,\omega^s \wedge \omega^t$). In this basis the metric will have the form

$$g_{\mu\nu} = \begin{bmatrix} 0 & 1 & 0 & 0 \\ 1 & g & 0 & 0 \\ 0 & 0 & 1 & 0 \\ 0 & 0 & 0 & 1 \end{bmatrix} \tag{10.22}$$

where $g(t)$ is a function which is zero at $t = t_0$. Thus

$$ds^2 = 2\sigma^0\sigma^1 + g(t)(\sigma^1)^2 + (\sigma^2)^2 + (\sigma^3)^2 \,. \tag{10.23}$$

The $\{\sigma^\mu\}$ system is not unique since t is not unique.

In this system the "zero" direction is lightlike. In Chapter 5 we proved the existence of such a basis, and in Chapter 6 we showed how we could choose the metric properties in the "zero" direction so that (10.23) is valid. When $g(t)$ is positive, we may make a transformation of coordinates and recover the synchronous system, but when $g(t)$ becomes negative $H(t)$ has a timelike direction.

The Einstein Equations

It is a straightforward, if tedious, process to compute the Einstein equations in this basis. It turns out that the R_{11} field equation, which is a constraint equation, is sufficient for our purpose. It is a relation between Ω, β, and their derivatives, and w, p, and u^μ (at least at t_0 when $g = 0$). If $C^s{}_{si} = 0$, however, this equation implies $(w+p) < 0$ at t_0, which is impossible.

Consequently to show that a singularity exists it is sufficient to calculate R_{11} at the moment when $g = 0$. In our basis

$$\begin{aligned} R_{11} &= R^0{}_{101} + R^2{}_{121} + R^3{}_{131} \\ &= gR^0{}_{001} + R^0{}_{212} + gR^1{}_{212} + R^0{}_{313} + gR^1{}_{313} \end{aligned}$$

and at $g = 0$

$$\begin{aligned} R_{11} &= R^0{}_{212} + R^0{}_{313} \\ &= (d\sigma^0{}_2)_{12} + (d\sigma^0{}_3)_{13} + (\sigma^0{}_s \wedge \sigma^s{}_2)_{12} + (\sigma^0{}_s \wedge \sigma^s{}_3)_{13} \end{aligned}$$

where $(\;)_{12}$ indicates that the coefficient of $\sigma^1 \wedge \sigma^2$ is to be taken. The explicit form of $\sigma^\mu{}_\nu$ is found by solving the equations $d\sigma^\mu = -\sigma^\mu{}_\nu \wedge \sigma^\nu$ and $dg_{\mu\nu} = g_{\mu\alpha}\sigma^\alpha{}_\nu + g_{\nu\alpha}\sigma^\alpha{}_\mu$ (remember that $dg_{\mu\nu} \neq 0$ because of the function $g(t)$).

The result for R_{11} is

$$R_{11} = -\tfrac{1}{2}(d^2{}_{13} + d^3{}_{12})^2 - (d^2{}_{12})^2 - (d^3{}_{13})^2 - \tfrac{1}{2}\dot{g}(d^s{}_{1s}) \qquad (10.24)$$

$$(\text{at } g = 0)\,.$$

If the group G is such that $C^s{}_{is} = 0$, then $d^s{}_{is} = 0$ and $R_{11} \leq 0$. The property $C^s_{is} = 0$ is shown by the groups of Types I, VIII, and IX, but not by the group of Bianchi Type V (see below).

Singularity

The preceding subsection showed that if $C^s{}_{is} = 0$ then $R_{11} \leq 0$ at the time when g vanishes and H is lightlike. We have, however, from Einstein's equations

$$R_{11} = (w+p)u_1{}^2 \geq 0 \text{ everywhere} \tag{10.25}$$

($p < -w$ is excluded on physical grounds). This inequality implies either $u_1{}^2 = 0$ (impossible, because at $t = t_0$, $g_{\mu\nu} u^\mu u^\nu = -1$ implies $u_0 u_1 = -1 - u_2{}^2 - u_3^2$) or $(w+p) = 0$ (vacuum). In a vacuum Type IX model with β diagonal, we find that the R_{11} equation reads

$$R_{11} = -\frac{1}{2}[e^{2\beta_1} - e^{2\beta_2}]e^{4\Omega} = 0 ,$$

or $\beta_1 = \beta_2$ (T-NUT-M!).

In the general non-vacuum model $(w+p) > 0$ strictly so (10.24) and (10.25) are contradictory. This contradiction means a non-singular model cannot contain a perfect fluid. Therefore a fluid-filled, spatially homogeneous cosmological model with an isometry group whose structure constants obey $C^s{}_{is} = 0$ must be singular.

This singularity appears precisely at the time when the volume of H(t) vanishes, that is, when the synchronous system breaks down. We shall later give an alternate proof of singularity for $p = 0$ in Type IX models where we show that the matter density becomes infinite at $t = t_0$. Further insight concerning the existence of the singularity and its structure results from the Hamiltonian discussion of the next chapter.

A Type V Model in Which Incompleteness is not Accompanied by a Matter Singularity

We have seen that all spatially homogeneous, fluid-filled models are incomplete. Some are also singular, but by no means all (witness T-NUT-M). Can a fluid-filled homogeneous model be non-singular even if incomplete? If $C^s{}_{is} \neq 0$, our proof of singularity breaks down and some spatially homogeneous models may not have a singularity at $t = t_0$ despite their incompleteness. Shepley (1969a) has described a Type V model in which there exists no barrier to evolution of the H(t) at $t = t_0$. He did this by showing that the equations split into two groups, propagation and constraint, and that the constraint equations can all be satisfied at t_0, so H(t) may evolve past t_0 without trouble. The matter is non-rotating and thus the model is singular somewhere, but it is useful to emphasize the difference between "singular" and "incomplete."

Brill's Electromagnetic Model

T-NUT-M is a vacuum Type IX model which is incomplete yet non-singular. A fluid-filled Type IX model is singular. Brill (1964) has given a non-singular Type IX model containing an electromagnetic field but no fluid. Brill's model is incomplete, as is T-NUT-M.

The field equations for a model containing only an electromagnetic field are (Weber, 1961)

$$R_{\mu\nu} = \frac{1}{2}\left(F_{\mu\sigma}F_{\nu}{}^{\sigma} - \frac{1}{4}F_{\sigma\tau}F^{\sigma\tau}g_{\mu\nu}\right), \tag{10.26}$$

where $F_{\mu\nu}$ is the electromagnetic field tensor. Because $R_{00} > 0$ in (10.2) the volume of a spacelike H(t) will collapse, that is $\Omega \to \infty$ in a finite interval of proper time. Thus H(t) must approach a lightlike hypersurface at some time t_0 if no singularity appears. In the σ^μ basis of (10.23) $R_{11} \leq 0$ (10.24). Although no fluid-filled model is compatible with the inequality, in Brill's model we have

$$R_{11} = \frac{1}{2}F_{1s}F_1{}^s = \frac{1}{2}[(F_{12})^2 + (F_{13})^2] \quad (\text{at } t = t_0). \tag{10.27}$$

This form of R_{11} is consistent with $R_{11} = 0$ if $F_{12} = F_{13} = 0$ at t_0, and we may continue H(t) through t_0.

Brill's universe is a manifold similar to T-NUT-M but filled with an electromagnetic field. Though we have shown why it is mathematically non-singular, it is incomplete. If it contained any amount of fluid it would be singular.

The Barrier to Classical Evolution

As shown above, some spatially homogeneous models are singular, some not. All non-static models are at least incomplete, and all fluid models known are singular. This singularity is a type of barrier. It prevents evolution of the model beyond t_0, when the spacelike invariant hypersurface H(t) would become lightlike if no singularity appeared.

Any classical calculation of the time evolution of the geometry grinds to a halt as it comes up to this stage of evolution. In this sense an "obstacle" may be said to block further prediction of the geometry beyond a certain lightlike hypersurface. This use of the word "obstacle" should not be taken to mean that the physics stops here; only that *classical* physics stops here. This is the place for the quantum form of general relativity. There is no obstacle – except lack of wisdom – to calculating the time evolution of the quantum state of the geometry. The "obstacle" is an obstacle only in this sense, that the classical predicting machinery of relativity cannot penetrate it.

What if we drop the spatial homogeneity postulate, while retaining the classical form of the theory? The works of Hawking, Geroch, and Penrose show that incompleteness, at least, is a property of any model reasonably close to a spatially homogeneous one, even though symmetry is dropped as a postulate. Whether it is also necessary to have a mathematical singularity as in a Type IX model is an open question.

Non-homogeneous, complete models containing matter do exist. The cylindrically symmetric universe of Maitra (1966) is stationary and singularity-free because it is complete. In addition, the Gödel universe has positive matter density and positive (if unduly large) pressure. Here there is a *three-dimensional invariance group (a subgroup of the full five-dimensional invariance group* of isometries), but the three-dimensional invariant hypersurfaces are always timelike.

Maitra's model and Gödel's model are unsatisfactory as cosmological models. Neither contains a region which can be identified with our neighborhood in the real universe. Moreover, any model which can serve as a practical and realistic cosmological model does have at least the property of being incomplete (Hawking and Ellis, 1965). All explicitly known models which can serve as cosmological models are mathematically singular – not merely incomplete (although some are less singular than others, see Collins, 1974). The existence of a singularity, its physical meaning, and the effect of quantum mechanics on it, all are objects of current thinking.

11. HAMILTONIAN COSMOLOGY

Река времен в своем стремленьи
Уносит все дела людей
И топит в пропасти забвенья
Народы, царства и царей.

– GAVRIIL ROMANOVICH DERZHAVIN

11.1. Realistic and Approximate Cosmography

The previous chapters dealt mainly with mathematical notions. Here we begin a series of four chapters on more physical questions appropriate

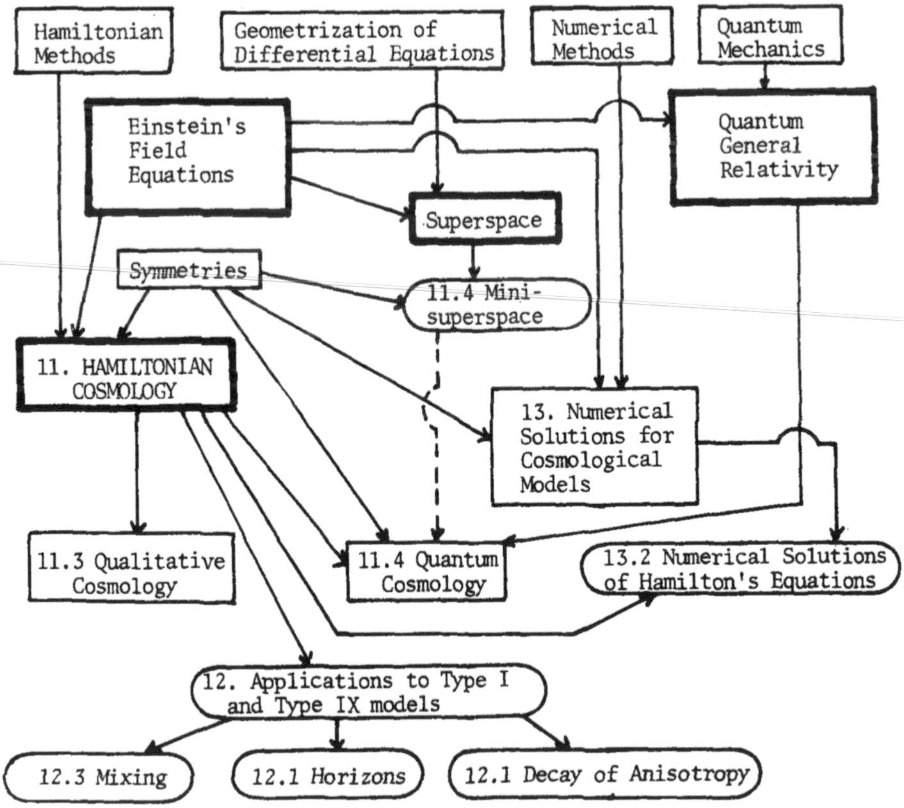

Fig. 11.1. Flow Chart for Chapters 11, 12, 13.

to the general spatially homogeneous cosmological model. The portrayal of the real universe by a homogeneous model allows very complex problems to be treated; at the same time, the high symmetry of the model makes these problems tractable. Figure 11.1 is a flow chart for Chapters 11, 12, and 13.

For the portraits we will paint of the universe (different portraits to emphasize different features), we accept the theory of general relativity without a cosmological constant. In our discussion of the FRW universe in Chapter 4, and in a discussion of the existence of singularities, we find that a dust-filled model (pressure assumed to be zero) is adequate for the description of certain interesting features. These features, in particular the qualitative behavior of the universe near a singularity, arise in a model where the curvature satisfies Einstein's equations for dust:

$$R_{\mu\nu} - \frac{1}{2} Rg_{\mu\nu} = \rho u_\mu u_\nu \,. \tag{11.1}$$

Here ρ is the matter density and u_μ the components of the covariant velocity field in some convenient basis.

Cosmological Studies with Spatially Homogeneous Models

In a spatially homogeneous model, space-time is filled with a one-parameter set of invariant hypersurfaces H(t). Spatial homogeneity means that the metric on each H(t) is described in terms of constants. As t changes, the H(t) are said to "evolve," and the metric becomes a set of functions of t only. We are able to exhibit this fact by using a set of differential forms ω^i (i = 1, 2, 3, and labels three one-forms) to express the three-dimensional metric of H(t). The ω^i obey

$$d\omega^i = \frac{1}{2} C^i{}_{st}\omega^s \wedge \omega^t \,. \tag{11.2}$$

The $C^i{}_{st}$ are the structure constants of the group of isometries. The full four-metric is obtained by taking the curl of t, dt, to be the fourth differential form used, $\omega^0 = dt$. The metric is given by

$$ds^2 = g_{\mu\nu}\omega^\mu\omega^\nu . \tag{11.3}$$

The spatial homogeneity is expressed by the statement that the functions $g_{\mu\nu}$ are functions of t *only*:

$$g_{\mu\nu} = g_{\mu\nu}(t) .$$

Cosmologies of Types I, V, and IX

It is desirable to look at spatially homogeneous models because of the enormous simplification in the equations which they involve: Instead of partial differential equations, ordinary differential equations appear (the danger, of course, is that *any* simplification of this kind is too much; destroying the essential randomness of nature may produce singularities which would not occur in less special situations).

However, of the nine types of three-dimensional groups, three stand out. The first is the simplest, T_3, (Bianchi Type I). All directions in an invariant hypersurface are infinite, and all Type I models possess singularities. T_3 is the spatial-homogeneity group of the "flat" FRW models.

The second especially interesting group is $SO(3, R)$ (Bianchi Type IX). The Type IX cosmologies are very general (seven adjustible parameters even if the pressure p is taken to be zero) and therefore have been studied extensively. The structure constants are

$$C^i{}_{jk} = \varepsilon_{ijk} \quad \text{or} \quad C^1{}_{23} = 1 \textit{ et cyc} .$$

The Type IX cosmologies, which include the closed FRW models, all have true singularities. The invariant hypersurfaces are closed and may be taken as copies of the three-sphere S^3.

The third group, Bianchi Type V, is the spatial-homogeneity group of the open FRW metrics. This group generates spacelike sections with constant negative curvature. Type V models have been moderately well studied, but we shall not treat them thoroughly here (see Matzner, 1969).

We shall discuss Type IX cosmologies here more thoroughly than other types, since these models have been and are studied abundantly. These models allow not only expansion but also rotation and shear, and in general are anisotropic. All of these models do have singular points, as we have shown and shall show again.

11.2. Hamiltonian Cosmology – The Homogeneous Cosmology As a Bouncing Particle

The application of Hamiltonian methods to cosmology grew out of the study of spatially homogeneous models. It was noted by Misner (1968) that the R_{00} field equation gives a first integral for the R_{ij} field equations. In consequence, R_{00} can be used to construct a Lagrangian whose variation gives the R_{ij} equations. This construction is especially useful in the case of spatially homogeneous models, in which the invariant hypersurfaces $H(t)$ can be parametrized by some set of parameters $p_a(t)$ $(a = 1, \cdots, n)$. The field equations resemble Lagrange's equations for a particle moving in an n-dimensional space. In this case, the problem of determining the metric becomes similar to a particle problem, which has long been studied, namely that of a particle bouncing around in a potential well.

Once one has a Lagrangian formulation, it is useful to reformulate the problem using a Hamiltonian. This reformulation was done by Misner (1969b) who used the Hamiltonian method for general relativity due to Arnowitt, Deser, and Misner (ADM) (1962). Misner's approach has since been used successfully by many authors to study homogeneous cosmologies (for a review, see Ryan 1972d). This approach has also lead to the idea of *quantum cosmology* and *minisuperspace* which we shall discuss briefly. We shall also discuss objections to the use of Misner's Hamiltonian formulation for certain Bianchi types (those whose group structure constants obey $C^s{}_{is} \neq 0$).

The ADM Formulation

The ADM formulation is a procedure for reducing the Einstein action, $I = \int R\sqrt{-g}\,d^4x$, to canonical form. One begins by writing the action in the first-order Palatini (1919) form in which the $\Gamma^{\mu}{}_{\nu\alpha}$ and the $g_{\mu\nu}$ are varied independently. This action is then reparametrized by introducing quantities

$$N = (-{}^4g^{00})^{-\frac{1}{2}}, \quad N_i = {}^4g_{0i}, g_{ij}, \pi^{ij}, \quad C^0, \text{ and } C^i, \tag{11.4}$$

where the superscript ^{4} denotes a four-dimensional geometrical object, and the superscript ^{3} will be used for objects on selected three-dimensional hypersurfaces. The g_{ij} $(i, j = 1, 2, 3)$ are the ij components of the metric $(g_{ij} = {}^4g_{ij})$ and the π^{ij} are defined in terms of the ${}^4g_{\mu\nu}$ and the ${}^4\Gamma^{\alpha}{}_{\mu\nu}$, while C^0 and C^i are algebraic combinations of the π^{ij}, the g_{ij} and their derivatives. ADM show that the Einstein action (from which the field equations can be derived) reduces to

$$I = (16\pi)^{-1} \int \left[\pi^{ij}\left(\frac{\partial g_{ij}}{\partial t}\right) - NC^0 - N_i C^i \right] d^4x \tag{11.5}$$

(we have discarded a total divergence). The π^{ij}, g_{ij}, N and N_i are to be varied separately. Varying N and N_i gives $C^0 = 0$ and $C^i = 0$, a set of constraints on g_{ij} and π^{ij} (in fact they are the R_{00} and R_{0i} Einstein equations). While we single out x^0 as t in (11.5), the action is still invariant under changes of all four coordinates and is completely general up to this point.

The novelty of the procedure of ADM is that they reduce (11.5) to "basic" variables, essentially two of the g_{ij} and their conjugate π^{ij}. They do this reduction by choosing four of the twelve g_{ij} and π^{ij} as coordinates (intrinsic coordinates) and by solving the four equations $C^0 = 0$, $C^i = 0$ to eliminate four more. This prescription is rather vague because each case needs careful study to see which variables should be chosen as coordinates and which should be eliminated by solving the constraints.

We write the final version of the action as

$$I = (16\pi)^{-1} \int \pi^{ij} \frac{\partial g_{ij}}{\partial t} d^4x , \qquad (11.6)$$

subject to the constraints $C^0 = 0$, $C^i = 0$. The C's are

$$C^0 = -(g)^{\frac{1}{2}} \left\{ {}^3R + g^{-1} \left[\frac{1}{2} (\pi^k{}_k) - \pi^{ij} \pi_{ij} \right] \right\} , \qquad (11.7a)$$

$$C^i = -2\pi^{ij}{}_{|j} , \qquad (11.7b)$$

where $g \equiv \det(g_{ij})$, ${}^3R$ is the scalar curvature of $t = const.$ surfaces, indices are raised and lowered by means of g_{ij}, and $|$ means covariant differentiation on $t = const.$ surfaces.

Equations (11.7a) and (11.7b) are vacuum equations. For some cosmological questions it will be necessary to consider models with matter. To include matter we need an action of the form $I = \int [(-{}^4g)^{\frac{1}{2}} R + \mathcal{L}_M] d^4x$, where the Lagrangian density $\mathcal{L}_M$ satisfies

$$\delta \int \mathcal{L}_M dx = -\int T_{\mu\nu} (-{}^4g)^{\frac{1}{2}} \delta g^{\mu\nu} d^4x . \qquad (11.8)$$

Once we have such an action we must break up $\mathcal{L}_M$ into terms such as $p\,dq$, $N\mathcal{L}^0{}_M$ and $N_i\mathcal{L}^i{}_M$, the first of these introducing new independent variables connected with matter and the second two modifying the constraints $C^0 = 0$, $C^i = 0$ to read $C^{0\prime} = C^0 + \mathcal{L}^0{}_M = 0$, $C^{i\prime} = C^i + \mathcal{L}^i{}_M = 0$.

Such a Lagrangian density exists for electromagnetic fields (see Hughston and Jacobs, 1970) and for fluids (Schutz, 1971). In certain fluid cases in spatially homogeneous models it is also possible to construct a Lagrangian density $\mathcal{L}_M = N\mathcal{L}_M + N_i\mathcal{L}^i{}_M$, where $\mathcal{L}^0{}_M$ and $\mathcal{L}^i{}_M$ are functions of metric variables and constants of motion (see Ryan, 1972d). This Lagrangian density introduces no independent matter variables and merely serves to modify the constraint equations.

11.3. Application to Homogeneous Cosmologies

The ADM formulation is very useful when applied to Class A Bianchi-type cosmologies and to the Kantowski-Sachs universe. In these cases we write the metric as

$$ds^2 = -dt^2 + R_0^2 e^{-2\Omega} e^{2\beta}{}_{ij}\omega^i\omega^j , \qquad (11.9)$$

where β is a 3×3 symmetric matrix and Ω a scalar, both being functions of time only. R_0 is a constant included for convenience in choosing units. For Bianchi-type universes the ω^i are invariant forms which obey $d\omega^i = \frac{1}{2}C^i{}_{jk}\omega^j \wedge \omega^k$, with the $C^i{}_{jk}$ the structure coefficients of the particular group under consideration. For the Kantowski-Sachs model $\omega^1 = dr$, $\omega^2 = d\theta$, $\omega^3 = \sin\theta d\phi$ for coordinates r, θ, ϕ.

Whenever Ω is a monotonic function of t, we can choose Ω as our time coordinate. This choice represents the first step of the ADM procedure, that is, choosing a function of the g_{ij} and π^{ij} as a coordinate; in this case $\Omega = -\frac{1}{6}\ln[\det(g_{ij})]$. The metric now becomes

$$ds^2 = -(N^2 + N_iN^i)d\Omega^2 + 2N_i d\Omega\omega^i + R_0^2 e^{-2\Omega} e^{2\beta(\Omega)}{}_{ij}\omega^i\omega^j , \qquad (11.10)$$

where $N = N(\Omega)$, $N_i = N_i(\Omega)$.

Inserting this metric into (11.6) we find

$$I = (16\pi)^{-1}\int 2[e^{\beta}{}_{is}\pi^s{}_t e^{-\beta}{}_{tj}\,đ\beta_{ij} - (\pi^k{}_k)d\Omega]\,d^3x . \qquad (11.11)$$

where $đ\beta = \frac{1}{2}[de^{\beta}e^{-\beta} + e^{-\beta}de^{\beta}]$.

Because we are considering homogeneous universes we can integrate over the space variables in (11.11) and eliminate them. Following Misner (1969b) we rescale the differential forms which appear in the metric or choose a subset of the manifold to make $\int d^3x = (4\pi)^2$. This integration leads to

$$I = 2\pi\int [e^{\beta}{}_{is}\pi^s{}_t e^{-\beta}{}_{tj}\,đ\beta_{ij} - (\pi^k{}_k)d\Omega] . \qquad (11.12)$$

We now define the matrix p_{ij} as

$$p_{ij} = 2\pi (e^{\beta}{}_{is}\pi^{s}{}_{t}e^{-\beta}{}_{tj} - \tfrac{1}{3}\delta_{ij}\pi^{\ell}{}_{\ell}) \tag{11.13}$$

and proceed to parametrize β_{ij} and p_{ij} in order to reduce the first term in the integrand of (11.12) to the form $p_A dq_A$. We write

$$\beta = e^{-\psi\kappa_3} e^{-\theta\kappa_1} e^{-\phi\kappa_3} \beta_d e^{\phi\kappa_3} e^{\theta\kappa_1} e^{\psi\kappa_3}, \tag{11.14}$$

where

$$\kappa_3 = \begin{bmatrix} 0 & 1 & 0 \\ -1 & 0 & 0 \\ 0 & 0 & 0 \end{bmatrix}, \qquad \kappa_1 = \begin{bmatrix} 0 & 0 & 0 \\ 0 & 0 & 1 \\ 0 & -1 & 0 \end{bmatrix} \tag{11.15}$$

and

$$\beta_d = \mathrm{diag}(\beta_+ + \sqrt{3}\beta_-, \beta_+ - \sqrt{3}\beta_-, -2\beta_+). \tag{11.16}$$

The conjugate variables are defined by

$$\begin{aligned} 6p_{ij} = e^{-\psi\kappa_3} e^{-\theta\kappa_1} e^{-\phi\kappa_3} \Big\{ & \alpha_1 p_+ + \alpha_2 p_- + \alpha_3 \frac{3p_\psi}{\sinh(2\sqrt{3}\beta_-)} \\ & + \alpha_4 \frac{3(p_\phi \sin\psi - p_\psi \cos\theta \sin\psi + p_\theta \cos\psi \sin\theta)}{\sin\theta \sinh(3\beta_+ + \sqrt{3}\beta_-)} \\ & + \alpha_5 \frac{3(p_\theta \sin^2\psi \sin\theta - p_\phi \sin\psi \cos\psi + p_\psi \cos\psi \sin\psi \cos\theta)}{\sin\psi \sin\theta \sinh(3\beta_+ - \sqrt{3}\beta_-)} \Big\} e^{\phi\kappa_3} e^{\theta\kappa_1} e^{\psi\kappa_3}, \end{aligned} \tag{11.17}$$

with $\alpha_1 = \mathrm{diag}(1,1,-2)$, $\alpha_2 = \mathrm{diag}(\sqrt{3}, -\sqrt{3}, 0)$, $\alpha_3 = \begin{bmatrix} 0 & 1 & 0 \\ 1 & 0 & 0 \\ 0 & 0 & 0 \end{bmatrix}$, $\alpha_4 = \begin{bmatrix} 0 & 0 & 1 \\ 0 & 0 & 0 \\ 1 & 0 & 0 \end{bmatrix}$, $\alpha_5 = \begin{bmatrix} 0 & 0 & 0 \\ 0 & 0 & 1 \\ 0 & 1 & 0 \end{bmatrix}$. This parametrization makes our action take the form

$$I = \int [p_+ d\beta_+ + p_- d\beta_- + p_\theta d\theta + p_\phi d\phi + p_\psi d\psi - H d\Omega] \tag{11.18}$$

where $H = 2\pi(\pi^k{}_k)$.

We can obtain H as a function of $\beta_\pm, p_\pm, p_\theta, p_\psi, p_\phi, \theta, \psi, \phi$, and Ω by solving $C^0 = 0$ (this is the second step of the ADM procedure). We find

$$H^2 = 6p_{ij}p_{ij} - 24\pi^2 g^3R . \tag{11.19}$$

where $p_{ij}p_{ij}$ is a quadratic form in the variables $\beta_+, \beta_-, \theta, \phi, \psi$ and their conjugate momenta, and g^3R is a function of these variables and Ω.

At this point we have not completed the ADM reduction. We still have to solve

$$C^i(p_\pm, p_\theta, p_\phi, p_\psi, \beta_\pm, \theta, \psi, \phi, \Omega) = 0 \tag{11.20}$$

and eliminate three of the variables. In practice it is sometimes more useful to leave these three equations unsolved, and think of (11.18) as an action subject to the three constraints (11.20).

To give the metric completely we need only specify N and N_i. We cannot specify these functions in the general case arbitrarily because the specifications of N and N_i and of coordinate choices are the same. In the cases we will consider we may choose N_i arbitrarily but must compute N with the coordinate choices we have made above. The simplest choice for Hamiltonian cosmology is

$$N_i = 0 .$$

Our choice of Ω as time implies that

$$N = H^{-1}e^{-3\Omega}(12\pi R_0^{\,3}) , \tag{11.21}$$

where all the variables in H must be solved for as functions of Ω. The choice $N_i = 0$ is not the only possible one. Ryan (1972d) discusses other choices and their meanings; we shall not consider them here.

The only other idea we need to consider to apply Hamiltonian methods to homogeneous cosmologies is that of matter – we need a matter Lagrangian. We shall consider fluid models as examples, and as was mentioned, Ryan (1972d) gives a matter Lagrangian for a fluid in Bianchi-type universes which is valid for universes of Ellis-MacCallum Class A (including

Bianchi Types I, II, VI_{-1}, VII_0, VIII, IX, see Table 6.2). A special case is the Lagrangian for dust $(p=0)$:

$$\mathcal{L}_M = -2\mu N(1+R_0^{-2}e^{2\Omega}e^{-2\beta}{}_{ij}u_i u_j)^{\frac{1}{2}} - 2N_i\mu u_j g^{ij} \,, \tag{11.22a}$$

For a fluid with $p = kw$ and $u_i = 0$,

$$\mathcal{L}_M = -2\mu N R_0^{-3k} e^{3k\Omega} \,. \tag{11.22b}$$

In these equations u_i are space components of the fluid velocity (chosen so that the geodesic equations are solved as extra equations) and μ is a constant of motion derived from the fluid equations defined by (ρ = density):

$$\mu = N\rho u^0 R_0^{\,3} e^{-3\Omega} \,. \tag{11.23}$$

With this Lagrangian we can complete the Hamiltonian formulation for dust-filled models. Notice that it is of the form $\mathcal{L}_M = N\mathcal{L}^0{}_M + N_i\mathcal{L}^i{}_M$ discussed above, so the addition of matter leaves (11.18) unchanged but with the Hamiltonian (11.19) rewritten as

$$H^2 = H^2_{\text{vac.}} - 24\pi^2 g^{\frac{1}{2}} \mathcal{L}^0{}_M \tag{11.24}$$

and the constraints $C^i = 0$ modified to imply

$$\pi^{ij}{}_{|j} = \frac{1}{2}\mathcal{L}^i{}_M \,. \tag{11.25}$$

The variables $\beta_\pm, \theta, \phi, \psi$, and their conjugates $p_\pm, p_\theta, p_\phi, p_\psi$ remain the only independent variables. They are solved for as functions of Ω provided the u_i are also solved for as functions of Ω by use of the auxiliary geodesic equations. The matter density ρ is determined from the constant of motion μ.

Application of the Hamiltonian Formulation

The Hamiltonian formulation has never been fully investigated for all possible Bianchi types, but Jacobs and Hughston (1970) have applied the method to vacuum models in which the matrix β is diagonal. In this case the action reduces to

$$I = \int p_+ d\beta_+ + p_- d\beta_- - H d\Omega \,, \tag{11.26}$$

with

$$H^2 = p_+{}^2 + p_-{}^2 - 24\pi^2 g^3R \,, \tag{11.27}$$

where g^3R is a function of Ω, β_+ and β_-. This Hamiltonian is the same as that of a particle (the *universe point*) moving in two dimensions on the $\beta_+\beta_-$-plane, with g^3R acting as a potential (a time-dependent potential because it is a function of Ω). In fact, we can let $g^3R = -R_0{}^4 e^{-4\Omega}(V-1)$, where $V = V(\beta_+, \beta_-)$. It turns out that the potentials $V(\beta_+, \beta_-)$ for all Bianchi types have exponentially steep walls in $\beta_+\beta_-$-space. Table 11.1 gives V for all of the Bianchi types with β diagonal. It also gives the constraints $C^i = 0$ reduced to statements about p_+ and p_-. Figure 11.2 shows the walls associated with $V(\beta_+, \beta_-)$ for all nine Bianchi types.

MacCallum and Taub (1972) have raised questions about the validity of this Hamiltonian approach for certain Bianchi types. They object to variations with homogeneity imposed from the start. By not making such an assumption they obtain results for models of Ellis-MacCallum Class B ($C^s{}_{is} \neq 0$) which are different from the Jacobs-Hughston picture as outlined above. Table 11.1 and Figure 11.2 describe the potentials even for the Class B cases correctly. Unfortunately the Einstein equations in some of the Class B cases are not completely described by equations derived in the Hamiltonian method using the listed potentials. These cases are still being studied, but see Ryan (1974).

Table 11.1. The Potentials in Hamiltonian Cosmology for all Diagonal Bianchi Type Models.

In general the potential is $V = 1 - \frac{2}{3} R_0^{-2} e^{-2\Omega}\, {}^3R$, and for diagonal $\boldsymbol{\beta}$, ${}^3R$ is:

$$ {}^3R = -\frac{1}{2} R_0^{-2} e^{2\Omega} [e^{2\beta_{ij}} e^{2\beta_{kl}} m^{ik} m^{jl} - 2m^*_{ij} e^{-2\beta_{ij}} + 12 a_i a_j e^{-2\beta_{ij}}] , $$

where m_{ij} and a_i are defined in Table 6.2, and m^*_{ij} is the classical adjoint of the matrix m_{ij}. Some of the Class B cases, Types III, IV, $VI_{h\neq -1}$, $VII_{h\neq 0}$, are not correctly described by the Hamiltonian method, although these potentials are nonetheless useful (Ryan, 1974). For Class A as well as diagonal Type V models, the Hamiltonian method is correct.

Bianchi Type	Potential	Constraint
I	None	None
II	$1 + \frac{1}{3} e^{4(\beta_+ + \sqrt{3}\beta_-)}$	None
III	$1 + \frac{4}{3} e^{4\beta_+}$	$3p_+ - \sqrt{3} p_- = 0$
IV	$1 + \frac{1}{2} e^{4\beta_+} [8 + \frac{2}{3} e^{4\sqrt{3}\beta_-}]$	$p_+ = 0$
V	$1 + 4e^{4\beta_+}$	$p_+ = 0$
VI	$1 + \frac{4}{3} [1 + h + h^2] e^{4\beta_+}$	$p_- = \sqrt{3} \left[\frac{h+1}{h-1}\right] p_+$
VII	$1 + \frac{2}{3} e^{4\beta_+} [\cosh(4\sqrt{3}\beta_-) + 2h^2 - 1]$	None (h=0) $p_- = \sqrt{3} p_+ (h \neq 0)$
VIII	$1 + \frac{2}{3} e^{4\beta_+} [\cosh(4\sqrt{3}\beta_-) - 1]$ $+ \frac{1}{3} e^{-8\beta_+} + \frac{4}{3} e^{-2\beta_+} \cosh(2\sqrt{3}\beta_-)$	None
IX	$1 + \frac{2}{3} e^{4\beta_+} [\cosh(4\sqrt{3}\beta_-) - 1]$ $+ \frac{1}{3} e^{-8\beta_+} - \frac{4}{3} e^{-2\beta_+} \cosh(2\sqrt{3}\beta_-)$	None

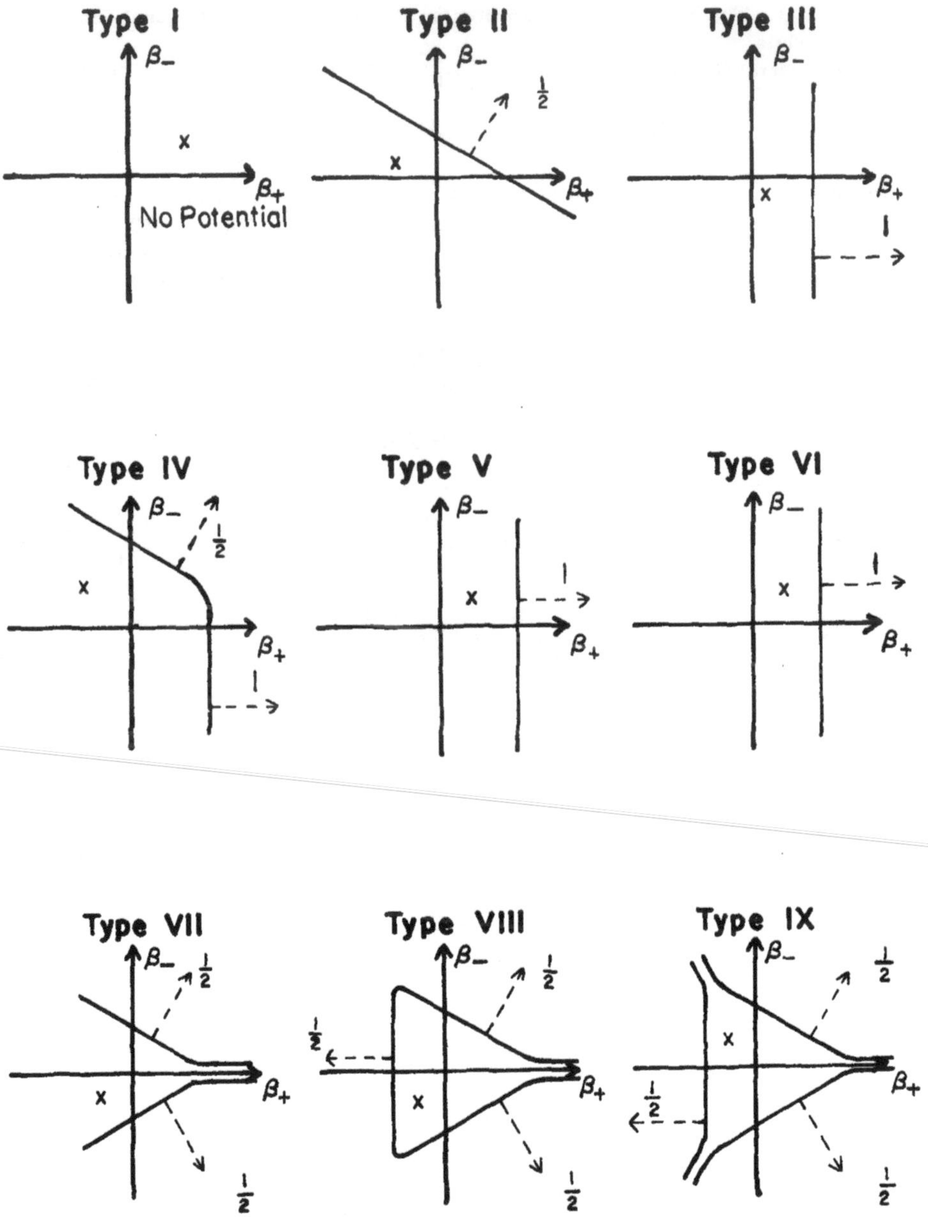

Fig. 11.2. The Potentials for Bianchi-Type Models. An equipotential is shown for each Bianchi type, and the dashed arrows give the velocity of the wall associated with the potential. The symbol x marks the position of a generic universe point. See Table 11.1.

Qualitative Cosmology

The Hamiltonian for Bianchi-type and Kantowski-Sachs models resembles a particle Hamiltonian with exponential potentials. This form leads to an interesting method of approximate solution for these universes. Because of the steepness of the potentials we may replace them by infinitely hard walls to a good approximation. These walls lie along the equipotentials shown in Figure 11.2 for each of the different diagonal Bianchi types. Far from the walls the universe point will move in a straight line. It will bounce off the walls from time to time. With straight-line motion between wall encounters and a derived set of "bounce laws" one can build up a solution in the form of a diagram which matches the behavior of a homogeneous universe quite well. We shall give examples of this procedure in the next chapter. Of course, there are some special situations in some models in which no bounces occur. In these cases the detailed shapes of the walls must be accounted for, and special care must be taken. Collins and Stewart (1971) and Collins (1971) have studied many properties of Bianchi-type universes by means of a similar qualitative method.

As a final note, the pictorial qualitative solution is, of course, equivalent to a series of analytic solutions: The straight-line portions of the pictorial solution correspond to an analytic solution characterized by a set of parameters. The bounce laws predict changes in these parameters which occur suddenly from time to time. This approach has been taken by Belinskii, Khalatnikov, and Lifshitz (1970).

11.4. Quantum Cosmology and Minisuperspace

Quantum cosmology should be the application of the quantum theory of gravity to the cosmological problem. That is, one should examine the equations of quantized general relativity to find solutions that become the Friedmann-Robertson-Walker (or some more complicated cosmological solution) metrics as $e^{-\Omega} \to \infty$. The obstacle to this scheme is that quantum gravity is not well enough understood to make it possible to find such solutions. DeWitt (1967a,b,c) and Misner (1969b) have studied compromise models in which homogeneity is imposed before the gravitational field is quantized. In practice this is what is known as quantum cosmology. It can be hoped that the models which result will at least have some of the more important features of a real quantum solution.

Quantum gravitation deals with wave functions whose domain is the set of positive-definite metrics (or geometries) on three-dimensional hypersurfaces (Wheeler, 1968). This domain set is called *superspace*. The subset of all homogeneous three-geometries, as it is applied to cosmology and the problem of quantum cosmology, is called *minisuperspace* (Misner, 1972). We will briefly discuss this concept below.

Application of the Quantum Principle

The action (11.18) with the Hamiltonian of (11.24) has the form of a particle action. We can quantize this action in the usual way by replacing various quantities by operators

$$H \to -i\partial/\partial\Omega,\ p_{\pm} \to -i\partial/\partial\beta_{\pm},\ p_{\phi} \to -i\partial/\partial\phi,\ p_{\psi} \to -i\partial/\partial\psi,\ p_{\theta} \to -i\partial/\partial\theta\,.$$

These operators are used to find a wave function $\Psi(\Omega,\beta_{\pm},\phi,\psi,\theta)$ as a solution of a differential equation. There are three difficulties with this program: (1) The Hamiltonian is explicitly time-dependent (Ω-dependent); (2) H is a square-root Hamiltonian; and (3) We must find some way of handling the constraints $C^i = 0$. The first of these difficulties is merely computational – one usually does not encounter a time-dependent Hamiltonian in elementary quantum mechanics. The second difficulty is more fundamental, but methods for handling square-root Hamiltonians exist. The third difficulty is basic, and we should discuss it briefly before studying Type I and diagonal and symmetric Type IX universes as examples.

There are two different ways of handling the constraints $C^i = 0$, and the arguments apply also to the important constraint $C^0 \approx 0$ which we have solved above. In fact, the ADM method obtains (11.24) by solving

$C^0 = 0$ classically to give our expression for H. This procedure is the spirit of the ADM method – we must solve the constraints classically to arrive at the "true" action for the gravitational field before quantizing. The other method (the Dirac method) for quantizing the action (11.18) is to retain both $C^0 = 0$ and $C^i = 0$ as operator equations. This method was used by DeWitt (1967a,b,c) to study the quantum behavior of FRW universes. These two methods lead to different equations for quantum universes, and we shall use a symmetric Type IX model later to illustrate this difference.

Consider now a Type I vacuum universe, in which

$$H = (p_+{}^2 + p_-{}^2)^{\frac{1}{2}} . \tag{11.28}$$

There are three methods of quantizing a square-root Hamiltonian such as this one: (i) the square-root method of Schweber, Bethe, and DeHoffmann (1955); (ii) the Dirac (1947) method of linearization; and (iii) the Schrödinger (1926)-Klein (1927)-Gordon (1926) (SKG) method. We shall not discuss the first method, which involves spectral techniques. We will only mention that the Dirac method leads to a linear two-component spinor equation, and at present we lack an experimental quantity to associate with the spinor components (see Ryan, 1972d). The SKG method seems to be best, even though we have the usual problem of possibly non-positive-definite probability densities.

The SKG method works with H^2 and leads to

$$-\frac{\partial^2\Psi}{\partial\Omega^2} + \frac{\partial^2\Psi}{\partial\beta_+{}^2} + \frac{\partial^2\Psi}{\partial\beta_-{}^2} = 0 \tag{11.29}$$

for the wave function $\Psi(\Omega, \beta_\pm)$. This equation has solutions

$$\Psi = Ae^{i(p_+\beta_+ + p_-\beta_- - E\Omega)} \quad (A, p_\pm, E \text{ constants with}$$
$$E = \pm(p_+{}^2 + p_-{}^2)^{\frac{1}{2}}) . \tag{11.30}$$

The positive and negative signs for E correspond to expanding and contracting universes respectively. A surprising feature of this solution is that there is no evidence of quantum mechanics causing the universe to avoid a singularity. If we make up a wave packet from the functions Ψ, we find that this wave packet marches sedately out to $\Omega = \infty$ without any tendency to avoid the singularity.

Minisuperspace

Superspace (Wheeler, 1968) is the set of all three-geometries. General relativity can be thought of as the development of a three-geometry in "time," the meaning of time depending on how one breaks four-dimensional spacetime into time and space. General relativity can be thought of, then, as the study of tracks in superspace, different tracks corresponding to different evolutions of metrics and to different choices of "time." Superspace is a very valuable concept for homogeneous cosmologies because it reduces from an infinite dimensional space to a finite dimensional minisuperspace, when homogeneity is imposed. In fact, for the Bianchi-type universes discussed in this chapter, superspace is $\beta_+, \beta_-, \phi, \psi, \theta, \Omega$-space.

It can be shown (Misner, 1972) that the constraint $C^0 = 0$ is equivalent to an equation of motion in minisuperspace for g^{ij}, the three-metric for a homogeneneous cosmology. In superspace (ij) represents one tensorial index, and the equation of motion is

$$\dot{g}^{(ij)}{}_{;(kl)}{}^{(kl)} = \frac{1}{2}\mathcal{R}^{;(ij)}, \quad \text{with } \dot{g}^{(ij)} \equiv dg^{(ij)}/d\lambda \tag{11.31}$$

for some affine path parameter λ. $\mathcal{R}$ is the scalar curvature computed from g^{ij}. The covariant derivative is taken with respect to the metric of superspace $\mathcal{G}_{(ij)(kl)}$,

$$\mathcal{G}_{(ij)(kl)} = (g_{ik}g_{jl} + g_{jk}g_{il} - 2g_{ij}g_{kl}) . \tag{11.32}$$

Table 11.2. The Misner Metrics of Minisuperspace.

A) For any Bianchi-type model:

$$ds^2 = 24[-d\Omega^2 + d\beta_{ij}\, d\beta_{ij}]$$

B) For diagonal Bianchi-types:

$$ds^2 = 24[-d\Omega^2 + d\beta_+{}^2 + d\beta_-{}^2]$$

C) For Bianchi-types with one off-diagonal term ($\boldsymbol{\beta} = e^{-\psi\kappa_3}\boldsymbol{\beta}_d e^{\psi\kappa_3}$):

$$ds^2 = 24[-d\Omega^2 + d\beta_+{}^2 + d\beta_-{}^2 + \tfrac{1}{3}\sinh^2(2\sqrt{3}\beta_-)\, d\psi^2]$$

D) For Bianchi-types with $\beta = e^{-\theta\kappa_1} e^{-\phi\kappa_3}\beta_d e^{\theta\kappa_3} e^{\phi\kappa_1}$ and for general Bianchi-types:

$$ds^2 = 24[-d\Omega^2 + d\beta_+{}^2 + d\beta_-{}^2 + \tfrac{1}{3}\sinh^2(2\sqrt{3}\beta_-)(\sigma^3)^2 + \tfrac{1}{3}\sinh^2(3\beta_+ + \sqrt{3}\beta_-)(\sigma^2)^2 + \tfrac{1}{3}\sinh^2(3\beta_+ - \sqrt{3}\beta_-)(\sigma^1)^2]$$

where: $\sigma^1 = \sin\theta\, d\phi$ and $\sigma^1 = \sin\psi\, d\theta - \cos\psi \sin\theta\, d\phi$

$\sigma^2 = d\theta$ $\quad\quad\sigma^2 = \cos\psi\, d\theta - \sin\psi \sin\theta\, d\phi$

$\sigma^3 = \cos\theta\, d\phi$ $\quad\quad\sigma^3 = -(d\psi + \cos\theta\, d\phi)$

respectively. The matrices κ_3 and κ_1 are:

$$\kappa_3 = \begin{bmatrix} 0 & 1 & 0 \\ -1 & 0 & 0 \\ 0 & 0 & 0 \end{bmatrix}, \quad \kappa_1 = \begin{bmatrix} 0 & 0 & 0 \\ 0 & 0 & 1 \\ 0 & -1 & 0 \end{bmatrix}.$$

Note that (11.31) is very nearly a geodesic equation, and DeWitt (1967a,b,c) and Gowdy (1970) have proposed new metrics for superspace to replace (11.32) for which (11.31) becomes a geodesic equation. In Table 11.2 we list the Misner metrics of minisuperspace corresponding to (11.32) for Bianchi-type metrics.

Superspace gives us an idea of how to proceed to parametrize homogeneous cosmologies. Notice that the quantum-mechanical equations we have obtained above contain terms in, say, $p_\pm{}^2$ which have been changed directly to $\partial^2/\partial\beta_\pm{}^2$. We ignored the problem of factor ordering in H. The ADM method, in which all of the constraints are solved before quantizing provides an unambiguous factor ordering, at least for the models considered above. Another method of consistently determining a factor ordering is to write the equation for Ψ as a covariant Laplace-Beltrami operator in superspace applied to the wave function Ψ plus "potential" terms. The Laplace-Beltrami operator in superspace depends on the superspace metric. In the next chapter we will see that factor ordering becomes more important in non-diagonal Type IX models. If we use Misner's (1972) metric (11.32) we find quantum mechanical equations for Type I which are the same as (11.29). If we use DeWitt's metric $\sqrt{g}\mathcal{G}_{(ij)(kl)}$ we find that $-\partial^2\Psi/\partial\Omega^2$ is replaced by $-\partial^2\Psi/\partial\Omega^2 - \frac{3}{2}\partial\Psi/\partial\Omega$, but otherwise the equations remain the same.

12. TYPE I MODELS AND TYPE IX MODELS – THE SIMPLEST AND THE MOST INTERESTING

We must use time as a tool, not as a couch
– JOHN KENNEDY

Type I models are the simplest of anisotropic models, but already illustrate some of the intriquing features of all Bianchi-type models, particularly in their Hamiltonian formulation. Type IX models illustrate the full range of problems encountered in classical and quantum cosmology.

12.1. Type I Models

In Chapter 11 we defined the quantities β_+, β_-, and Ω, the general Type I metric being

$$ds^2 = -dt^2 + e^{-2\Omega}\Big[e^{2(\beta_+ + \sqrt{3}\beta_-)}(dx^1)^2 + e^{2(\beta_+ - \sqrt{3}\beta_-)}(dx^2)^2 + e^{-4\beta_+}(dx^3)^2\Big].$$

From (11.19) and (11.22b), with appropriate resealing of μ and R_0, we find

$$H^2 = p_+{}^2 + p_-{}^2 + \mu e^{-3(1+k)\Omega}\ , \tag{12.1}$$

where Ω plays the role of time in H, and p_+ and p_- are momenta conjugate to β_+ and β_- respectively. As a technical aside we mention that this form of H assumes that the matrix β is diagonal. The diagonal assumption, however, does not restrict the generality of the fluid Type I model. The constant k comes from the solution of

$$T^{\mu\nu}{}_{;\nu} = 0 \ \text{ for } \ T_{\mu\nu} = \frac{1}{2}(w+p)u_\mu u_\nu + pg_{\mu\nu},\ p = kw \tag{12.2}$$

which implies

$$w = \mu e^{3(1+k)\Omega}\ . \tag{12.3}$$

From the form of H, we see that

$$p_+ = \text{const.}, \quad p_- = \text{const.}$$

Therefore the universe point moves in a straight line across the $\beta_+\beta_-$-plane with velocity

$$[(d\beta_+/d\Omega)^2 + (d\beta_-/d\Omega)^2]^{\frac{1}{2}} = (p_+^{\,2} + p_-^{\,2})^{\frac{1}{2}} H^{-1} . \tag{12.4}$$

Notice that $H \to (p_+^{\,2} + p_-^{\,2})^{\frac{1}{2}} = \text{const.}$ as $\Omega \to \infty$ (near the singularity).

To complete the description of the fluid-filled model we need only $\Omega(t)$. From (11.21), we have

$$d\Omega/dt = \frac{1}{12\pi} R_0^{\,-3} H e^{3\Omega} . \tag{12.5}$$

Near the singularity where H becomes a constant this equation is particularly easy to integrate. The asymptotic form of $\Omega(t)$ shows that a true singularity exists in every Type I model since the physical observable w blows up in a finite amount of proper time t.

The Decay of Anisotropy

The Hamiltonian description of the evolution of Type I models allows us to show directly that the observed anisotropy of the universe decreases as one moves from the singularity toward the present. By "observed anisotropy" we mean anisotropy in the Hubble constant, which is not the same in all directions. In an anisotropic model the redshifts of galaxies and the temperature of the 3K black-body radiation are anisotropic.

Misner (1968) has shown that this anisotropy is described by $n_i \sigma_{ij} n_j$, where n_i is the direction vector of an observation and where

$$\sigma_{ij} = \frac{1}{2}\left[e^{-\beta}{}_{si} \frac{d}{dt} e^{\beta}{}_{sj} + e^{-\beta}{}_{sj} \frac{d}{dt} e^{\beta}{}_{si}\right] . \tag{12.6}$$

An especially convenient measure of the root-mean-square anisotropy is $\mathrm{tr}(\sigma^2) = \sigma_{ij}\sigma_{ij}$. For diagonal Type I universes we find

$$\begin{aligned} \mathrm{tr}(\sigma^2) &= 6[(d\beta_+/d\Omega)^2 + (d\beta_-/d\Omega)^2](d\Omega/dt)^2 \\ &= 6(12\pi R_0^3)^{-2}(p_+^2 + p_-^2)e^{6\Omega} \propto e^{6\Omega} . \end{aligned} \tag{12.7}$$

Thus as Ω decreases from ∞ (remember Ω decreases monotonically as t increases) anisotropy decays. Decay of anisotropy of some sort may, in fact, be the reason for the presently observed isotropy of the 3K black-body radiation.

Horizons

Because of the finite age of the universe, light can reach an observer typically only from a limited amount of matter. The farthest distance an observer can see in a given direction is called the *horizon* distance. The existence of a horizon thus depends on the existence of a singularity, and this concept gives important insight concerning the singularity and astrophysical processes. For example, the spacelike sections of a Type IX model are each finite in diameter, and it is possible for certain horizons to be larger than this diameter. A model with such large horizons is called "mixing" for each material object can influence every other piece of matter. These models will be discussed later in this chapter.

Horizons are most easily defined and computed in a Type I model, but in these mixing does not occur. Let n_i be a given spatial direction (that is, a triplet of numbers such that $n_1^2 + n_2^2 + n_3^2 = 1$; n_i is a direction in an orthonormal frame of the $t = \text{const}$ hypersurface). We will denote by t_1 the time coordinate of the observer and by t_0 the time of the initial singularity (the horizon, of course, depends only on $t_1 - t_0$ and not on the spatial position x^i of the observer, due to the spatial homogeneity). We draw a lightlike line in direction n_i from the observer, backward in time until it meets the singularity.

Let the observer, for convenience, be at $x^i = 0$. The lightlike line obeys the differential law

$$-dt^2 + e^{-2\Omega} e^{2\beta}_{ij} dx^i dx^j = 0, \quad \text{where } \frac{dx^i}{dt} \propto n_i \text{ at } t = t_1 . \tag{12.8}$$

We now assume β to be diagonal and n_i to be an eigendirection: That is, $n_i = 1$ for some given i, the other components being zero. The light-like line equation is

$$-dt^2 + e^{-2\Omega} e^{2\beta_{ii}} (dx^i)^2 = 0 \quad \text{(no sum on } i\text{)} \tag{12.9}$$

and the coordinate of the horizon is

$$x^i{}_H(t_1) = \int_{t_0}^{t_1} e^{\Omega} e^{-\beta_{ii}} dt \,. \tag{12.10}$$

This result gives the coordinate of the farthest point that can be seen. At time t_1, this point is at proper distance $X^i{}_H(t_1)$ from the observer, where

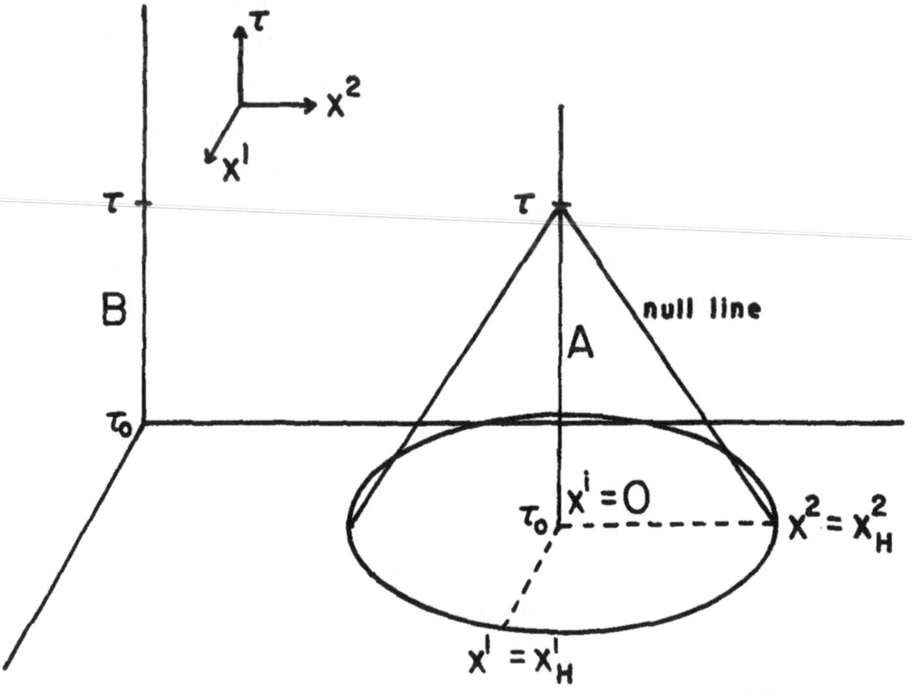

Fig. 12.1. Horizons. The horizon length for an observer A at time τ is the maximum spatial distance he can see in a given direction. The time τ_0 corresponds to the initial singularity (the "Big Bang"). Because the model is spatially homogeneous, the horizon lengths in i^{th} direction for observers A and B are equal if the values of τ are the same. $x^i{}_H$ is a function of τ, however.

$$X^i_H(t_1) = e^{-\Omega}e^{\beta_{ii}} \int_{t_0}^{t_1} e^{\Omega}e^{-\beta_{ii}} dt . \tag{12.11}$$

The horizon size may be infinite. In certain models there is one and only one direction with an infinite value of X^i_H. These models are highly special; in the generic model all horizon sizes are finite.

If the horizon size at time t_1 is finite, then $X^i_H(t_1) \to 0$ as $t_1 \to t_0$. This important result shows that near the singularity, in the general Type I model, a piece of matter can be influenced only by its nearest neighbors. At the present time, because of the expansion of the horizon, an observer is continually being influenced by new stars. This effect could be important in modern astrology (Figure 12.1).

12.2. Bianchi Type IX Universes in the Hamiltonian Approach

Aside from the general Bianchi Type I models, which for the fluid-filled case exhibits unbroken straight-line motion, the most thoroughly studied Bianchi-type model is of Type IX. Type IX models may be broken down into subclasses of varying degrees of complexity. The simplest are those in which the β_{ij} of (11.9) is diagonal (the *diagonal* case) studied by Misner (1969a,b). Slightly more complicated is the *symmetric* or *non-tumbling* case in which β has exactly one off-diagonal element. The Hamiltonian form for its Einstein equations was considered by Ryan (1971a). The most complex case is the *general* case in which is a general 3×3 matrix. The Hamiltonian formulation for this case was also given by Ryan (1971b).

The Diagonal Case

We can make β diagonal by choosing $\phi = 0$, $\theta = \pi/2$, $\psi = 0$ in (11.18). This choice reduces the action (11.18) to

$$I = \quad p_+ d\beta_+ + p_- d\beta_- - H d\Omega , \tag{12.12}$$

with

$$H^2 = p_+{}^2 + p_-{}^2 + e^{-4\Omega}(V(\beta_+, \beta_-) - 1) \tag{12.13}$$

in vacuum. In this case we can generally ignore matter for large enough Ω (that is, near the singularity): The matter terms in the Hamiltonian do not depend on β_+, β_- because $\mathcal{L}^i{}_M = 0$ and $\mathcal{L}^0{}_M = \text{const}$. Because $\pi^{ij}{}_{|j} = 0$ the constraints $C^{i'} = 0$ are identically satisfied. V as a function of β_+ and β_- is given in Table 11.1. The equipotentials of V are shown in Figure 12.2.

The complicated form of V implies that it will be almost impossible to find an analytic solution for $\beta_+(\Omega)$, $\beta_-(\Omega)$ which is valid everywhere.

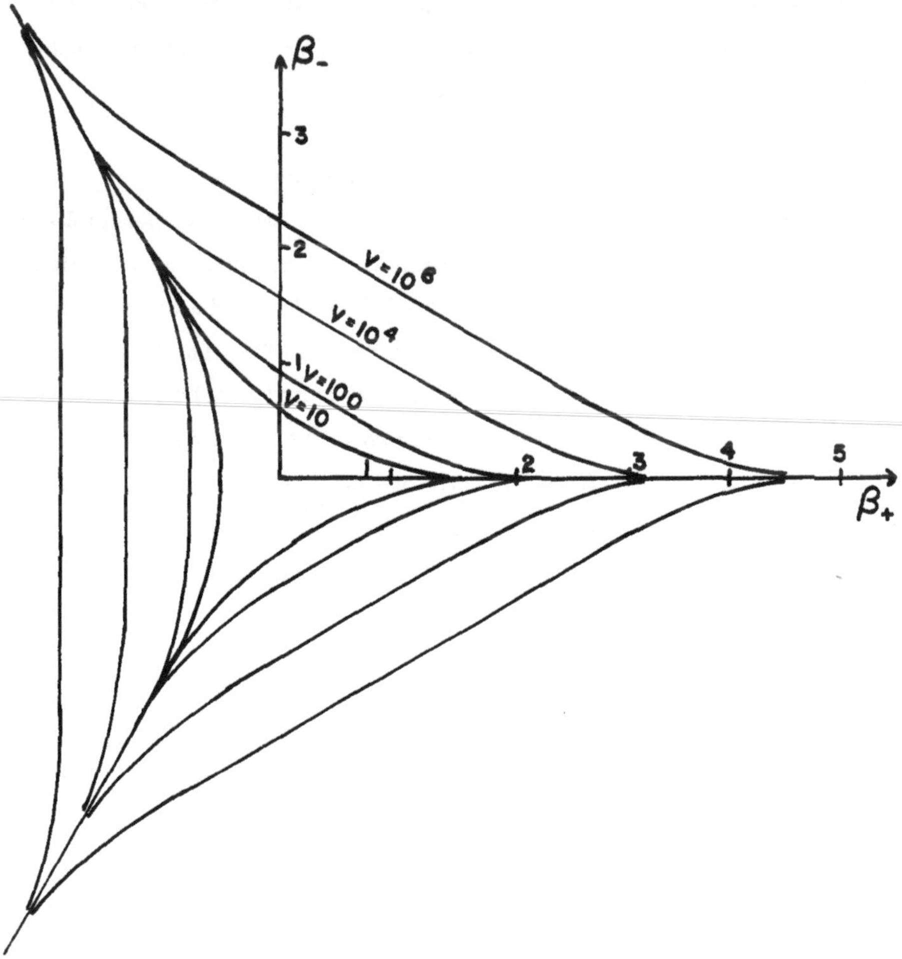

Fig. 12.2. The Potential $V(\beta_+, \beta_-)$ for Type IX Models

However, the technique of qualitative cosmology allows us to display an approximate solution in the form of a diagram. The hard-wall approximation implies that the universe point moves in a triangular well with unit velocity (since $(d\beta_+/d\Omega)^2 + (d\beta_-/d\Omega)^2 \approx 1$ far from the walls). This well is shown in Figure 11.2. Because $V(\beta_+, \beta_-)$ is multiplied by $e^{-4\Omega}$, the size of this well changes with time and Misner (1969b) has shown that the walls expand isotropically with velocity $d\beta_{wall}/d\Omega \approx \frac{1}{2}$. Thus the universe point will catch up with the wall if its angle of incidence is not too large. Generically the motion will be that of a particle bouncing around in a triangular box.

We now need a set of "bounce laws" for the reflection of the universe point from the walls. Because the walls are moving in Ω-time, reflection is not specular. Instead the angle of incidence θ_{in} is related to the angle of reflection θ_{out} by

$$\sin(\theta_{out}) = \frac{3\sin(\theta_{in})}{5 - 4\cos(\theta_{in})} . \tag{12.14}$$

When the universe point moves through the body of the triangle H is a constant, but this constant changes suddenly during a bounce. The incident H, H_{in} is related to H after reflection, H_{out}, by

$$H_{in}\sin(\theta_{in}) = H_{out}\sin(\theta_{out}) . \tag{12.15}$$

It is now possible to draw a qualitative picture of β_+, β_- as a function of the time Ω. The picture makes use of the straight-line-constant-speed property of the motion of the universe point between bounces. It also makes use of the bounce laws. An example is shown in Figure 12.3.

A situation where the qualitative solution is not sufficient to give us all the important information is during a "mixing bounce." ("Mixing" will be discussed below.) This type of bounce occurs when the universe point moves almost directly into one of the corner channels. In this case the universe point begins to oscillate rapidly between the two walls of the channel, finally reversing direction and moving out of the channel. The mixing bounce will be studied in more detail in below.

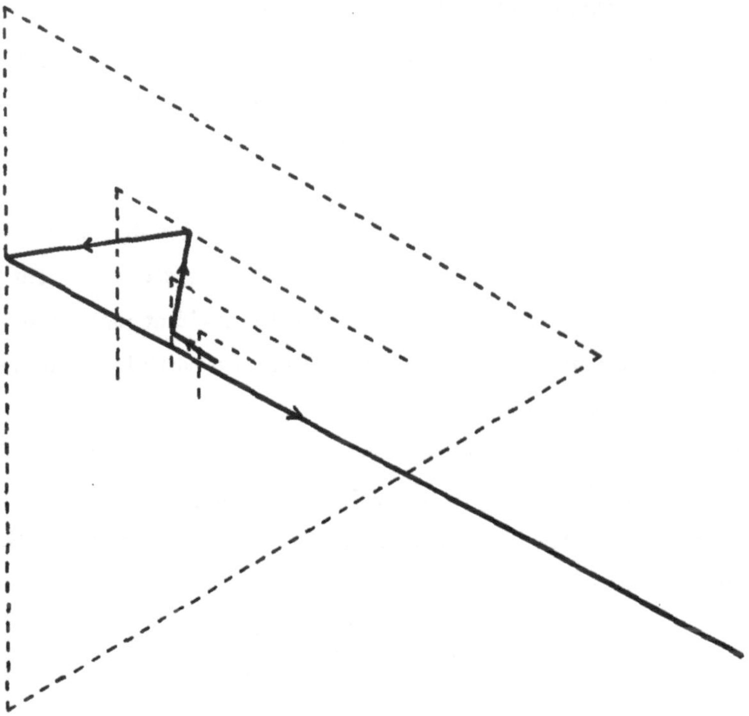

Fig. 12.3. A Series of Bounces of the Universe Point in a Type IX Potential. The dashed lines show the expanding potential wall at various times, and the solid line is the track of the universe point.

The Symmetric or Non-Tumbling Case

In the simplest case with rotating matter β has precisely one off-diagonal entry. The terms "symmetric" (Gödel, 1950) and "non-tumbling" (Matzner, Shepley, Warren, 1970) are both used for this case. In terms of the parameters of (11.18), we have $\phi = 0$, $\theta = \pi/2$, and ψ is allowed to vary. In this model we must allow matter, and here we let this matter be a pressureless fluid $(T_{\mu\nu} = \rho u_\mu u_\nu)$. The equation $T^{\mu\nu}{}_{;\nu} = 0$ can be shown to be satisfied if

$$u_1 = u_2 = 0, \quad u_3 = \overline{C} = \text{const.}$$

and

$$\bar{\mu} = \text{const.} = N\rho u^0 e^{-3\Omega} R_0{}^3 \tag{12.16}$$

in the invariant frame $\{d\Omega, \sigma^i\}$. Using the metric of the symmetric case and the constants $\bar{\mu}$ and $\bar{C}$ we find

$$\mathfrak{L}_M = -2\bar{\mu}(1+R_0^{-2}\bar{C}^2 e^{2\Omega} e^{2\beta_+})^{\frac{1}{2}} N - 2N_i \bar{\mu}\bar{C}R_0^{-2} e^{2\Omega} e^{2\beta_+} . \quad (12.17)$$

If we let $36\pi^2 R_0^4 = 1$ and define $\mu = 48\pi^2\bar{\mu}$, $C = 32\pi^2 \frac{\mu}{\bar{\mu}} \bar{C}$ then the action (11.18) reduces to

$$I = \int p_+ d\beta_+ + p_- d\beta_- + p_\phi d\phi - H d\Omega , \quad (12.18)$$

with

$$\begin{aligned} H^2 = p_+^2 + p_-^2 + \frac{3p_\psi^2}{\sinh^2(2\sqrt{3}\beta_-)} + e^{-4\Omega}(V(\beta_+, \beta_-) - 1) \\ + \mu e^{-3\Omega}(1 + 4C^2 e^{2\Omega} e^{4\beta_+})^{\frac{1}{2}} , \end{aligned} \quad (12.19)$$

where $V(\beta_+, \beta_-)$ is the *same* as that given in Figure (12.2) for the diagonal case.

When we compute $\pi^{ij}{}_{|j}$ for our metric we find that the constraint $C^{i'} = 0$ implies

$$p_\psi = \mu C . \quad (12.20)$$

This constant value of p_ψ is consistent with the field equations since H is cyclic in ψ. We replace p_ψ by μC and arrive at a Hamiltonian for $\beta_\pm$. The equation for ψ is then given by

$$\dot{\psi} = \partial H / \partial p_\psi |_{p_\psi = \mu C} , \quad (12.21)$$

and so ψ may be found as a function of Ω once β_+ and β_- are given as functions of Ω.

We can now construct a diagrammatic solution by means of the wall approximation. The additional potential terms in H add additional walls. The potential $3p_\psi^2/\sinh(2\sqrt{3}\beta_-)$ is called the *centrifugal* potential, and the wall it generates is the centrifugal wall. $\mu e^{-3\Omega}(1 + 4C^2 e^{2\Omega} e^{2\beta_+})^{\frac{1}{2}}$ is

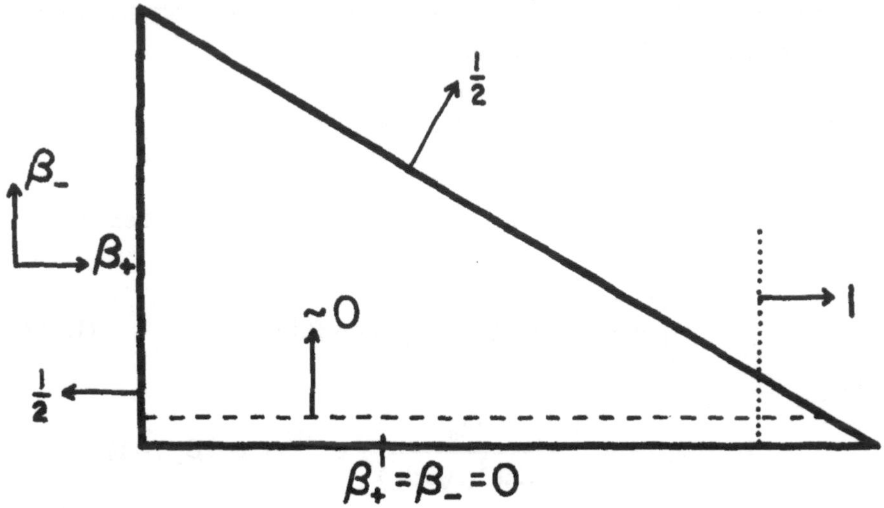

Fig. 12.4. The Walls Associated with the Potentials in a Symmetric or Non-Tumbling Type IX Model. The solid line is the gravitation wall; the dotted line is the rotation wall; and the dashed line is the centrifugal wall. The arrows attached to the walls show their respective velocities.

the *rotation* potential, and the wall it generates is the rotation wall. The positions and velocities of these walls are shown in Figure 12.4. It is interesting to note that β_- is most naturally interpreted as a radial coordinate in the parametrization and therefore is strictly positive. When we find ψ by integrating (12.21), we see that ψ will only change during bounces off the centrifugal wall. The diagrammatic solution has been used by Ryan (1971a) to show that the change in ψ during a bounce goes to zero as we approach the singularity. (Belinskii, Lifshitz and Khalatnikov, 1971a, have used the analytic approach to show the same thing.) More detail about the symmetric case can be found in Ryan (1971a, 1972d).

The General Case

The details of the derivation of the Hamiltonian form of the Einstein equations in the general case are found in Ryan (1972d). We shall give the final results and briefly discuss the diagrammatic solution for the case of a pressureless fluid.

It can be shown that the Euclidean sum

$$\sum_{i=1}^{3} u_i u_i \equiv \bar{C}^2 \tag{12.22}$$

is a constant of the motion (see Chapter 13). Because of this constant we can parametrize the u_i by use of the parameters γ, λ in the form (κ_1, κ_3 are the constant matrices of 11.15):

$$u_i = \bar{C}(e^{-\psi\kappa_3} e^{-\theta\kappa_1} e^{-\phi\kappa_3})_{ij} n^j , \tag{12.23}$$

where

$$n^j = (\sin\gamma \sin\lambda, \sin\gamma \cos\lambda, \cos\gamma) , \tag{12.24}$$

with γ and λ functions of Ω, and $\bar{C}$ a constant. The quantity

$$\bar{\mu} = N\rho u^0 e^{-3\Omega} R_0^{\,3} \tag{12.25}$$

is still a constant in the general case. We rescale $R_0, \bar{\mu}, \bar{C}: 36\pi^2 R_0^{\,4} \to 1$, $48\pi^2 \bar{\mu} \to \mu$, and $32\pi^2 \bar{\mu}\bar{C} \to \mu C$. We arrive at the following Hamiltonian for $\beta_\pm$:

$$\begin{aligned} H^2 = p_+^{\,2} + p_-^{\,2} &+ \frac{3(\mu C)^2 \cos^2\gamma}{\sinh^2(2\sqrt{3}\beta_-)} + \frac{3(\mu C)^2 \sin^2\gamma \cos^2\lambda}{\sinh(3\beta_+ + \sqrt{3}\beta_-)} + \frac{3(\mu C)^2 \sin^2\gamma \sin^2\lambda}{\sinh(3\beta_+ - \sqrt{3}\beta_-)} \\ &+ e^{-4\Omega}(V-1) + e^{-3\Omega}\mu\Big(1 + (2C)^2 e^{2\Omega}\Big[\sin^2\gamma \sin^2\lambda\, e^{-2(\beta_+ + \sqrt{3}\beta_-)} \\ &\quad \sin^2\gamma \cos^2\lambda\, e^{-2(\beta_+ - \sqrt{3}\beta_-)} + \cos^2\gamma\, e^{4\beta_+}\Big]\Big)^{\frac{1}{2}} , \end{aligned} \tag{12.26}$$

where γ and λ are to be treated as known functions of Ω in Hamilton's equations and where $V(\beta_+, \beta_-)$ is again the same function of $\beta_\pm$ as in the diagonal and symmetric cases. We also have the supplementary equations

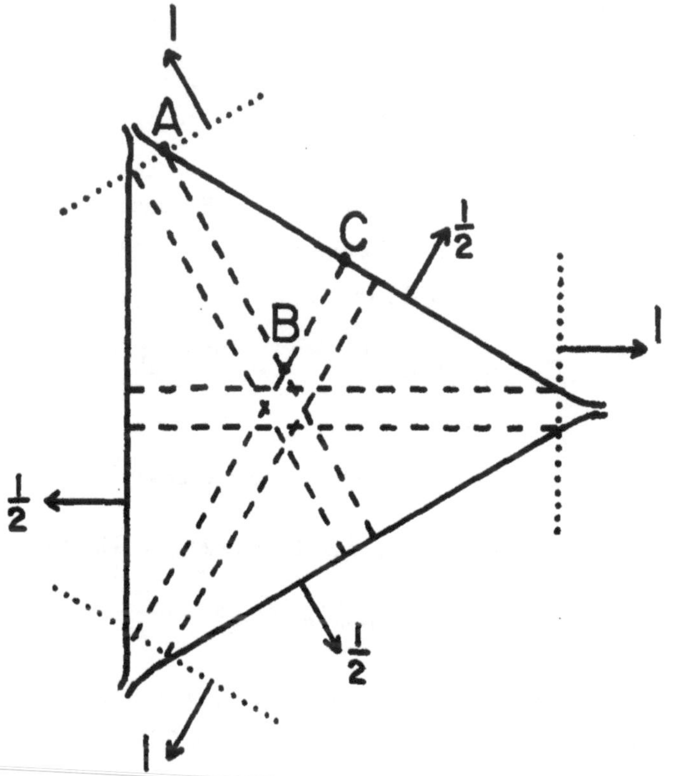

Fig. 12.5. The Walls Associated with the Potentials in the General Type IX Model. The solid line is the gravitation wall; the dotted lines the rotation walls; and the dashed lines the centrifugal walls.

$$\dot{\lambda} = d\lambda/d\Omega = \frac{8Ce^{-\Omega}}{H}\,\frac{\cos\gamma\, e^{(\beta_+ + \sqrt{3}\beta_-)}\sinh(3\beta_+ - \sqrt{3}\beta_-)}{\sqrt{(1+D)}}$$

$$+\frac{8Ce^{-\Omega}}{H}\,\frac{\cos\gamma\,\sin^2\lambda\,\sinh(2\sqrt{3}\beta_-)}{\sqrt{(1+D)}} - \frac{3\mu C}{H}\,\frac{\cos\gamma}{\sinh^2(2\sqrt{3}\beta_-)}$$

$$-\frac{3\mu C}{H}\cos\gamma\,\sin\lambda\,\cos\lambda\{\sinh^{-2}(3\beta_+ - \sqrt{3}\beta_-) - \sinh^{-2}(3\beta_+ + 3\beta_-)\} \qquad (12.27)$$

$$\dot{\gamma} = d\gamma/d\Omega = \sin\gamma\left\{\frac{8Ce^{-\Omega}}{H}\,\frac{\cos\lambda\sin\lambda e^{-2\beta_+}\sinh(2\sqrt{3}\beta_-)}{\sqrt{(1+D)}} + \right.$$

$$\left. + \frac{3\mu C}{H}\,\frac{\cos^2\lambda}{\sinh^2(\beta_+ + \sqrt{3}\beta_-)} + \frac{3\mu C}{H}\,\frac{\sin^2\lambda}{\sinh^2(3\beta_+ - \sqrt{3}\beta_-)}\right\}$$

where

$$D = \bar{C}^{-2} u_i e^{-2\beta} d_{ij} u_j \,.$$

Once $\beta_\pm$, γ, and λ have been found as functions of Ω, Hamilton's equations for ϕ, ψ, and θ may be solved by simple integration.

The walls associated with the various potentials in 12.26 are shown in Figure 12.5. The walls are gravitational, centrifugal and rotational, coming from the potentials analogous to those of the same names in the symmetric case. Because the corresponding potential terms contain γ and λ, the centrifugal and rotation walls shown in Figure 12.5 appear and disappear. We can think of the rotation walls as "flaps" which cover the channels at the corners of the triangle and the centrifugal walls, which pass through the center of the diagram, as becoming "transparent" or "opaque." It has been shown (see Belinskii, Khalatnikov, and Ryan, 1971) that γ and λ approach constants near the singularity. This behavior freezes the centrifugal and rotation walls and traps the universe point in one of the regions similar to the triangle ABC in Figure 12.5.

Interpretation of the Qualitative Diagrams; The Existence of a True Singularity

Diagrammatic solutions most directly show many of the most important facets of the behavior of various universes. For Type IX models they show that a true singularity exists and that the character of this singularity is similar in the diagonal, symmetric, and general cases. In particular, rotation has little effect on the singularity.

We use Hamiltonian techniques and the equation

$$\frac{d\Omega}{dt} = \frac{1}{N} = -(12\pi R_0^3)^{-1} H e^{3\Omega} \,, \tag{12.28}$$

to prove the existence of a singularity. First, we show that the break-down envisaged in Chapter 10, where $|\dot{\Omega}|$ became infinite is also accompanied by an $\Omega = \infty$ singularity in Type I and Type IX universes. In fact,

H decreases as Ω becomes large. We take H positive and thus assume the universe is expanding. Equation (10.6) indicates that H never goes through zero before $\dot{\Omega}$ becomes infinite toward the past (that is, toward increasing Ω). With these restrictions, let us look at Type I fluid models. There, for a fluid model with $p = kw\ (k \leq 1)$,

$$H^2 = p_+{}^2 + p_-{}^2 + e^{-3(1-k)\Omega}, \quad p_{\pm} \text{ constants}.$$

As we go back toward the singularity, Ω increases, and H becomes a constant. Thus when $\dot{\Omega}$ is infinite, Ω is infinite, and (10.14) shows that $n \to \infty$.

For Type IX universes, the problem becomes more complicated. In the diagonal case with $p = kw$,

$$H^2 = p_+{}^2 + p_-{}^2 + e^{-4\Omega}(V-1) + \mu e^{-3(1-k)\Omega}, \quad V = V(\beta_+, \beta_-) \geq 0 .$$

Consider $\dot{H} = \partial H/\partial\Omega$ which yields

$$2H\dot{H} = -4e^{-4\Omega}V - 3(1-k)\mu e^{3(1-k)\Omega} + 4e^{-4\Omega} .$$

H is thus a decreasing function of Ω once Ω is large enough (observation shows Ω is sufficiently large at present in the real universe). This decrease of H implies that $\dot{\Omega}$ becomes infinite only when Ω does. Again a mathematical singularity results because of (11.23).

Since we do not have a matter Lagrangian for more complicated forms of matter, we can only treat dust models in the symmetric and general cases. In the symmetric Type IX case H is given by (12.19) and we find

$$2H\dot{H} = -4e^{-4\Omega}(V-1) - \mu e^{-3\Omega}(3 + 8C^2e^{2\Omega}e^{4\beta_+})^{\frac{1}{2}}(1 + 4C^2e^{2\Omega}e^{4\beta_+})^{-\frac{1}{2}} .$$

The same argument as in the diagonal case obtains: H always decreases. The general case is even more complicated. However, at large Ω, γ and λ in (12.26) become constant. We may use essentially the same argument as in the symmetric case, and the result is the same: Ω becomes infinite at a finite value of t, namely at the value when $\dot{\Omega}$ becomes infinite.

In order to show that $\Omega = \infty$ is a true singularity in the symmetric and general, dust-filled, Type IX cases, let us look at (12.25). This equation gives us the physically observable matter density:

$$\rho = \bar{\mu} R_0^{-3} e^{3\Omega} [1 + e^{2\Omega} e^{-2\beta}_{ij} u_i u_j]^{-\frac{1}{2}} . \tag{12.29}$$

Diagrammatic arguments show that because of (12.22) the square root term in (12.29) is dominated by the $e^{2\Omega}$ term. Hence $\rho \sim e^{2\Omega}$, and the infinite value of ρ at $\Omega = \infty$ indicates a true singularity.

As the singularity nears, the universe point bounces within its well. This oscillatory approach to the singularity must be studied in more detail than can be obtained from the qualitative discussion we have given here. In particular, the phenomenon of mixing, to be discussed in the next section requires such a more thorough presentation.

12.3. Decay of Anisotropy in Type IX Models – Mixing

The behavior of anisotropy is much more interesting and complicated in a Bianchi Type IX model than in Type I models. Not only does anisotropy decay, but even if the matter is postulated to be non-rotating, "mixing" may occur. Misner's mixmaster model (Misner, 1969a) is the result.

Non-Rotating Type IX Models

As in the Type I case we discuss anisotropy within the Hamiltonian formulation for non-rotating Type IX models. The most general Type IX model may be handled similarly. From the previous subsection we find that the Hamiltonian H which governs non-rotating, dust-filled, Type IX universes is given by

$$H^2 = p_+{}^2 + p_-{}^2 + e^{-4\Omega}(V(\beta_+, \beta_-) - 1) + \mu e^{-3\Omega} , \tag{12.30}$$

where $V(\beta_+, \beta_-)$ is displayed in Figure 12.2. The general behavior of the universe point under this Hamiltonian was discussed qualitatively in the previous subsection. As before anisotropy is measured by $\mathrm{tr}(\sigma^2)$ which is again

$$\mathrm{tr}(\sigma^2) = \frac{1}{24\pi^2} R_0^{-6}(p_+^{\,2} + p_-^{\,2}) e^{6\Omega} . \tag{12.31}$$

In this case $(p_+^{\,2} + p_-^{\,2})$ is not a constant. However, it is effectively constant while the universe point moves across the center of the triangular potential. It changes suddenly when the universe point bounces off the walls. Consequently $(p_+^{\,2} + p_-^{\,2})$ changes very slowly in comparison to the $e^{6\Omega}$ *term in* $\mathrm{tr}(\sigma^2)$, and on the average anisotropy decays as Ω decreases.

Notice that as Ω decreases and t increases, anisotropy is lowered even more strongly due to the shrinking of the triangular potential. This shrinking forces the universe point into a smaller and smaller region near $\beta_+ = \beta_- = 0$. The decay of anisotropy is further enhanced if the additional postulate is accepted that energy is transported from one part of the universe to another along lightlike geodesics. Such transport of energy is termed "neutrino viscosity," for it involves a hypothetical lightlike particle which can pass through a large amount of matter before depositing its energy.

The Phenomenon of Mixing

Insight is gained by considering certain special paths in the plane as examples. The trivial path $\beta_+ = \beta_- = 0$ for all t corresponds to the isotropic, FRW model. A path which follows the line $\beta_+ = 0$ corresponds to the Taub-NUT-Misner universe discussed previously. In that model, distance measured along one axis of the $t = $ constant sections becomes zero as $\Omega \to \infty$. Such a singularity is called a "pancake" singularity in contrast to the complete collapse which occurs in the isotropic model.

We saw in Chapter 8 that some geodesics traverse the T-NUT-M space sections an infinite number of times as the Misner interface is approached. Of course, the T-NUT-M model is a vacuum model, and we are here dealing with a dust-filled model. In a dust-filled model, the Misner interface disappears and is replaced by a singularity. However, if the anisotropy of the model follows a path along the $\beta_+ = 0$ axis, it too has a pancake type of singularity. It too contains geodesics which wrap around space an infinite number of times near the singularity.

When a null geodesic can wrap around the universe, there is the possibility of communication over cosmic distances by photons or neutrinos or even shock waves. This transfer of energy may be a way of ironing out inhomogeneities and is called *mixing*. We used the term *horizon* to denote the boundary of cosmic matter visible to a given observer, and therefore, mixing corresponds to a sufficiently large horizon in all directions.

In the above case of a pancake singularity, not all directions are directions of mixing. Consider now the most general Type IX model in which the anisotropy path does not exactly follow the $\beta_+ = 0$ line in the $\beta_+\beta_-$-plane. The general behavior of the universe point in this case was discussed above. If the universe point moves for a sufficient time approximately parallel to the $\beta_+ = 0$ line there is still one direction of mixing. Chitre (1972a) has shown that this direction will always change. In fact, even if the universe point begins to leave the center by one of the corner channels, it will eventually emerge much as a charged particle from a converging magnetic field ("magnetic mirror," Misner, 1969a). Chitre further showed that in approximately 2% of all diagonal Type IX models, the change of mixing direction occurs often enough that complete mixing occurs (but see MacCallum, 1971a). Misner called a model with mixing in all directions the *Mixmaster Universe*.

Chaotic Cosmology

One of the most interesting new ideas in cosmology is the concept of "chaotic cosmology." As given by Misner (1968) the idea is that it is no accident that the universe is highly isotropic and homogeneous now. It is postulated that all universes, no matter how anisotropic and inhomogeneous at early times, eventually, through some process, become homogeneous and isotropic as time goes by. The program for proving this idea was begun by Misner (1967a, 1968, 1969a,b), who investigated the decay of anisotropy in Type I and Type IX models. He proposed the mixmaster universe to illustrate processes by which isotropization and homogenization could come about. The next step in such a program would be the study of inhomogeneous cosmologies to see if they actually tend toward homogeneity.

The entire concept has received a number of hard blows, including objections by Stewart (1968), Doroshkevich, Zel'dovich and Novikov (1967a), and Collins and Stewart (1971), concerning the possible amount of decay of anisotropy. The low probability of mixing in Type IX universes is another problem. Despite these objections the original idea is so persuasive that it is very difficult to discard. Recently Collins and Hawking (1973a) have proposed that in any universe in which galaxies can form (and hence life develop) anisotropy and inhomogeneity will decay to observed levels as the universe approaches the present.

12.4. Quantum Type IX Cosmologies

In Chapter 11 we discussed the basic concepts of quantum cosmology and minisuperspace. It is when these ideas are applied to Type IX cosmologies that we see the full range of problems encountered in quantized Bianchi-type universes.

If we apply the procedure of Chapter 11 to a diagonal Type IX universe, we find the vacuum Schrodinger-Klein-Gordon (SKG) equation is

$$-\frac{\partial^2\Psi}{\partial\Omega^2}+\frac{\partial^2\Psi}{\partial\beta_-{}^2}+\frac{\partial^2\Psi}{\partial\beta_-{}^2}+e^{-4\Omega}(V(\beta_+,\beta_-)-1)\Psi=0\,. \qquad (12.32)$$

This equation is only complicated by the fact that the "potential" term is explicitly Ω-dependent. If we approximate $V(\beta_\pm)$ by a triangular potential with infinitely hard walls, these walls are expanding, and the expanding well problem is difficult to solve exactly. Misner (1969b) has pointed out that if the solution is similar to the solution for an expanding one-dimensional square well (Zapolsky, 1970, in Ryan, 1972d), then the energy levels of the triangular box should obey

$$E_n \sim |n|\,\Omega^{-1}\;, \qquad (12.33)$$

where $|n|$ is some combination of the two quantum numbers which determine the energy level. This value of the energy plus the classical adiabatic *invariant* (Misner, 1969b) $H\Omega \sim \text{const.}$ can be used to imply that on the average the quantum number of a wave packet remains fixed. Misner (1969b) interprets this result to say that the universe, if it is classical now (and if it is diagonal Type IX), does not become more quantum-mechanical as we go back toward the singularity and that therefore quantum mechanics cannot affect the singularity.

With the symmetric case we see, for the first time, the problem that the space constraints become important. The quantized analogue of (12.19) is

$$-\frac{\partial^2\Psi}{\partial\Omega^2}+\frac{\partial^2\Psi}{\partial\beta_+{}^2}+\frac{\partial^2\Psi}{\partial\beta_-{}^2}+3(\sinh 2\sqrt{3}\beta_-)^{-2}\frac{\partial^2\Psi}{\partial\psi^2}+e^{-4\Omega}(V-1)\Psi$$
$$+\mu e^{-3\Omega}(1+4C^2e^{2\Omega}e^{4\beta_+})^{\frac{1}{2}}\Psi=0\,. \tag{12.34}$$

where we assume μ and C are c-numbers. The constraint $p_\psi = \mu C$ becomes $\mu C\Psi = -i\frac{\partial\Psi}{\partial\psi}$ in the Dirac method of handling constraint. Substituting the solution to this equation into (12.34) we find that we have the same equation as we would if we had solved the constraint $p_\psi = \mu C$ before we had quantized the ADM method. A problem arises in the Dirac method, however, when one requires that $(\beta_\pm, \psi + \pi/2)$ represent the same universe as $(\beta_\pm, \psi)$. In that case we find $\mu C = 4n$, n an integer. This result is disturbing because we assumed μ and C were c-numbers. We have not, however, quantized the matter field, but assumed it to be classical. We can hope that if we were to quantize the matter field, thus making μ and C quantum numbers, the difficulties with the Dirac method would disappear.

Another possible method for ordering factors in the quantum-mechanical equations for Bianchi-type universes was mentioned in Chapter 11, the use of the superspace metric to write the derivative part of the equations as a covariant Laplace-Beltrami operator applied to Ψ. The DeWitt metric in the diagonal Type IX case only trivially modifies the $\partial^2\Psi/\partial\Omega^2$ term, and the Misner metric gives (12.34). In the symmetric case, however, even the Misner metric leads to an equation different from (12.34). This new equation is

$$-\frac{\partial^2\Psi}{\partial\Omega^2}+\frac{\partial^2\Psi}{\partial\beta_+{}^2}+\frac{1}{\sinh(2\sqrt{3}\beta_-)}\frac{\partial}{\partial\beta_-}\left[\sinh(2\sqrt{3}\beta_-)\frac{\partial\Psi}{\partial\beta_-}\right]+e^{-4\Omega}(V-1)\Psi$$

$$+\mu e^{-3\Omega}(1+4C^2e^{2\Omega}e^{4\beta_+})^{\frac{1}{2}}\Psi=\frac{-3(\mu C)^2\Psi}{\sinh^2(2\sqrt{3}\beta_-)}\ . \tag{12.35}$$

As was mentioned in Chapter 11, not enough is known about quantum gravitation to choose among the various alternative methods of quantization.

13. NUMERICAL TECHNIQUES

Ο βιος ανθρωποις λογισμου και αριθμου δειται πανυ

– EPICHARMUS

Except in the case of the FRW models and simple Bianchi Type I cosmologies, we cannot expect to find exact solutions for homogeneous cosmologies (even though qualitative solutions can be found by means of the Hamiltonian techniques of Chapter 11). Inhomogeneous models are even harder to handle. Numerical analysis of homogeneous cosmologies is therefore a necessity – not only to give us exact solutions (to act as checks of our qualitative solutions) but as a testing ground for general numerical studies in cosmology. In this chapter we present solutions for Bianchi Type IX universes as examples of numerical techniques in cosmology.

13.1. General Techniques – Initial Data

It is most convenient to compute in an orthonormal basis $\{dt, \sigma^i\}$ using equations for a 3×3 matrix B whose elements are functions of time t only. Different Bianchi types are specified by the structure constraints of the spatial homogeneity group, $C^i{}_{jk}$. We presume that the orthonormal-synchronous basis never breaks down (until a singularity is reached), so that it may be used for computation.

The Einstein equation for Bianchi-type universes are a set of ten coupled *ordinary* differential equations ideally suited for numerical solutions. Behr (1965a) and others have studied this problem. In order to solve the equations we specify initial data on a $t = t_0$ hypersurface $H(t_0)$. These data are limited by four *constraint* (C) equations which they must satisfy. The remaining six *propagation* (P) equations allow one to compute the metric at points off the initial surface. Because the P equations are second order, the initial data consist of the intrinsic metric of H and the first time derivative of this 3-metric – or equivalents of these quantities.

Initial Data

In the synchronous basis, the metric is of the form

$$ds^2 = -dt^2 + g_{st}(t)\omega^s\omega^t, \qquad d\omega^i = \frac{1}{2} C^i{}_{st}\omega^s \wedge \omega^t. \tag{13.1}$$

We define the four one-forms σ^μ by

$$\sigma^i = b_{it}\omega^t, \quad \sigma^0 = dt, \text{ with } b_{is}b_{sj} = g_{ij} \tag{13.2}$$

where $B = (b_{ij})$ is the square root of $G = (g_{ij})$. The metric is now in orthonormal form; $ds^2 = \eta_{\mu\nu}\sigma^\mu\sigma^\nu$.

As we saw in Chapter 2, what really counts in computing properties of a manifold is not the metric components but both the metric components and the structure coefficients. In the form of the metric given by (13.1), the structure coefficients are constant, and the metric may be truly said to be represented by the matrix function $G(t)$. Since $B(t)$ is determined by $G(t)$, B may be regarded as taking the place of the metric on the spacelike sections $H(t)$.

We derived the affine connection forms and the curvature tensor in Chapter 9. There the matrix functions $K(t)$ and $L(t)$ were defined by

$$K = \dot{B}B^{-1}; \quad L = \frac{1}{2}(K + K^T) = (\ell_{ij}). \tag{13.3}$$

Remember that $L(t)$ is the second fundamental form of the invariant hypersurface $H(t)$ in the $\{\sigma^\mu\}$ basis. In addition, if B and L are given at a time t_0, they determine $K(t_0)$. We may regard $L(t_0)$ as equivalent to the first time-derivative of the metric at the hypersurface $H(t_0)$.

The Einstein equations are second order, so the metric and its first derivative at t_0 serve as initial data. Equivalently, we may specify $B(t_0)$ and $L(t_0)$ initially and use the field equations to find $B(t)$ for other times.

Constraint and Propagation Equations

From the expressions of Chapter 9 we may find the P and C equations. The P equations involve the six (ij) components of the Ricci tensor:

$$R_{ij}[=\dot{\ell}_{ij}+\text{func}\,(L,B)] = (w+p)u_i u_j + \frac{1}{2}(w-p)\delta_{ij}, \ (i,j=1,2,3)\,. \quad (13.4)$$

This form results from the fact that the matrix M in (9.14) can be expressed as a function of B and L. Thus, if $B(t_0)$, $L(t_0)$ are given and if w, p, and u_i are known functions of B and L, the P equations give $\dot{L}(t_0)$. $\dot{B}(t_0)$ is computed from $B(t_0)$ and $L(t_0)$ by use of (9.15). Now, $B(t_0+\Delta t)$ and $L(t_0+\Delta t)$ may be calculated. In this way we may watch the universe evolve.

But we must first find w, p, and u_μ as functions of B and L. To do so we use the C equations. Three of these involve the (0i) components of the Ricci tensor, and the fourth involves the previously defined quantity S:

$$\begin{aligned} R_{0i}\,[&=\ell_{st}\,d^t{}_{si}+\ell_{ti}\,d^s{}_{ts}] = (w+p)u_0 u_i\,. \\ S &= \frac{1}{2}\,[R_{0i}+R_{11}+R_{22}+R_{33}] = wu_0^2 + p(u_0^2-1)\,. \end{aligned} \quad (13.5)$$

These equations do not involve $\dot{L}$ and hence act as constraints which $L(t_0)$ and $B(t_0)$ must satisfy: A choice of $w(t_0)$, $p(t_0)$, and $u_\mu(t_0)$ limits our choice of $L(t_0)$ and $B(t_0)$. If the model were a vacuum model, $w=p=0$, these equations explicitly restrict the permissible assignment of values for $B(t_0)$ and $L(t_0)$.

Alternately, when $B(t_0)$ and $L(t_0)$ are specified, we can compute $w(t_0)$, $p(t_0)$, and $u_\mu(t_0)$ using the C equations. These four equations are sufficient for this computation since $u_\mu u^\mu = -1$ and since we specify p as a function of w by an equation of state $p = p(w)$.

As an example, we impose the "dust" equation of state $p = 0$, so that $w = \rho$. Moreover, we shall consider the one particular group, SO(3, R) (Bianchi Type IX). The methods, of course, will apply to any group and any equation of state.

Given $B(t_0)$, $L(t_0)$, we calculate the Ricci components $R_{0i}(t_0)$ and $S(t_0)$. The values of $\rho(t_0)$ and $u_\mu(t_0)$ are then determined by the formulae

$$\rho = S - \sum_{s=1}^{3} (R_{0s})^2 S^{-1} ,$$

$$u_i = R_{0i}\left(S^2 - \sum_{s=1}^{3} R_{0s}{}^2\right)^{-\frac{1}{2}} , \tag{13.6}$$

$$u_0 = (1 + u_1{}^2 + u_2{}^2 + u_3{}^2)^{\frac{1}{2}} .$$

(The sign of u_0 is arbitrarily taken as $+$.) As we see, there are two conditions which must be satisfied in a dust-filled universe (or for that matter, in a fluid-filled model)

$$S > 0, \quad S^2 - \sum R_{0i}{}^2 > 0 . \tag{13.7}$$

In addition, of course, the matrix $B(t_0)$ should be positive definite (have three positive eigenvalues). Provided these inequalities hold, $B(t_0)$ and $L(t_0)$ serve as appropriate initial data, and given $B(t_0)$ and $L(t_0)$, $\rho(t_0)$ and $u_\mu(t_0)$ may be calculated, and the P equations employed as outlined above.

13.2. Numerical Bianchi Type IX Models

There are two possible procedures in the generation of a numerical model. In both, the equation of state $p(w)$ and the spatial homogeneity group must be specified in advance of the selection of initial data. We have made this specification by setting $p = 0$ and requiring the structure constants $C^i{}_{jk}$ to be those of the Type IX group, $C^i{}_{jk} = \varepsilon_{ijk}$.

The two procedures involve methods of selecting initial data. In the first, $\rho(t_0)$, $u_\mu(t_0)$, or other fluid parameters are chosen. A specification of $B(t_0)$ and $L(t_0)$ which conforms to the C equations must then be found. The second procedure is to choose $B(t_0)$ and $L(t_0)$, using (13.6) to compute $\rho(t_0)$ and $u_\mu(t_0)$. We follow this second method.

Our computer program is schematically illustrated in Figure 13.1. The input data consist of two 3×3 matrices, $B(t_0)$, $L(t_0)$. One of these matrices, that used as $B(t_0)$, is positive definite. We compute $K(t_0)$ from $B(t_0)$ and $L(t_0)$, and from this find $\dot{B}(t_0)$. From $B(t_0)$, we calcu-

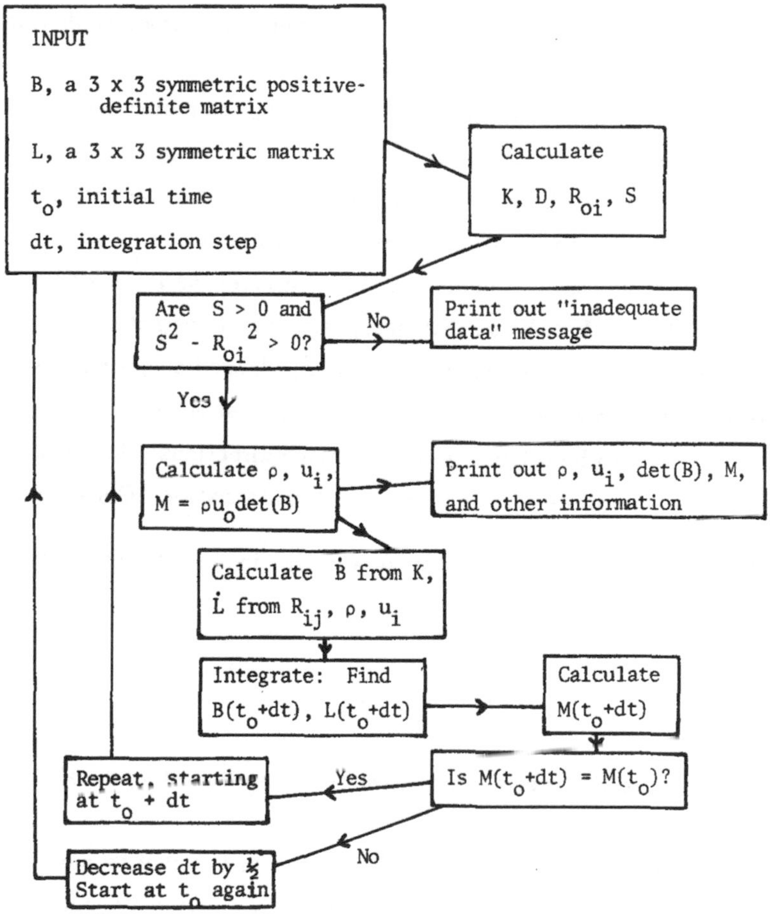

Fig. 13.1. Computer Flow Chart

late $D(t_0) = (d^i{}_{jk}(t_0))$. From $K(t_0)$ and $D(t_0)$, we then calculate $R_{0i}(t_0)$ and $S(t_0)$. The inequalities (13.7) are checked, and if they are satisfied, $\rho(t_0)$ and $u_\mu(t_0)$ are calculated. Finally, the values of $\dot{L}(t_0)$ are found from (13.4). A small step forward is taken by numerical integration and values of the fundamental quantities are found at the new time $t_0 + \Delta t$. Knowing $B(t_0 + \Delta t)$ and $L(t_0 + \Delta t)$, we repeat the process.

The program also can internally determine the size of the integration step. This determination makes use of an important constant of the motion derived from (10.9), the conservation law. In a Type IX universe, $M = \rho u_0 \det(B) = \rho u_0 e^{-3\Omega} R_0{}^3 = \text{const}$. At each step, $M(t+\Delta t)$ is computed and compared with $M(t)$. If there is a change, the program shortens the integration step.

In certain rotating models, the size of the integration step may be determined using a second constant of the motion. If ξ is a Killing vector field (one of the three which are present because of the $SO(3,R)$ symmetry), $u \cdot \xi$ is a constant along the matter paths since u is a geodesic. This fact was proved in Section 8.3. We label the three Killing vector fields ξ_i, $i = 1, 2, 3$, so that $u \cdot \xi_i$ are three constants. These constants are not themselves useful, for the ξ_i depend on position within each given $H(t)$. However, at one point in the spacetime manifold, the ξ_i may be chosen to be aligned with three fixed directions. At other points, the ξ_i are rotated by an orthogonal transformation. Consequently, $C^2 \equiv \sum_{i=1}^{3} (u \cdot \xi_i)^2$, which is constant, is a fairly simple function of B and L. C^2 is given explicitly by

$$C^2 = u_i b_{is} b_{sj} u_j \,, \tag{13.8}$$

where $u^i = u_i$ are the spatial components of u in the orthonormal $\{\sigma^\mu\}$ basis.

In certain models, C^2 vanishes identically, and cannot be monitored to see if the integration step need be changed. In practice the constancy of M is used to check the usefulness of a specified size of the integration step and C^2 is monitored as a check. A further check consists of verifying that all matter variables exhibit the same time development characteristics when a given integration is reperformed using a smaller integration step.

The details of the program are only interesting to the computer himself so we will not give them here. The Runge-Kutta method of integration is used, along with double-precision computation where needed.

Classification and Number of Type IX Homogeneous Universes

It is especially convenient to express the initial data in the parametrization of the Hamiltonian formulation of Chapter 11. From that chapter we find we use the quantities $\beta_{\pm}$, $p_{\pm}$, ϕ, ψ, θ, p_ϕ, p_ψ, p_θ, R_0, Ω, μ, u_1, u_2, u_3. (These quantities may readily be combined to give B and L.) These quantities are redundant, however. We may diagonalize β_{ij} at any one time (t_0), so ϕ and ψ may be taken to be 0 and θ to be $\pi/2$ without loss of generality. H_1 and u_1, u_2, u_3 are defined in terms of the other variables by the constraints. R_0 may be chosen as the specification of length units, and μ chosen as the specification of mass unit, so the physical state of the universe does not depend on them. Consequently, as initial conditions we must specify $\beta_{\pm}$, $p_{\pm}$, p_ϕ, p_ψ, p_θ and Ω. However, this choice corresponds to an entire family of initial conditions, because a universe with $\beta_{\pm}$, $p_{\pm}$, p_ϕ, p_ψ, p_θ at one value of Ω is the *same* as one with $p_{\pm}'$, $\beta_{\pm}'$, p_ϕ', p_ψ', p_θ' at another value of Ω. There is thus a seven-parameter family of initial conditions for general Type IX universes.

Within the general class of Type IX universes there are several subclasses most easily defined in terms of initial data restrictions:

1) The FRW $k = +1$ model. The initial data have the form $B(t_0) = (B\delta_{ij})$, $L(t_0) = (L\delta_{ij})$ or equivalently, $\beta(t_0) = 0$, $p(t_0) = 0$, $\Omega(t_0) = \Omega =$ const. These forms are preserved as t varies.

2) The diagonal case, $B(t_0) = \mathrm{diag}(B_1, B_2, B_3)$, $L(t_0) = \mathrm{diag}(L_1, L_2, L_3)$. Equivalently, $\beta_\pm$ and $p_\pm$ are specified, as is Ω. $p_\phi = p_\psi = p_\theta = 0$. Again these forms are preserved as t varies.

3) The T-NUT-M-like case is the diagonal case in which $B_1 = B_2$ and $L_1 = L_2$ or $\beta_- = p_- = 0$.

4) The symmetric case (also called the non-tumbling case) has diagonal $B(t_0)$ as without loss of generality do all cases. $L(t_0)$ has one off-diagonal element. As t varies one off-diagonal element of B becomes non-zero, but the other two off-diagonal elements of B and L remain zero. Exactly one spatial component of the velocity u is non-zero in this case. In terms of the Hamiltonian parameters, $\beta_\pm$, $p_\pm$, p_ϕ and Ω are specified, and $p_\psi = p_\theta = 0$.

5) The time-symmetric case: $B(t) = B(-t)$. This case is defined by $L(t_0) = 0$.

6) The pseudo-time-symmetric case: The matter variables (ρ, u^μ) are time-symmetric, but $B(t)$ need not be. This case is defined by $B(t_0)$ diagonal and $L(t_0)$ with only one component (and that a non-diagonal component) non-zero. The Hamiltonian momentum parameters $p_\pm$, p_ψ, p_θ are all zero at $t = t_0$, but p_ϕ is non-zero.

7) The general case: $B(t_0)$ is diagonal, but $L(t_0)$ is any matrix. The Hamiltonian parameters $\beta_\pm$, $p_\pm$, p_ϕ, p_ψ, p_θ, Ω are all arbitrary. Models in the last four classes may "mix" in the sense of Chapter 12.

Numerical Examples

It would be impossible to give an exhaustive set of numerical solutions for even the diagonal case, so great is the number of possible initial conditions. Hamiltonian techniques and qualitative cosmology can be used to sort out those initial conditions which are most interesting. One problem in displaying results is that cosmic time t moves too quickly, and a large amount of computer time is spent covering only a small part of the life of the universe. This is another advantage of the Hamiltonian formulation: Ω-time moves much more slowly near the singularity. However, t-time is very useful near a turn-around, det(B) = maximum, where Ω-time breaks down.

The figures below give several examples of numerical solutions for various initial conditions and special cases. The diagonal case has been studied numerically by Okerson (1969) using the Lagrangian formulation and Ω-time, but we shall not give any figures showing its behavior. Figures given in Matzner, Shepley, and Warren (1970) in which the general case is studied in t-time near turnaround for various initial conditions will not be given here. We include figures which show the behavior of the symmetric case in the Hamiltonian formulation, studied in Ω-time. This is work of Moser, Matzner, and Ryan (1973).

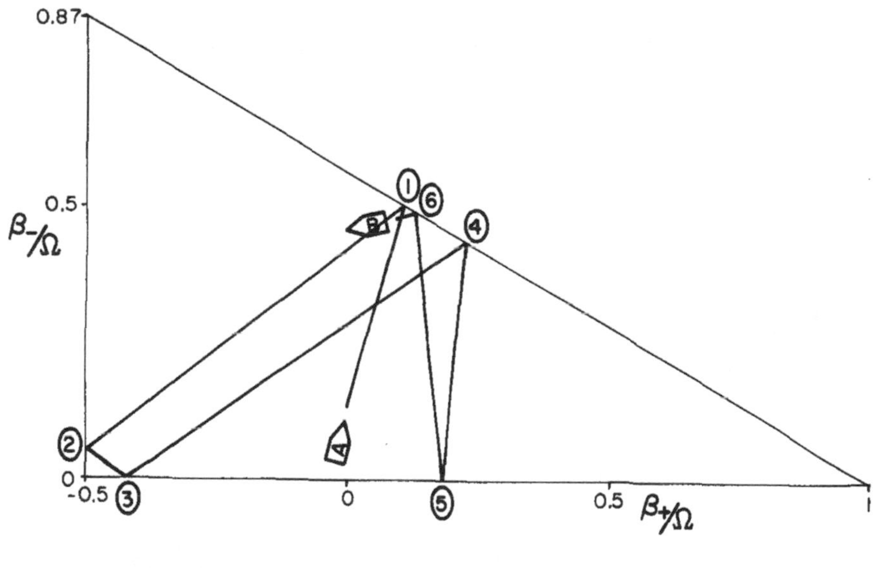

Fig. 13.2. The results of numerical solution of Hamilton's equations for a symmetric Type IX model. The triangle shown is that of Figure 12.4. The positions of the walls and of the universe point have been divided by Ω to keep the walls static and to confine the motion of the universe point to the fixed triangle. This transformation preserves straight-line motion of the universe point, but distorts angles, so bounce laws cannot be checked directly against this figure. For $\mu=2$, $C = 10^{-4}$ (μ, C as in Chapter 12) this figure starts the universe point at a generic point A moving in a randomly chosen direction at $\Omega = 15$ and follows it through six bounces to end at point B at $\Omega = 10^3$.

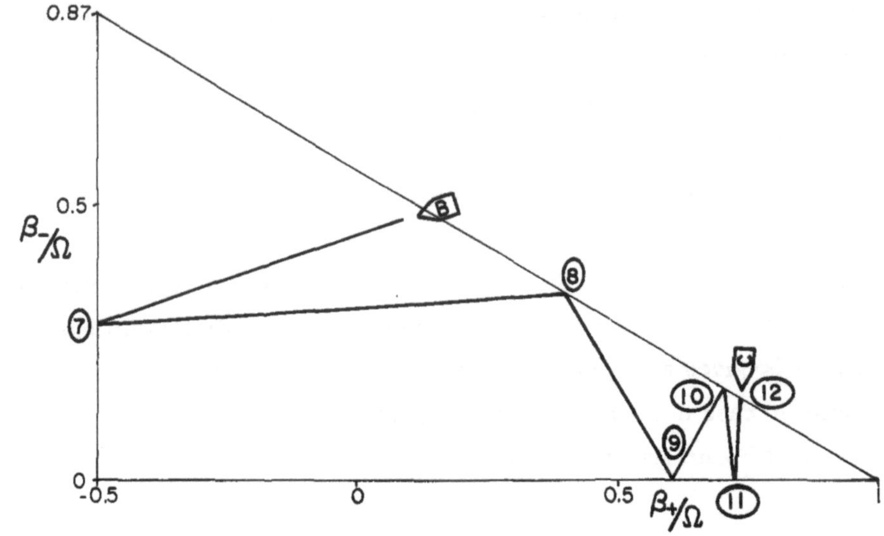

Fig. 13.3. This figure is a continuation of Figure 13.2. The universe point begins at point B (B of Figure 13.2) at $\Omega = 10^3$ and makes six bounces ending at point C.

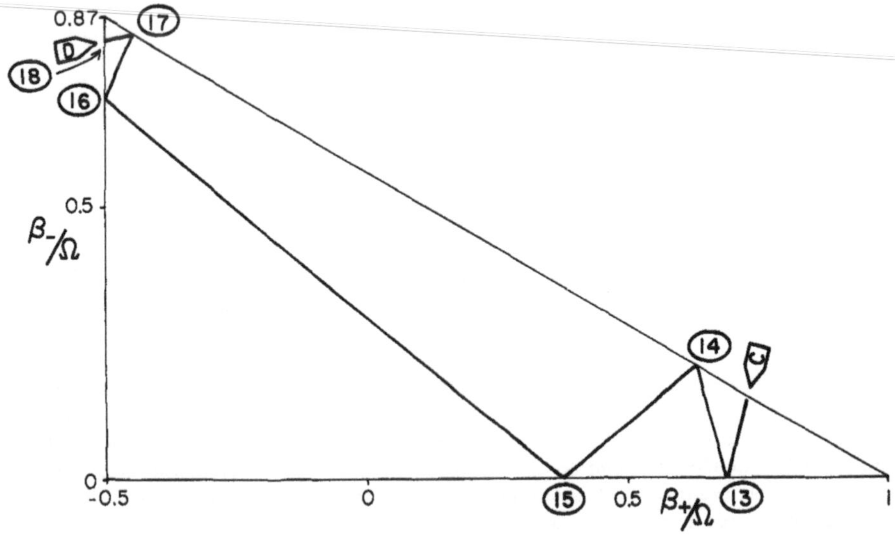

Fig. 13.4. A continuation of Figure 13.2. The universe point begins at C, makes six bounces, ending at point D at $\Omega = 10^6$.

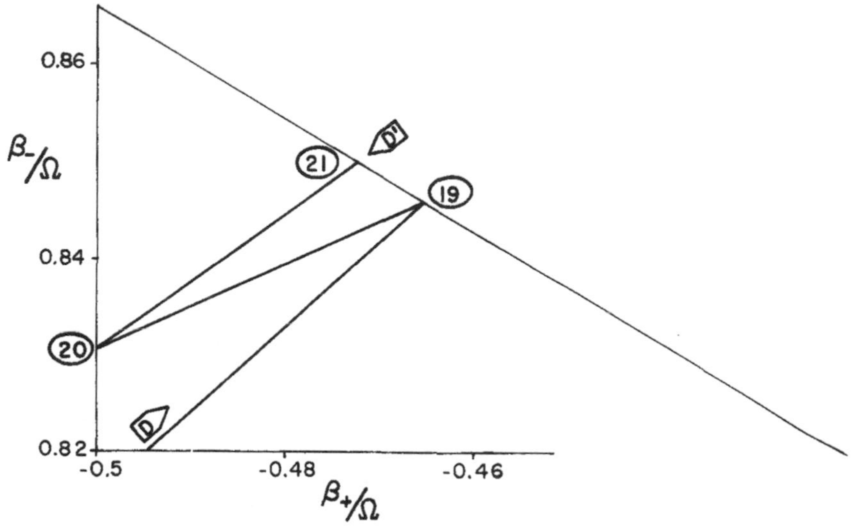

Fig. 13.5. A continuation of Figure 13.4. The universe point begins at D, makes three bounces, ending at D′ at $\Omega = 2.10^6$. Note that the scale of the plot is expanded to prevent loss of detail.

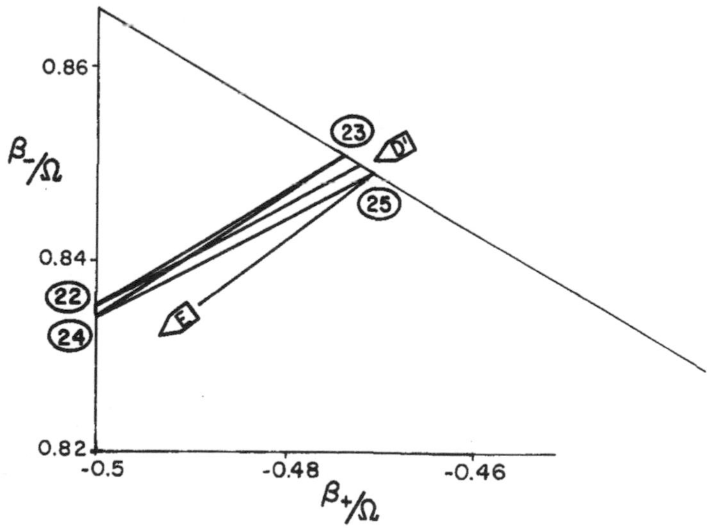

Fig. 13.6. A continuation of Figure 13.5. The universe point begins at D′, makes four bounces, ending at E at $\Omega = 3.3.10^6$.

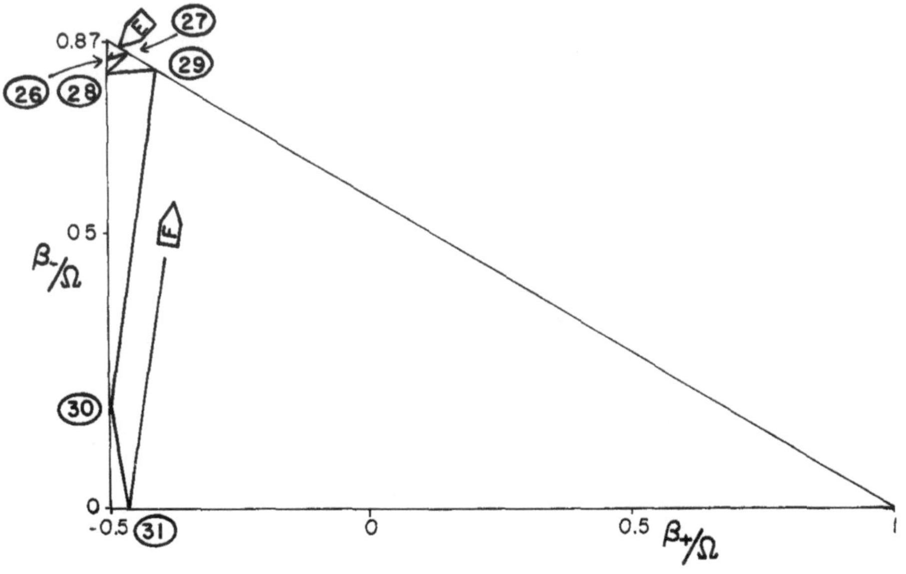

Fig. 13.7. A continuation of Figure 13.6. The universe point begins at E, makes six bounces, ending at F at $\Omega = 10^7$.

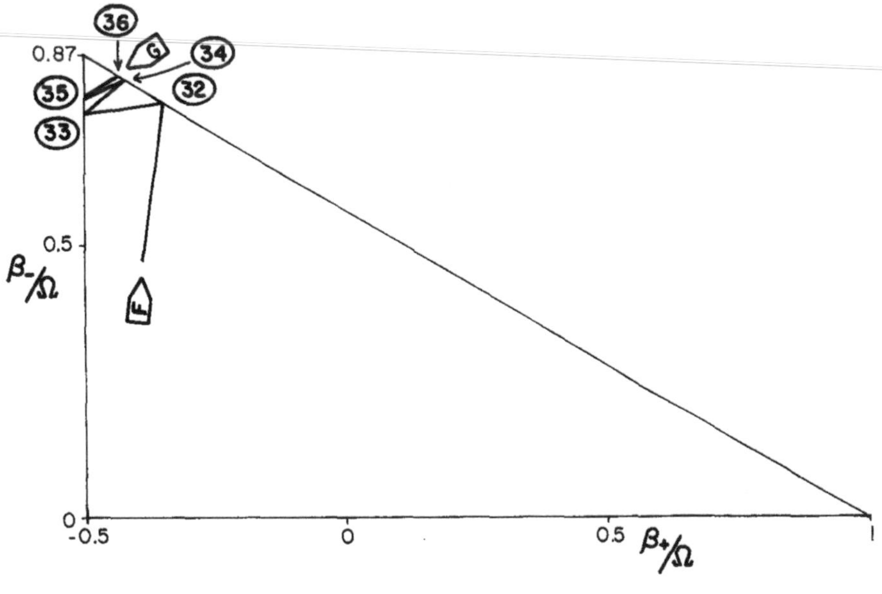

Fig. 13.8. A continuation of Figure 13.7. The universe point begins at F, makes five bounces, ending at G at $\Omega = 2.5 \cdot 10^8$.

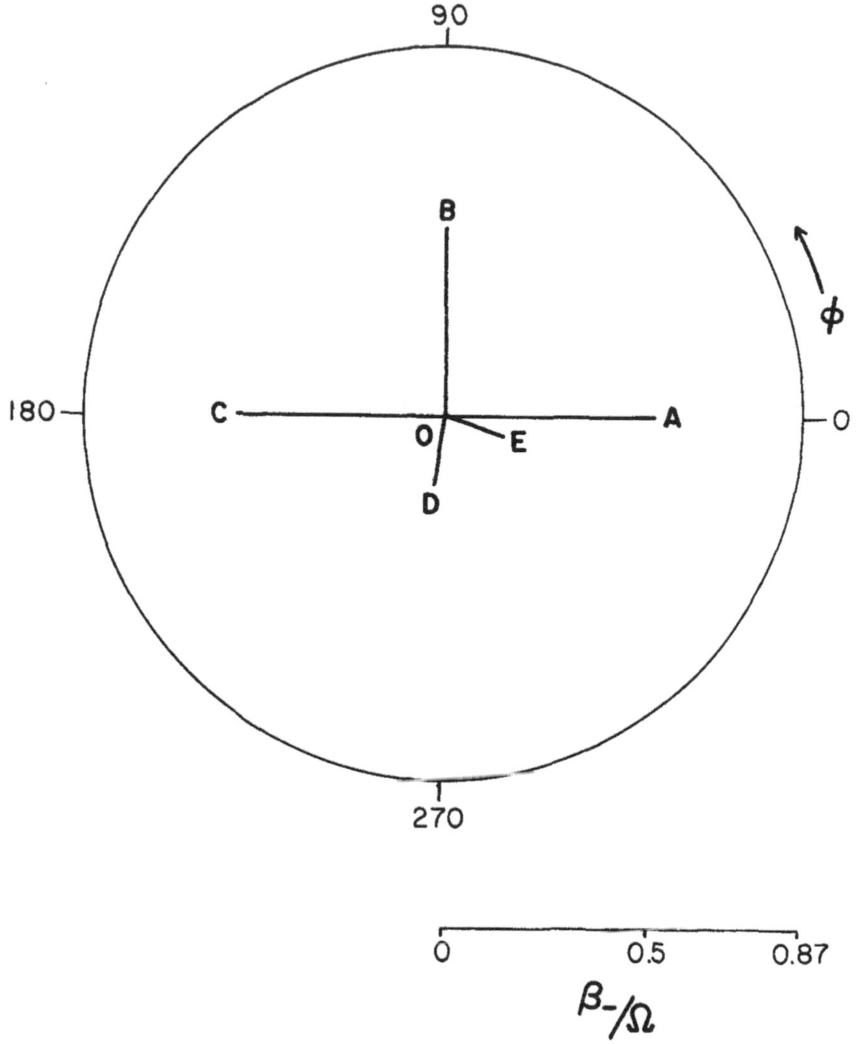

Fig. 13.9. The motion of the universe point in the angle ϕ. The motion is projected onto the plane $\beta_+/\Omega = -\frac{1}{2}$ in β_+, β_-, ϕ-space. The β_+-axis is marked by 0. The universe point begins at A and follows the path A0-0B-B0-0C-C0-0D-D0-0E, ending at E at $\Omega = 1.1.10^4$. Bounces 3, 5, 9, 11 of the previous figures are shown. The varying lengths of the paths correspond to bounces against different parts of the gravitation potential triangle.

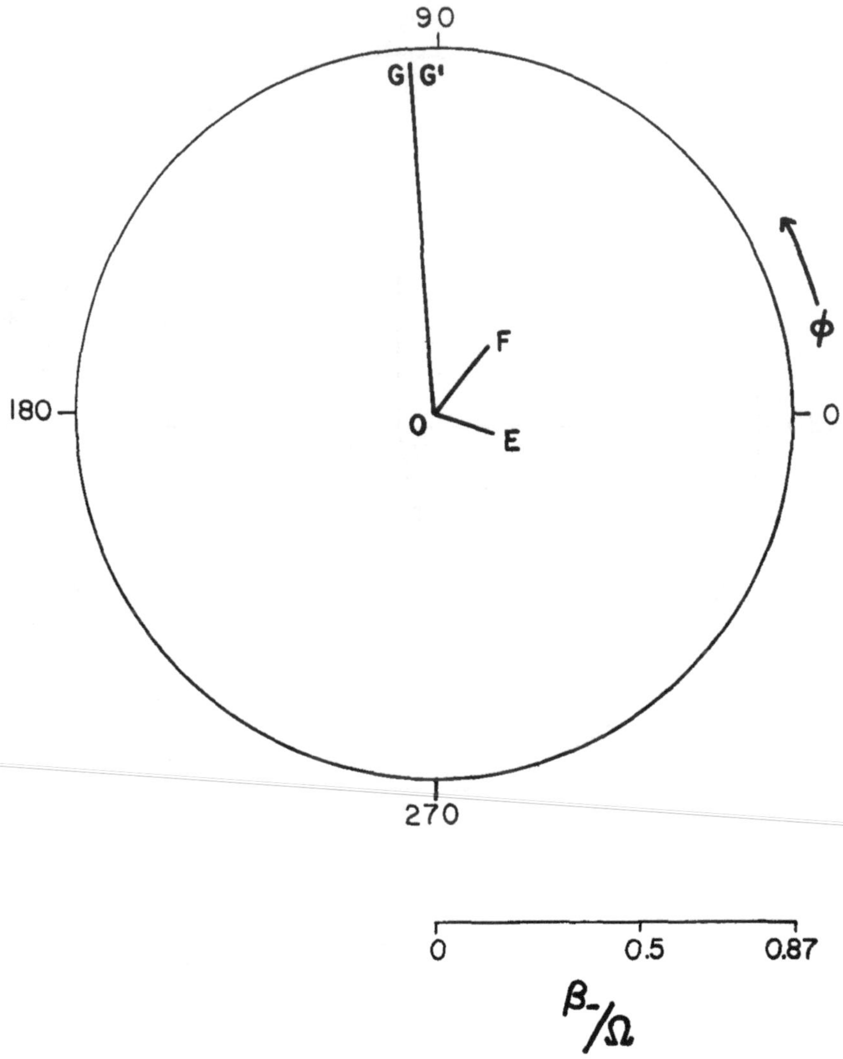

Fig. 13.10. Bounces 13, 15, and 31. The universe point begins at E (the E of Figure 13.9) and follows the path E0-0F-F0-0G-G0-0G′. Note that the change in ϕ during the bounce G0G′ is very nearly zero.

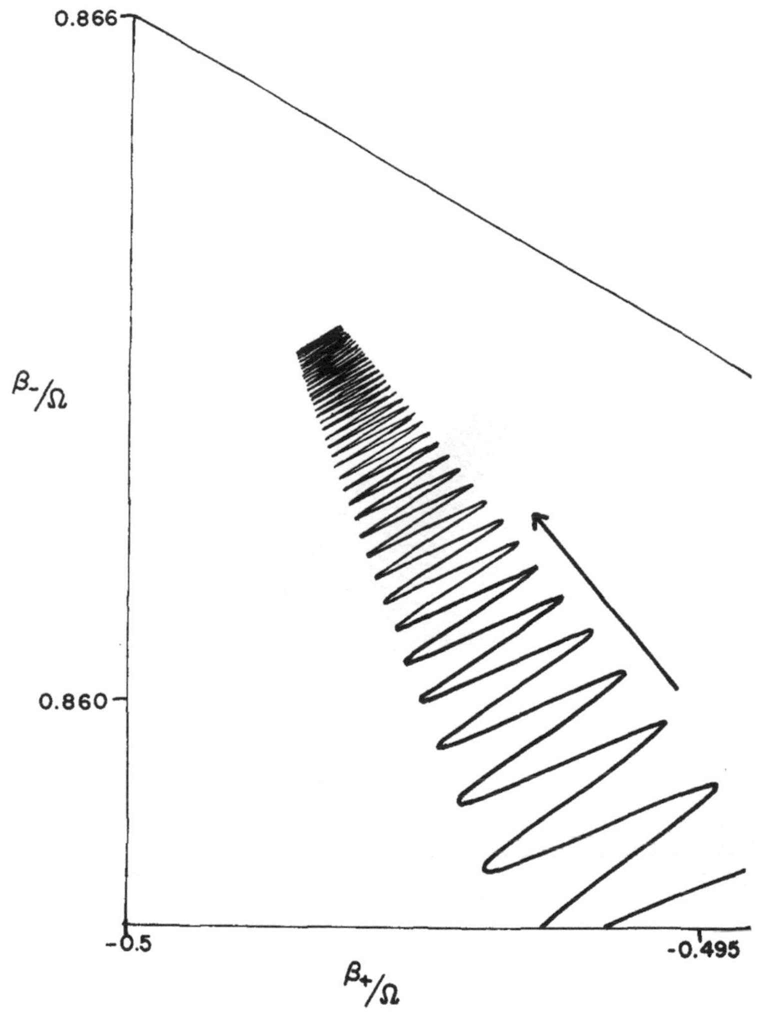

Fig. 13.11. A Mixing Bounce. The universe point is sent into the corner channel nearly directly. The general motion is in the direction of the arrow.

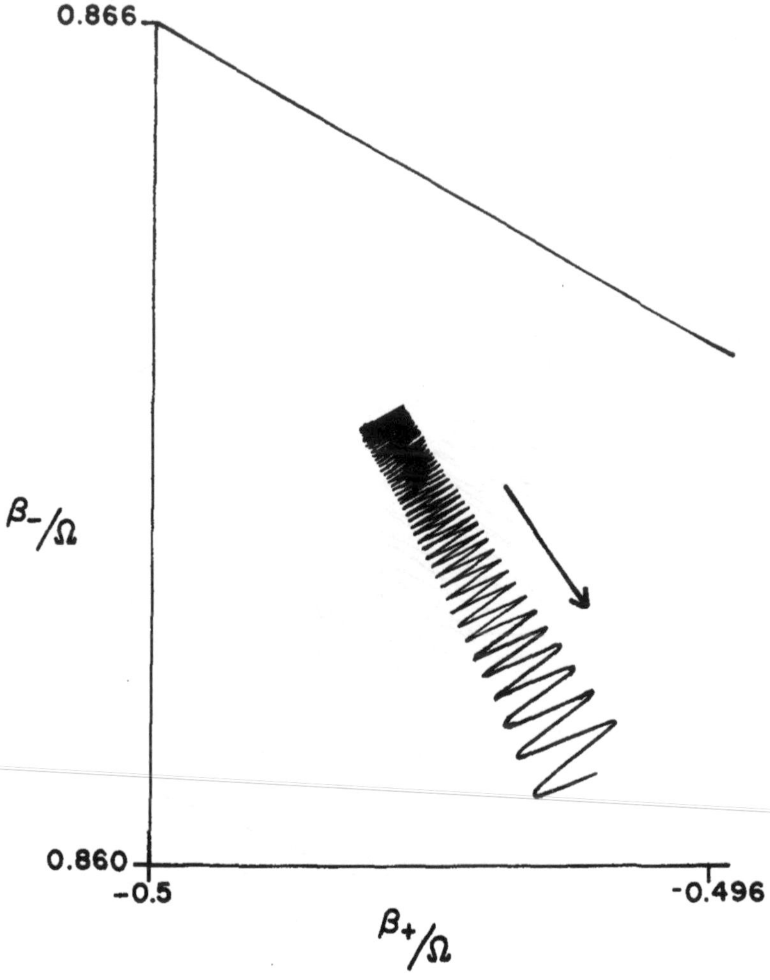

Fig. 13.12. A continuation of Figure 13.11. This figure shows the exit from the mixing bounce. The motion is in the direction of the arrow.

14. ASTROPHYSICAL STUDIES IN ANISOTROPIC TYPE I MODELS

In tauros Libyci ruunt Leones;
Non sunt papilionibus molesti
– MARTIAL

14.1. Large Scale Magnetic Fields

Spatial homogeneity without isotropy represents a compromise between simplicity and generality. The existence of anisotropy in the model allows a theoretical discussion of many vital effects, and we shall discuss three in this chapter. They are the effects of primordial magnetic fields, neutrino viscosity and kinetic theory, and perturbation theory as it pertains to the formation of galaxies. Our discussion will be brief, principally dealing with setting up equations and with basic properties, to serve as a program for further research. We will deal with Type I models in this chapter: models with anisotropy, but at the same time simple enough to allow explicit examples to be exhibited.

Electromagnetic Fields

The models we have previously described were mostly perfect fluid models. When a substantial large-scale electromagnetic field exists, the energy-momentum tensor must be augmented by the energy-momentum tensor of the field:

$$T^{M}{}_{\mu\nu} = F_{\mu\sigma}F_{\nu}{}^{\sigma} - \frac{1}{4}F_{\sigma\tau}F^{\sigma\tau}g_{\mu\nu} , \tag{14.1}$$

where the $F_{\mu\nu}$ are the components of the electromagnetic field tensor. This tensor is antisymmetric $F_{\mu\nu} = -F_{\nu\mu}$, so the electromagnetic field F is a two-form, and a very elegant treatment of Maxwell's theory can be given in the language of forms. Half of the Maxwell equations can be written

$$d\left(\frac{1}{2} F_{\mu\nu} \sigma^{\mu} \wedge \sigma^{\nu}\right) = 0 \tag{14.2}$$

if $\{\sigma^{\mu}\}$ is a basis for the one-forms. The other Maxwell equations are most conveniently written in component form:

$$F^{\mu\sigma}{}_{;\sigma} = J^{\mu} \tag{14.3}$$

where the J^{μ} are the components of the current density vector field associated with cosmic matter.

In spatially-homogeneous models there is a cosmic time t which measures proper time in the direction orthogonal to a given homogeneous hypersurface. Further, the metric can be expressed in the orthonormal synchronous basis $\{\sigma^{\mu}\}$ of Chapter 9, $ds^2 = \eta_{\mu\nu} \sigma^{\mu} \sigma^{\nu}$. With respect to an observer moving along a t-line, the components of F are defined in terms of the components of the electric field E_i and of the magnetic field B_i $(i=1,2,3)$ as $F_{0i} = E_i$; $F_{12} = B_3$ *et cyc.*

A Simple Primordial Magnetic Field

Suppose now that the model we deal with contains neutral matter (on the average), with a velocity field u orthogonal to homogeneous hypersurfaces. We further suppose that of the electromagnetic field components, only the component B_3 is non-zero. These are drastic simplifications but the behavior of these simple models is nonetheless most provocative.

The primordial magnetic field is assumed to be pointing solely in the "three" direction. The stress-energy tensor is found to be diagonal, with components (in the $\{\sigma^{\mu}\}$ basis where $\sigma^0 = dt$ and B_3 is a function of t only)

$$T^{M}{}_{\mu\nu} = \operatorname{diag}\left(\frac{1}{2} B_3^2, \frac{1}{2} B_3^2, \frac{1}{2} B_3^2, -\frac{1}{2} B_3^2\right). \tag{14.4}$$

The (00) component of the stress-energy tensor is positive and therefore contributes to the collapse to a singularity as described in Chapter 10. There is a negative stress, however, in the (33) component, and this term affects the detailed dynamics of the model near the singularity. In particular, a large scale magnetic field requires some anisotropy.

Time Development of the Field

The time development of the magnetic field is governed by Maxwell's equations. The first of these equations (14.2) says

$$\begin{aligned} d\left(\frac{1}{2} F_{\mu\nu}\sigma^\mu \wedge \sigma^\nu\right) &= d(B_3\sigma^1 \wedge \sigma^2) = 0 \\ &= \dot{B}_3\, dt \wedge \sigma^1 \wedge \sigma^2 + B_3\,(d\sigma^1 \wedge \sigma^2 - \sigma^1 \wedge d\sigma^2)\,. \end{aligned} \tag{14.5}$$

We have made use of the fact that $B_3 = B_3(t)$ and have written dt for σ^0. From (9.8) we see that $d\sigma^i = k_{is} dt \wedge \sigma^s + \frac{1}{2} d^i{}_{st}\sigma^s \wedge \sigma^t$. Consequently (14.5) implies, first, that $\dot{B}_3 + B_3(k_{11} + k_{22}) = 0$, and second that $k_{13} + k_{23} = d^1{}_{13} + d^2{}_{23} = 0$ if $B_3 \neq 0$. The first of these equations is the time-development equation for B_3, and the second is a constraint which must be satisfied by a spatially homogeneous model in order that it admit a magnetic field of the type described. The remaining Maxwell equation serves to define the current J.

A Type I Universe with a Magnetic Field

To see the effects of a magnetic field in a particularly simple case, consider a Type I model with diagonalized metric. In this model the matrix K is diagonal, and $d^i{}_{jk} = 0$. The basis forms are

$$\sigma^0 = dt, \quad \sigma^i = e^{-\Omega} e^{\beta}{}_{ii}\, dx^i \text{ (NS)}\,. \tag{14.6}$$

The curls of the σ^μ are therefore

$$d\sigma^0 = 0, \quad d\sigma^i = (\dot{\beta}_{ii} - \dot{\Omega})\, dt \wedge \sigma^i \text{ (NS)}\,. \tag{14.7}$$

The Maxwell equation which determines the time-development of B_3 then is written with $\beta = \mathrm{diag}\,(\beta_+ + \sqrt{3}\beta_-, \beta_+ - \sqrt{3}\beta_-, -2\beta_+)$. In consequence, $B_3 = Ce^{2(\Omega-\beta_+)}$, $C = \text{const}$. As it happens, the current vector field J vanishes in this case, and the magnetic field is sourceless.

The full field equations include a fluid stress-energy tensor as well as $T^M{}_{\mu\nu}$. If the fluid is dust, the source term for the Einstein field equations is

$$T_{\mu\nu} = \rho u_\mu u_\nu + T^M{}_{\mu\nu} \,. \tag{14.8}$$

As in the purely fluid case discussed earlier, the fluid velocity field is $u^\mu = (1,0,0,0)$ or $u_\mu \sigma^\mu = -\sigma^0$. The field equations are not independent. The conservation law, $T^{\mu\sigma}{}_{;\sigma} = (\rho u^\mu u^\sigma)_{;\sigma} + T^{M\mu\sigma}{}_{;\sigma} = 0$, serves to make one of them a consequence of the others. The conservation law is satisfied automatically for $T^M{}_{\mu\nu}$ because of the Maxwell equations: $T^{M\mu\sigma}{}_{;\sigma} = 0$; so $(\rho u^\mu u^\sigma)_{;\sigma} = 0$. The dust part of the conservation law can be integrated to yield $\rho = Me^{3\Omega}$, $M = \text{const}$.

The immediate consequence of the difference of the R_{22} and R_{33} equations is that $\dot{\beta}_- = De^{\Omega+2\beta_+}$, $D = \text{const}$. It is possible, and of course, simpler, to consider the case $\beta_- = 0$, so that $D = 0$. We now make this assumption, noting that it is not possible to have the case $\beta_+ = \beta_- = 0$ (isotropy) unless $C = B_3 = 0$ also. The general solution of the field equations is given implicitly as a function of t by

$$t - t_0 = \frac{2}{3E^2}\left(Ee^{\beta_+-\Omega} - C^2\right)\left[Ee^{\beta_+-\Omega} - \tfrac{1}{2}C^2\right]^{\frac{1}{2}};$$

$$e^{-(2\beta_++\Omega)} = ke^{\Omega-\beta_+}\left[Ee^{\beta_+-\Omega} - \tfrac{1}{2}C^2\right]^{\frac{1}{2}} \tag{14.9}$$

$$+ \frac{M}{3E^2}e^{\Omega-\beta_+}\left[E^2e^{2(\beta_+-\Omega)} + 2e^{\beta_+-\Omega}EC^2 - 2C^4\right];$$

$$E,\ k,\ t_0 \ \text{constants},\ E > 0\,.$$

It is seen from the above solution that the singularity is a "pancake" singularity, for $e^{-\Omega+\beta_+}$ remains finite when $e^{-\Omega-2\beta_+} \to 0$, when $C \neq 0$, that is, in the presence of a magnetic field. In this model the presence of the magnetic field requires at least a minimal amount of anisotropy and affects the nature of the singularity.

We have seen that the singularity $(\Omega = \infty)$ is not prevented, however. That the time when $\Omega = \infty$ is a true singularity is shown by the fact that the fluid energy density ρ is infinite at that time. The magnetic energy density is not necessarily infinite then.

Change of the Type of Singularity Because of the Magnetic Field

The effect of a magnetic field – on cosmologies, on collapsing stars, on ultra dense materials – is a subject of current study. The direction this study is taking is shown by the Hamiltonian work of Hughston and Jacobs (1970). Even at the primitive level of the simple solution given above, the effects of negative and positive stresses in the magnetic field can be seen.

The specialty and simplicity of the solution does not make it the less interesting, for the effect of the magnetic field was to prevent a collapse in two of the three spatial directions. In Chapter 10 we saw that Brill's model, an SO(3, R)-homogeneous model, contained a magnetic field – but no fluid – and remained nonsingular in the same sense that the T-NUT-M model is non-singular: mathematically non-singular.

What then is the role of magnetism in preventing or altering a singularity? How does the magnetic effect change when the Bianchi type of the spatial homogeneity group is changed? What other effects have yet to be discovered? These and other questions are the beginnings of a potentially very fruitful line of inquiry.

14.2. Kinetic Theory and Neutrino Viscosity

We have treated several types of stress-energy tensors in our cosmological models. In addition to the vacuum tensor, $T_{\mu\nu} = 0$, these types included perfect fluids, electromagnetic fields, and combinations of fluids and electromagnetic fields. These cases were treated in a phenomenological manner; in the fluid cases, for example, it was assumed that an equation of state was given.

The form of the stress-energy tensor may also be derived from kinetic theory with cosmic matter assumed to be composed of particles which interact by colliding (cross sections measured in laboratory experiments) or which interact with electromagnetic fields. We shall briefly describe the technique here and then indicate how neutrino viscosity – the transfer of energy by a postulated cosmic sea of neutrinos – is described by kinetic theory.

The Distribution Function

The particles in the cosmic gas are described by a function which tells how the particles are distributed in phase space. The typical case considered by a researcher in this area concerns a system of particles all having the same mass m. We use a coordinate system to describe a point in the spacetime manifold as x^μ $(\mu = 0,1,2,3)$. The momentum of a single particle is described by its contravariant components $k^\mu = mv^\mu$, restricted to lie on the "mass-shell" P_m; $(k^\mu) \in P_m \leftrightarrow k^\mu k^\nu g_{\mu\nu} = -m^2$. A point in phase space is described by the octet (x^μ, k^a), and phase space is seven-dimensional because of the restriction of k^μ to the mass-shell.

The distribution function f is defined by the requirement that the integral

$$\int_{P_m} f(x^\mu, k^a)\, dp_m \tag{14.10}$$

be the number density of particles at the point in M whose coordinates are x^μ. The integration is over the mass shell P_m in the space of contravariant vectors (the tangent space) at x^μ. The volume element dp_m is the volume element in P_m, to be described more fully below in the case $m = 0$. The components of the stress-energy tensor are formed from the distribution function:

$$T^{a\beta}(x^\mu) = \int_{P_m} f(x^\mu, k^\nu)\, k^a k^\beta\, dp_m\,. \tag{14.11}$$

This tensor field serves as the source term in Einstein's field equations.

Neutrinos in a Type I Universe

An important problem in relativistic kinetic theory is the effect to be expected from the existence of a copious number of neutrinos in the early stages of the universe. We note that (1) a neutrino travels on a lightlike geodesic between collisions (that is, it is a massless particle); (2) it travels a long distance before interacting (very small cross-section). Kinetic theory is then used to find a general form of the stress-energy tensor which reflects the long-distance transport of energy by a "viscosity" term. The field equations are then solved to yield, among other results, the effect on the anisotropy of the model.

Misner's (1968) treatment of neutrino viscosity in a Type I model is especially straightforward. His model has metric

$$ds^2 = -dt^2 + e^{-2\Omega}(e^{2\beta})_{ij}\, dx^i dx^j .$$

The fact that neutrinos are massless says that they travel on null geodesics between collisions. We write the momentum vector field as $k^\mu = dx^\mu/d\lambda$ where $x^\mu(\lambda)$ is the parametrized null geodesic path; therefore $k^\mu k_\mu = k^\mu{}_{;\nu} k^\nu = 0$. In consequence, $k_i = \text{const}$, and $k^i = e^{2\Omega} e^{-2\beta}{}_{is} k_s$, so that $k^0 = [e^{2\Omega} e^{-2\beta}{}_{st} k_s k_t]^{\frac{1}{2}}$.

Misner uses the smallness of the collision cross-section – indeed the fact that neutrinos are collisionless to a good approximation – to say that the conservation of energy law holds for each k^μ. Because the distribution function f describes the number of particles having momentum k^μ, this conservation law becomes

$$(fk^\mu)_{;\mu} = 0 . \tag{14.12}$$

(This equation along with the geodesic equation for k^μ, implies $T^{\mu\nu}{}_{;\nu} = 0$.)

In the case of a Type I model, f does not directly depend on x^i (i = 1,2,3). Equation (14.12) implies $fk^0 e^{-3\Omega} = \text{const}$. The "constant" in this equation is, in fact, an arbitrary function N of the three constants k_i:

$$f(x^\mu, k^\mu) = N(k_i)e^{3\Omega}(k^0)^{-1} = N(k_i)e^{2\Omega}[e^{-2\beta}{}_{st}k_s k_t]^{-\frac{1}{2}} . \qquad (14.13)$$

We continue to follow Misner in using the parameters k_i as coordinates on the mass-shell, P_m, momentum space. In all, the stress-energy tensor components $T^{\mu\nu}$ are given by

$$T^{00} = \int f(k^0)^2 d^3k = e^{4\Omega} \int N(k_i)[e^{-2\beta}{}_{st}k_s k_t]^{\frac{1}{2}} d^3k$$

$$T^{0i} = \int fk^0 k^i d^3k = e^{5\Omega} \int N(k_i)e^{-2\beta}{}_{is}k_s d^3k \qquad (14.14)$$

$$T^{ij} = \int fk^i k^j d^3k = e^{6\Omega} \int N(K_i)[e^{-2\beta}{}_{st}k_s k_t]^{\frac{1}{2}} e^{-2\beta}_{ia} k_a e^{-2\beta}_{jb} k_b d^3k .$$

The integration is over the three-space with coordinates k_1, k_2, k_3.

The Distribution Function

At this point a choice of $N(k_i)$ (equivalent to a choice of the form of the distribution function f), must be made. If the anisotropy matrix does not vanish, a realistic choice for $N(k_i)$ will in general yield an integral which can only be evaluated approximately. One property of $N(k_i)$ is needed to avoid a conflict with the field equations, however. This property is that $N(k_i)$ is an even function, and the potential conflict is due to the fact that R_{0i} always vanishes in a T_3-homogeneous model. We have

$$N(k_i) = N(-k_i) \rightarrow R_{0i} = T_{0i} = 0 .$$

Misner used a form of $N(k_i)$ that can be justified by physical arguments. He assumed that during the initial stages of the cosmos the neutrinos were in thermal equilibrium with other cosmic matter. The stress-energy tensor is then of the form of a $p = \frac{1}{3}w$ perfect fluid. When cosmic expansion caused the temperature to drop below a certain critical value T_0, the neutrinos became almost free. $N(k_i)$ is then fixed at

$$N(k_i) = a[e^{|k|/T_0} \pm 1]^{-1}, \; a = \text{const},$$
$$|k|^2 = (k_1)^2 + (k_2)^2 + (k_3)^2 . \tag{14.15}$$

Here a is a normalization constant. The sign is plus for fermions (such as neutrinos) and minus for bosons (such as photons or gravitons – the theory holds for these particles, too).

At the time of decoupling, when the temperature drops below T_0, the mean free path for neutrinos increases greatly. When the mean free path is neither very small nor very large, energy dissipation (transfer of energy from one part of the universe to another – neutrino viscosity) can take place with some degree of efficiency. Here the Boltzmann equation, which includes collisions, must be used to predict the detailed behavior of the distribution function.

Neutrino Viscosity

Misner, in lieu of solving the Boltzmann equation, approximates the effect of neutrino viscosity on the anisotropy of the universe. The traceless part of the stress-energy tensor (in the orthonormal tetrad $\{\sigma^\mu\}$) is

$$T_{ij} - \frac{1}{3} T_{ss} \delta_{ij} = -A\bar{\beta}_{ij} . \tag{14.16}$$

where A is some slowly varying function. The matrix $\bar{\boldsymbol{\beta}}$ depends on t only and is determined by approximating the effect of collisions.

The matrix $\bar{\boldsymbol{\beta}}$ is determined by a differential equation which includes the effect of the metric anisotropy matrix $\boldsymbol{\beta}$ as well as the effect of collisions. This equation is

$$\frac{d}{dt}\bar{\boldsymbol{\beta}} = \frac{d}{dt}\boldsymbol{\beta} - \frac{1}{t_c}\bar{\boldsymbol{\beta}} . \tag{14.17}$$

where t_c measures the mean collision time. At the time when the temperature drops to T_0, the time scale for change of $\boldsymbol{\beta}$ is short with respect to t_c, but t_c itself is not infinite. Collisions soon cause $\bar{\boldsymbol{\beta}}$ to reach the value $\bar{\boldsymbol{\beta}} = t_c \dot{\boldsymbol{\beta}}$.

Misner's treatment of the effect of the neutrino viscosity term therefore results in a negative term proportional to $\dot{\boldsymbol{\beta}}$ in the (ij) field equations $(i, j = 1,2,3)$. The traceless part of these equations becomes

$$(\dot{\beta}_{ij} e^{-3\Omega})^{\dot{}} = -\eta e^{-3\Omega} \dot{\beta}_{ij} \tag{14.18}$$

for some slowly varying (positive) function η whose explicit calculation requires a knowledge of t_c. The right side, proportional as it is to $\dot{\boldsymbol{\beta}}$, which in turn is proportional to the mean shear of cosmic matter, is the source of the nomenclature *viscosity*.

Detailed results depend on the explicit behavior of η as a function of the temperature of cosmic matter. For example, if η is constant, (14.18) implies

$$\dot{\beta}_{ij} \propto e^{-\eta t} e^{-3\Omega} .$$

No matter what the detailed form of η is, positive η, resulting from neutrino viscosity, causes more rapid decay of anisotropy than if no neutrino transfer of energy is possible.

The drop of $\dot{\beta}_{ij}$ to zero because of (14.18) measures the decay of anisotropy, because constant β_{ij} is equivalent to zero β_{ij} under a transformation of spatial coordinates x^i. Therefore we see, first, that anisotropy falls even in a perfect-fluid, expanding universe, and second, *that the decay of anisotropy is enhanced at the time when neutrinos first start to decouple from cosmic matter.* Present observations show that the real universe is isotropic: a summary of Misner's results is that the present isotropy does not necessarily mean that the universe was isotropic near its beginning (however, see Stewart, 1968).

14.3. Perturbation Theory and the Formation of Galaxies

Large scale magnetic fields can affect the type of singularity from which the universe expands, as we saw. Neutrino viscosity at an early epoch can cause an initial anisotropy to decrease rapidly. At a later cosmic epoch (it is thought), after both neutrino and electromagnetic radiation have become decoupled from cosmic matter, galaxies start to condense.

Perturbation theory is usually used to describe this condensation. The theory has not been completely successful, however. In an expanding model universe, a density perturbation typically grows too slowly for galaxies to reach their present size. Large amplitude perturbations must therefore be postulated to exist at a very early cosmic epoch if galaxies are to evolve to presently observed density variations (10^6 factor in relative density). Random or statistical fluctuations in cosmic density are too small to grow into galaxies in the 10^{10} years since the time of matter-radiation decoupling.

The Unperturbed Model

In anisotropic universes perturbation theory begins with a simple, spatially homogeneous model containing a smooth fluid. A concrete anisotropic model is the general Bianchi Type I cosmology. The metric is given in the coordinated basis $\{dt, dx^i\}$ by

$$\begin{aligned} ds^2 = -dt^2 + e^{-2\Omega}\Big[& e^{2(\beta_+ + \sqrt{3}\beta_-)}(dx^1)^2 + e^{2(\beta_+ - \sqrt{3}\beta_-)}(dx^2)^2 \\ & + e^{-4\beta_+}(dx^3)^2\Big] . \end{aligned} \tag{14.19}$$

The stress-energy tensor is that of a perfect fluid:

$$T^{\mu\nu} = (w+p)u^\mu u^\nu + pg^{\mu\nu} .$$

The pressure p is given in terms of the energy density w by the equation of state $p(w)$, and the fluid velocity field has the form $u^\mu = (1,0,0,0)$. The fluid has no rotation but does have shear, and the fluid velocity is geodesic. A second model, the general Type IX cosmology, can allow rotating matter. It is an especially interesting model in its own right, but the complexity of the general Type IX model makes the study of perturbations so difficult that only in special cases has such a study been performed.

The anisotropy functions $\beta_{\pm}$ in (14.19) have the time dependence

$$\beta_{\pm} = \int b_{\pm} e^{3\Omega} dt + c_{\pm} , \tag{14.20}$$

where $b_{\pm}$ and $c_{\pm}$ are constants. The $c_{\pm}$ are arbitrary and can be set zero through a coordinate transformation of the coordinates $x^i (i = 1,2,3)$. The expansion function Ω obeys the field equation $(\cdot = d/dt)$

$$3\dot{\Omega}^2 = -\frac{1}{2}(b_+^2 + b_-^2) e^{6\Omega} + w \tag{14.21}$$

where w is the energy density of the fluid. If the equation of state is $p = \sigma w$, $\sigma = \text{const}$, then w has the form

$$w = M e^{3(1+\sigma)\Omega}, \quad M = \text{const.} , \tag{14.22}$$

and (14.21) may then be integrated. The constant σ is less than or equal to one because of the requirement that the speed of sound $c_s = [dp/dw]^{\frac{1}{2}}$ be not greater than one. We then see the following fact: at $\Omega = \infty$ (this is the "Big Bang" singularity at $t = t_0$, where $e^{-3\Omega} = 0$), the anisotropy "energy" $(b_+^2 + b_-^2)$ dominates (or at least is of the same order of magnitude as) the energy density in affecting the expansion of the universe, unless of course the anisotropy strictly vanishes.

This latter case, the "flat" FRW model, will be discussed in detail, and we will later describe briefly the effects of anisotropy on a density perturbation. Its metric is a special case of (14.19):

$$ds^2 = -dt^2 + e^{-2\Omega}[(dx^1)^2 + (dx^2)^2 + (dx^3)^2] . \tag{14.23}$$

The interpretation of perturbation modes is especially straightforward in the isotropic model, and this interpretation is the point of departure for understanding perturbations in anisotropic models.

Functions Describing the Perturbation of a Flat FRW Model

To write the perturbed metric we first choose a basis $\{dt, dx^i\}$. One suitable basis uses a proper-time axis orthogonal to the three-space of coordinates x^i. In this basis $g_{00} = -1$, $g_{0i} = 0$, so that we have $\delta g_{00} = \delta g_{0i} = 0$. The perturbed fluid velocity has components in spatial directions. (An alternate choice is a comoving basis, in which $\delta g_{0\alpha} \neq 0$ but $\delta u^{\alpha} = 0$.)

We write the perturbed metric as

$$d\bar{s}^2 = -dt^2 + e^{-2\Omega}(\delta_{ij} + h_{ij})\,dx^i dx^j . \tag{14.24}$$

The metric perturbation h_{ij} in turn can be Fourier analyzed in the usual sense. This technique is applicable in the anisotropic Type I model also. We therefore consider the single "frequency" perturbation determined by the wave vector $k_i\,(i=1,2,3)$:

$$h_{ij} = \mu_{ij}(t)\,e^{ik_s x^s}, \qquad k_i = \text{const.} \tag{14.25}$$

The perturbation in the fluid variables is likewise Fourier analyzed. We write

$$\begin{aligned} \delta w &= W(t)\,e^{ik_s x^s}, \qquad \delta p = P(t)\,e^{ik_s x^s}, \\ \delta u_a &= iV_a(t)\,e^{ik_s x^s}, \qquad \delta u_0 = 0 . \end{aligned} \tag{14.26}$$

The last equation, $\delta u_0 = 0$, is a consequence of the conditions $\delta g_{0\alpha} = 0$.

The independent perturbation modes are now obtained by considering components of the matrix μ_{ij} singled out by the k_i vector. To project out terms orthogonal to k_i we define the projection operator $k_{ij} = \delta_{ij} - k_i k_j / k^2$, $k^2 = k_s k_s$. The matrix μ_{ij} is determined by the functions

$$\begin{aligned} \mu &\equiv \mu_{ss} , \\ \xi &\equiv \mu_{st} k_s k_t / k^2 , \\ \zeta_i &\equiv \mu_{st} k_s k_{it} , \\ \eta_{ij} &\equiv \left(k_{is} k_{it} - \tfrac{1}{2} k_{st} k_{ij}\right) \mu_{st} . \end{aligned} \tag{14.27}$$

We similarly break the three-vector V_i into a "parallel" and a "perpendicular" part; $A = V_s k_s$, $B_i = V_s k_{is}$.

Independent Perturbations

At this point we can start enumerating the independent perturbations. There are six functions μ_{ij}. The perturbed field equations fall into two classes. In the first class are the constraint equations. These equations allow us to express the five fluid variable perturbation functions W, P, A, B_i in terms of the metric perturbations and their first time derivatives.

To complete the specification of the five fluid perturbations it is usual to carry over the equation of state $p = p(w)$ into the perturbed model. The ratio of P to W is taken to be the same as dp/dw. This function in turn is taken to be the same function of time which is observed as the unperturbed model expands; $(P/W = dp/dw = \dot{p}/\dot{w})$. Such a perturbation is called "adiabatic."

The second class of equations include the six propagation equations, second order in time. When the fluid perturbation functions have been expressed in terms of μ_{ij} and $\dot{\mu}_{ij}$, these equations involve only the metric perturbation.

Because the propagation equations are second order, the solution is defined by two numbers for each of the six μ_{ij}. However, these twelve numbers, the perturbation parameters, do not represent twelve meaningfully different types of perturbations. Instead, four of these numbers may be set arbitrarily by means of gauge conditions.

Perturbation Equations in the Isotropic Case

The Fourier-transformed matter variables defined above are W, P, A, and B_i, defined above. The Fourier-transformed metric perturbation variables are $\mu, \xi, \zeta_i, \eta_{ij}$, defined by (14.27). The perturbation equations can be written in the form

$$e^{2\Omega}(\mu-\xi)k^2 - 2\dot{\Omega}\dot{\mu} = 2W \tag{14.28a}$$

$$\dot{\xi} - \dot{\mu} = -2(w+p)A/k^2 \tag{14.28b}$$

$$\dot{\zeta}_i = -2(w+p)B_i \tag{14.28c}$$

$$\ddot{\xi} - 3\dot{\Omega}\dot{\xi} - \dot{\Omega}\dot{\mu} + e^{2\Omega}(\mu-\xi)k^2 = W - P \tag{14.28d}$$

$$\ddot{\mu} - 2\dot{\Omega}\dot{\mu} = -W - 3P \tag{14.28e}$$

$$\ddot{\zeta}_i - 3\dot{\Omega}\dot{\zeta}_i = 0 \tag{14.28f}$$

$$\ddot{\eta}_{ij} - 3\dot{\Omega}\dot{\eta}_{ij} + e^{2\Omega}k^2\eta_{ij} = 0\ . \tag{14.28g}$$

Equations (14.28a) to (14.28c) involve only the first time-derivative of metric variables and are constraint equations. The remainder are propagation equations.

Equations (14.28c, f, g) reveal the interesting fact that the propagation equations of the individual components of the quantities ζ_i and η_{ij} are independent of any other perturbation mode. The η_{ij} are in turn independent of any matter variable, and they obey a wave-like equation. The $\dot{\zeta}_i$ are proportional to Ω^{-3} because of (14.28f), and in turn are proportional to the matter velocity variables B_i. A precise interpretation of these quantities is that the η_{ij} represent gravitational waves and the $\dot{\zeta}_i$ represent rotations. There are two independent η_{ij} (because $\eta_{ij}k_i = 0 = \eta_{ss}$) and four initial data values needed. There are two independent $\dot{\zeta}_i$ (because $\dot{\zeta}_i k_i = 0$) and two initial data values needed. We shall return to these quantities below and to the two extra initial data values needed to specify not only $\dot{\zeta}_i$ but also ζ_i.

The final perturbations, μ and ξ, obey coupled propagation equations. The constraint equations relate them to W (and thus to P) and to A, the "parallel" component of the perturbed velocity. Four initial data or perturbation parameters are needed to characterize μ and ξ.

Gauge Conditions

At this point we return to the full metric perturbation $\delta g_{\mu\nu}$. The specification $\delta g_{0\alpha} = 0$ does not exhaust all freedom of setting coordinate conditions, for there is still the freedom of positioning each $t = \text{const}$ hypersurface and of choosing coordinates within it. These freedoms are represented by infinitesimal coordinate transformations which preserve the conditions $g_{00} = -1$, $g_{0i} = 0$.

We define the infinitesimal coordinate transformation by

$$x^{\mu} \to x^{\mu} + \xi^{\mu} . \tag{14.29}$$

The change it produces in the metric is given by minus the Lie derivative of $g_{\mu\nu}$ with respect to $\xi^{\mu}: \delta g_{\mu\nu} = -(\mathcal{L}_{\xi} g)_{\mu\nu} = -(\xi_{\mu;\nu} + \xi_{\nu;\mu})$. The four conditions $\delta g_{0\alpha} = 0$ determine ξ^{μ} up to four functions of x^i only:

$$\begin{aligned} \xi_0 &= f_0(x^k) = -\xi^0 \\ \xi_i &= -f_{0,i}\, e^{-2\Omega} \int e^{2\Omega}\, dt + e^{-2\Omega} f_i(x^k) = e^{-2\Omega} \xi^i . \end{aligned} \tag{14.30}$$

The f_{μ} are Fourier analyzed to yield four constants F_{μ}:

$$f_0 = F_0 e^{ik_s x^s} , \quad f_i = iF_i e^{ik_s x^s} . \tag{14.31}$$

Then the effect of the ξ^{μ} on δg_{ij} is determined and from that the effect on the Fourier-analyzed metric perturbation variables.

The result is that a coordinate transformation is characterized by the F_0 and F_i of (14.31), and μ_{ij} is changed to

$$\mu_{ij} \to \mu_{ij} + \left[2k_i k_j \int e^{2\Omega} dt + 2\Omega \delta_{ij} \right] F + F_i k_j + F_j k_i . \tag{14.32}$$

In consequence, the metric variables enumerated in (14.27) are changed to

$$
\begin{aligned}
\mu &\to \mu + \left(2k^2 \int e^{2\Omega} dt + 6\Omega\right) F_0 + 2F_s k_s \\
\xi &\to \xi + \left(2k^2 \int e^{2\Omega} dt + 2\Omega\right) F_0 + 2F_s k_s \\
\zeta_i &\to \zeta_i + k^2 F_j k_{ji} \\
\eta_{ij} &\to \eta_{ij} \,.
\end{aligned}
\tag{14.33}
$$

We see that the η_{ij} are gauge-invariant. So are the $\dot{\zeta}_i$, but not the ζ_i. All six of the initial data parameters involved in the $\dot{\zeta}_i$ and η_{ij} are therefore physically meaningful and determine the gravitational wave and angular momentum content of the perturbation. The extra two pieces of initial data needed to determine ζ_i in addition to $\dot{\zeta}_i$ are physically meaningless, for they may be set arbitrarily by choice of the F_μ ("setting the gauge").

The two F_μ not involved in the ζ_i affect ξ and μ and through them the perturbation functions W, P, and A. Of the four initial data parameters involved, therefore, only two are physically meaningful. It is these perturbations which include sound waves (if $p \neq 0$). It is also these which include protogalaxies. It is these whose slow growth in an expanding universe does not permit a simpleminded, naive explanation of galaxy formation solely on the basis of statistical fluctuations condensing because of gravitational forces.

Density Perturbations in the Isotropic Dust Model

To obtain a feel for the growth rate of density perturbations, it is useful to consider the case of dust. In this case both the unperturbed and perturbed pressure vanish: $(p = P = 0)$. The unperturbed metric function $\Omega(t)$ is given by integrating (14.21), using (14.22) with $b_+^2 + b_-^2 = 0$, to yield

$$
e^{-\Omega} = Ct^{\frac{2}{3}}, \quad C = \text{const.} \tag{14.34}
$$

The gauge change effects shown in (14.33) for μ and ξ reduce to

$$\mu \to \mu - \left(Dt^{-\frac{1}{3}} + 4t^{-1}\right) F_0 + G$$
$$\xi \to \xi - \left(Dt^{-\frac{1}{3}} + \frac{4}{3}\,t^{-1}\right) F_0 + G\,, \tag{14.35}$$

F_0 and $G = 2F_s k_s$ are gauge constants, and D stands for $\frac{3}{2}C^{-2}k^2$. Examination of (14.35) shows that A may be made zero by choice of F_0. If B_i happen to be zero because the perturbation has zero rotation, this choice of F_0 is the same as choosing comoving coordinates. If the pressure doesn't vanish, the existence of sound waves prevents the elimination of A by this sort of infinitesimal coordinate transformation.

This choice of gauge is equivalent to setting $\xi = \mu + E$, $E = \text{const.}$ There is now only the one gauge freedom left $\mu \to \mu + G$, $\xi \to \xi + G$. The perturbation equations (14.28a, b, d, e) now reduce to the one equation

$$\ddot{\mu} + 2t^{-1}\dot{\mu} + \frac{1}{2}\,C^{-2}\,Ek^2\,t^{-\frac{4}{3}} = 0\,. \tag{14.36}$$

and W is given by

$$W = C^{-2}\,Ek^2\,t^{-\frac{4}{3}} + \frac{4}{3}\,t^{-1}\,\dot{\mu}\,. \tag{14.37}$$

The solution of (14.36) shows that $W = K_1 t^{-\frac{4}{3}} + K_2 t^{-3}$, ($K_1, K_2$ constants). The density contrast W/w is therefore:

$$\frac{W}{w} = L_1\,t^{\frac{2}{3}} + L_2\,t^{-1}, \quad L_1, L_2 \text{ constants.} \tag{14.38}$$

Thus the density contrast can grow, but at most as a power of time. If the pressure does not vanish, no striking change in the growth rate of the density contrast is seen.

Perturbation Modes in the Anisotropic Model

In the anisotropic model the perturbation is again represented by a 3×3 matrix h_{ij}. In the $\{dt, dx^i\}$ basis we have

$$\begin{aligned}\delta g_{00} &= \delta g_{0i} = 0\ , \\ g_{ij} + \delta g_{ij} &= e^{-2\Omega}\, e^{2\beta}{}_{ij} + \tfrac{1}{2}\, e^{-2\Omega}\, (e^{2\beta}{}_{ik} h_{kj} + e^{2\beta}{}_{jk} h_{ki})\ .\end{aligned} \tag{14.39}$$

Again the functions h_{ij} are Fourier analyzed, and a single frequency is selected for detailed examination: $h_{ij} = \mu_{ij}(t)\, e^{ik_s x^s}$. Finally the μ_{ij} matrix is separated into the functions $\mu, \xi, \zeta_i, \eta_{ij}$ defined by

$$\begin{aligned}&\mu = \mu_{ss}\ , \quad \xi = \mu_{st}\, k_s\, k_u\, g^{ut}/k^2\ , \\ &\zeta_i = \mu_{st}\, k_s\, k_{ti}\ , \quad \text{and} \quad \eta_{ij} = \left(k_{si}\, k_j - \tfrac{1}{2}\, g^{tb} k_{sb}\, g_{ic}\, k_{cj}\right) g_{sa}\, \mu_{at}\ .\end{aligned} \tag{14.40}$$

One additional function proves to be useful. We know that the anisotropy matrix β obeys $\dot{\beta}_{ij} = B_{ij}\, e^{3\Omega} (B_{ij} = \text{const}, B_{ss} = 0)$ (see 14.20). From the B_{ij} comes the definition $\theta = B_{ij}\mu_{ij}$, where θ is a linear combination of the functions already defined.

The analysis of the perturbations proceeds as in the isotropic case. The perturbation equations are written down, and either a numerical or an exact solution is sought. Again the separation into constraint and propagation equations is made. The constraint equations relate the perturbation functions to the fluid perturbation functions W, P, A, B_i defined earlier. The propagation equations are ordinary linear differential equations, second order in time, in μ_{ij}. The general solution involves 12 initial value parameters. As in the isotropic case, four of these parameters are affected by gauge transformations.

Independent Modes in the Anisotropic Case

It is discovered that the two ζ_i are changed by an additive constant under a gauge (infinitesimal coordinate) transformation. As in the isotropic case the $\dot{\zeta}_i$ are gauge invariant. Again as in the isotropic case, each $\dot{\zeta}_i$ is found to obey an equation which contains no other perturbation function. As in the isotropic case, $\dot{\zeta}_i$ has the interpretation that it is proportional to the rotation of the cosmic fluid. The simple, decoupled behavior of rotation in these models is due to the absence of rotation in the unperturbed model.

The η_{ij} functions are not decoupled from μ and ξ in the anisotropic case, however. The gravitational waves, in other words, are affected by the density perturbations. The analysis of these modes has recently been carried out in detail by Perko, Matzner, and Shepley (1972), a preliminary analysis having been carried out by Doroshkevich (1966) in an axially-symmetric, anisotropic model.

The analysis of density perturbations is simplified if k_i is an eigenvector of β_{ij} (that is, an eigenvector of B_{ij}). Thus, for example, we may choose coordinates so that β_{ij} is diagonal and k_i has a component only in the "three" direction. In this case one of the two independent components of η_{ij} does obey an equation independent of all other perturbation functions. This component is gauge invariant and does not affect the fluid variables. It represents one mode of gravitational waves.

The other mode of gravitational waves is coupled to the pressure perturbation P, the density perturbation W, and the "parallel" velocity perturbation component A. This coupling is due to the presence of fluid shear in the unperturbed model. It is here that the variable θ is useful. The second order differential equations for μ, ξ, and θ are all coupled. The general solution involves six parameters (initial data) of which two are affected by gauge transformations.

Effect of Gravitational Waves on a Density Perturbation

The effect of the coupling of gravitational radiation to density perturbations can be spectacular. In certain cases the density perturbation grows at a much faster rate as the model expands than in the isotropic case. Even in these cases, however, the perturbation grows as a power of t. There is no exponential growth such as Jeans (1929) predicted in a stationary cloud of gas.

There are two effects due to the anisotropy. The first is due to the different growth rates of the background metric in different directions. If the wave vector k_i is oriented in a direction with rapidly growing metric component, a slow growth rate of the density is observed. Alternatively, a direction of slow background expansion corresponds to rapidly growing density contrast.

The second effect is the coupling of the η_{ij}, "gravitational wave" mode, to the density perturbation because of the background anisotropy. Energy may be fed from a gravitational wave into a density perturbation or vice versa. This second effect could be of great importance in the theoretical discussion of galaxy formation should some independent estimate of initial cosmic gravitational wave density be obtainable.

The overall growth rate of a density perturbation – with all the effects of gravitational waves and anisotropy included – may be much higher or much lower than the corresponding growth rate in an isotropic universe. The upper limit is a density contrast growth rate of $W/w \propto t^{\frac{8}{3}}$ or less. This is an *upper* limit, and is a power-law rate of growth.

The Making of a Galaxy

The observed structure of galaxies is consistent with a formation time of about 10^8 years. In an isotropic universe, radiation and matter decouple when the temperature drops below about 3000 K, at perhaps 10^5 years after the Big Bang. Even if the large limit of growth rate in (14.38) obtains, galaxies cannot form in 10^8 years starting from a random perturbation at $t = 10^5$ years. The proof of the above statement is a naive one: a galaxy contains about $10^{70} \sim N$ particles (photons, etc.), and a random perturbation of galaxy size at $t = 10^5$ years is of relative magnitude $10^{-35} \sim N^{-\frac{1}{2}}$. In the limiting case of $t^{\frac{8}{3}}$ growth rate we thus have $W/w = 10^{-27}$ at $t = 10^8$ years.

In contrast to this small value of W/w, we expect that a value of W/w of about 1 is needed before non-linear effects take over to initiate rapid condensation. In fact, if W/w is about 1% at $t = 10^5$ years, then by 10^8 years, even at the low growth of $t^{\frac{2}{3}}$ in the isotropic case, (14.38), we have $W/w = 1$ at $t = 10^8$ years. Rapid condensation to the observed density contrast between galaxies and intergalactic space ($W/w \sim 10^6$) then ensues.

The effect of anisotropy is more than an enhancement of the growth rate, however. There is also an effect which may allow a random perturbation to commence at a time earlier than $t = 10^5$ years. In any model, a given observer can "see" all matter within a distance called the horizon size. A quick calculation shows that the horizon included a total mass of much more than one galaxy at $t = 10^5$ years in the isotropic model. In the anisotropic model, some directions have even larger horizons. If one postulates that a random perturbation occurs and starts to condense sooner when the horizon is larger, then even larger density contrasts are computed – none of the magnitude needed for galaxy formation, however.

Alternatively, one can postulate an initial perturbation which includes matter not seen by an observer at the center of the perturbation (not seen because light has not had time since the Big Bang to reach the observer from the edges of the perturbation). Peebles (1971a) makes such a postulate – a postulate of white noise perturbations – at early cosmic epochs. He concludes that it is quite reasonable that globular clusters result – conglomerations of perhaps 10^5 stars – in a reasonable time. Galaxies (10^{11} stars) then presumably form from these clusters.

Does Each Galaxy Contain the Imprint of the Big Bang?

Peebles' result is especially provocative in that it draws attention to the initial singularity. The initial perturbations he postulates do not have the $N^{-\frac{1}{2}}$ spectrum of random perturbations. Peebles suggests that the existence of galaxies shows the imprint – the structure – of the Big Bang itself.

Is Peebles' postulate of large initial perturbations necessary? We did not calculate horizon sizes in our anisotropic models because the horizon may be infinitely large in a more realistic model. Misner's mixmaster model has directions in which a given observer can send or receive signals completely around the universe – and these directions are continually changing. One can argue that in such a model the $W/w = 10^{-35}$ perturbation associated with a galactic mass occurs at a time much earlier than $t = 10^5$ years. Although the power law rate of growth varies drastically with direction in an anisotropic model (depending on local expansion rate) the continual shifting of directions in the Mixmaster model should result in an average power law $W/w \propto t^{\sigma}$, with $\sigma \sim 1$ (certainly $\sigma > 0$) (the beginnings of this calculation have been given by Hu and Regge, 1972). An early enough perturbation then can result in $W/w \sim 1$ by $t = 10^8$ years.

This is the unsolved problem of galaxy formation theory: Can a random perturbation result in a galaxy-sized condensation within about 10^8 years after the initial singularity (the Big Bang)? Or does each galaxy contain within itself a structure – the shape of the initial perturbation – left over from the Big Bang? Most arguments – those presented in this section – indicate the latter possibility, that to understand galaxies one must understand the initial singularity. And then one must again ask: Was the Big Bang a truly singular region or was it merely an epoch of rather dense matter – or was it an epoch when the laws of classical physics themselves were inoperable?

15. FINAL REMARKS: WHAT IS, WHAT IS NOT, AND WHAT SHOULD BE

The great tragedy of science – the slaying
of a beautiful hypothesis by an ugly fact
– THOMAS HENRY HUXLEY

15.1. A Potpourri of Cosmological Subjects

In a book of this kind one necessarily leaves out certain subjects. Instead of covering such subjects in detail we will here present them in outline form. Our personal preferences are clearly toward the mathematical end of this subject. What we have left out, therefore, are areas of primarily astrophysical or physical content. The general categories we will outline here include cosmogony and the physical universe as well as cosmological aspects of theories which compete with general relativity.

The Physical Universe

Relating mathematical forms given for cosmological models to the actual physical universe is one task of observation. The most distressing thing about the observations is the paucity of results which can allow one to distinguish first, among the various cosmological models in general relativity, and second, among competing cosmological or gravitational theories. An excellent review of observations up to the present time has been given by Peebles (1971b). The observations of most importance to cosmology include the existence and isotropy of the 3K black body radio background, the density of matter and radiation in our immediate neighborhood and at cosmological distances, and the red shift versus magnitude relation.

One of the earliest and most striking observations – the one that actually founded modern observational cosmology – was the detection of the red shift versus magnitude relation by Hubble (1936). Important recent criticisms of this most fundamental of observations have been reviewed by Burbidge (1971) and Segal (1972). In spite of the strength of many of these criticisms, we have accepted the Hubble law at least to the extent of discarding the model universes homogeneous both in space and in time. These models, of course, should be reexamined should it turn out that the red shift of distant galaxies does not indicate a universal expansion. Perhaps it will only be when the mystery of the quasar red shifts is solved that the validity of Hubble's observations will be verified.

The best current numerical value for the Hubble parameter is found in Sandage (1972, 73). Of course, it must be assumed that his value does give a true indication of expansion. The numerical value is

$$H = \frac{1}{R}\frac{dR}{dt} = 55 \pm 7 \frac{\text{km}}{\text{sec-Mpc}} = [(18 \pm 2)\times 10^9 \text{ years}]^{-1} . \tag{15.1}$$

Sandage also gives a value for the deceleration parameter q:

$$q = -\frac{1}{RH^2}\frac{d^2R}{dt^2} = 0.96 \pm 0.4 . \tag{15.2}$$

This value seems to indicate that the universe is closed, that is, a Type IX model. The error in q, however, is so large that no one model can be preferred over another.

It is unfortunate that the state of the observation of the deceleration parameter q is in such a primitive condition. If q really is the value that Sandage gives, then the universe is a Type IX model and therefore will recollapse eventually. If, on the other hand, the low observed value for the density of luminous matter is accepted, then the universe is a Type V model, and no recollapse will ever occur. Although apparently only of academic interest – for such recollapse will not occur for many dozens of billions of years in any event – a proof that our universe will turn around would be very provocative. If our universe is actually shown to be a Type IX model the old theories of a bouncing or cyclic universe spring immediately to mind. Perhaps it is true that some mechanism exists that will take the universe through the collapse phases predicted by the cyclic nature of the closed FRW model.

The measure of the total density of matter and radiation, even in our neighborhood of the cosmos, is in much cruder shape than the red shift – magnitude relation. Matter which is luminous has density about equal to (Peebles, 1971b).

$$\rho_L = 2 \times 10^{-31} \text{ gm/cm}^3 . \tag{15.3}$$

This value for the density is much lower than that needed to be consistent with a closed or Type IX model. The value of ρ_L, if indeed luminous matter represents a substantial portion of the total matter of the universe, would indicate an open or Type V cosmology and would therefore be inconsistent with Sandage's value of the deceleration parameter q. If there is a substantial amount of invisible matter (by substantial we mean hundreds of times more than ρ_L) it is certainly possible that the total matter density would be consistent with Sandage's value of q. Of course, people are continuing to look for such non-luminous matter (Peebles, 1971b).

Perhaps the most significant of all observations is the discovery of the 3K black body radiation. This radiation is taken by most to be an indication that the universe was once very hot and dense. The spectrum of this radiation seems to be close to a black body spectrum, with a temperature of 2.7K (Penzias and Wilson, 1965; Dicke, Peebles, Roll, and Wilkinson, 1965). Just as important as the spectrum and temperature of this radiation is its isotropy. One of the most remarkably precise measurements in cosmology shows this isotropy to be better than a part in 10^3 (Partridge and Wilkinson, 1967). Isotropy measurements directly from the red shift magnitude relation are not at all significant and perhaps can never be made as precise as the black body isotropy determination. The black body isotropy measurement, however, shows us that models which are presently isotropic are definitely to be preferred, for this radiation passes freely through matter in our cosmic neighborhood. It is an indication of the isotropy as far back as the cosmic epoch when the temperature was 3000 K (but not a definitive indication, see Misner, 1968; Collins and Hawking, 1973b). This temperature occurred only about 10^5 years after the initial singularity. The universe could have been enormously anisotropic before this epoch, and it is for this reason that we have studied such model universes.

Other observations include those of greater importance to cosmogony. The relative abundances of helium and other elements will be briefly discussed later. Still other observations have been proposed and indeed would be useful could they be carried out with any precision. The most striking of these is the suggestion that correlations among the orientation of galaxies be determined. Such a correlation could show that the rotation of the universe was significant during the formation of galaxies.

Cosmogony

Although the dictionary defines cosmogony as pertaining to the creation of the universe, most researchers mean by this term the theory of the creation of the chemical elements and of galaxies. The greatest success of this theory has been in the description of the relative abundances of heavy elements. Unfortunately for cosmology, this theory indicates that heavy elements are produced in stars and therefore pertain to stellar physics and not to cosmic physics. The abundance of the very light elements, however, in particular helium and deuterium, is affected significantly by the very early epochs of the cosmos. Observations of helium abundance are still in a very primitive state. What is desired is a measure of the primordal helium abundance, that is, the helium abundance at the time when galaxies and stars were just beginning to form. A typical result is that of Iben and Rood (1969) which gives a value of 25% for the helium abundance. This value is consistent with the temperature history of the simplest cosmologies, namely the FRW models (Peebles, 1966). The value, however, is recognized to be not definitive and other cosmologies are definitely not ruled out.

Perhaps of more significance than the helium abundance is the abundance of deuterium. Although its cosmic abundance is quite poorly known, its galactic abundance is better known (for example Rogerson and York, 1973). The cosmic abundance of deuterium would be a relatively sensitive indicator of the early cosmic temperature history (Reeves, Audouze, Fowler, and Schramm, 1973). The galactic deuterium figure, if it is accepted as the cosmic abundance, implies present cosmic matter density not different by more than an order of magnitude from the luminous matter density of about 2×10^{-31} gm/cm^3.

Most cosmogonic theories assume that the entire content of the universe is matter and not anti-matter. Whereas high energy processes can produce both matter and anti-matter, the fact remains that locally there seems to be no anti-matter. It is not known whether other stars, other galaxies, or other clusters of stars are anti-matter. Several theoretical models have been proposed in which matter and anti-matter coexist at an early stage and separate into pockets at later stages (in particular, see Omnes, 1969). These models do not seem to be consistent with observation, for it is known (see Steigman, 1973) that there are no gamma rays which would occur from the interaction ("leidenfrost") of matter and anti-matter in the nearest intergalactic regions. Many people believe observation rules out the most naive of the matter and anti-matter cosmologies.

An especially fascinating idea in cosmogony concerns the origin of the fundamental particles themselves from which the entire matter content of the universe is built up. This idea suggests that quantum processes produce particles from an initially empty universe. By an initially vacuum universe is meant one which contains only gravitational radiation at its start. By application of curved space quantum mechanics, Parker (1971, 1972a, b) has computed the particle production in various cosmological models. His results show that there is indeed particle production, but unfortunately as yet no precise numerical values for the total number of particles has come out. The basic problem with this type of theory is that particle production is not covariantly defined, and indeed it is possible to have particle production in a completely flat manifold. For example, in a manifold with metric

$$ds^2 = -dt^2 + f(t)\,dx^2 + dy^2 + dz^2$$

there is particle production, although if $f(t) = t^2$ the model is flat. Other earlier theories (e.g., Sexl and Urbantke, 1969) give some indication of the total number of particles which would be expected from a theory of this kind. This number is only very roughly the same as the observed 10^{80} value. Of course, any theory which predicts the complete production of observed matter from gravitational waves would result in a universe with equal numbers of baryons and anti-baryons, assuming of course that the fundamental laws of baryon production are indeed charge-reversal invariant in a curved spacetime.

At the very first few moments of the existence of the cosmos the matter is believed to be extremely hot and dense. A full detailed description of this epoch therefore requires knowledge of a realistic equation of state and knowledge of the realitivistic thermodynamics or statistical mechanics which govern the evolution of matter. Such a discussion, however, is more appropriate for stellar collapse problems than for cosmological applications for two reasons. First, unless the real universe is exactly isotropic or contains matter so stiff that the speed of sound is equal to the speed of light, then the effects of anisotropy will dominate any matter effects right at the initial stages of the universe. Second, the universe very quickly expands to the point where the matter density is sufficiently small that a highly relativistic equation of state is not necessary. For example, at the epoch when the temperature has dropped to about 3000 K the density of matter has fallen to about one gram/cm^3. It is at this epoch that galaxies are thought to form. Indeed a full theory of their formation would include the effects of a *non-relativistic*, realistic equation of state and of magnetic fields and of radiation (Field, 1974). It may, however, turn out that a heuristic treatment of the singularity is possible using a strongly relativistic equation of state of the type envisaged by Bahcall and Frautsche (1971), Hagedorn (1970), or Bowers and Zimmerman (1973).

Other Cosmology

The use of quantum field theoretic processes to explain particle production represents one extension of the ideas of the classical theory of general relativity. In cosmology certainly other theories of space and time have been and are being employed. Surprisingly, it is only recently that Newtonian cosmologies have been looked at (Milne, 1934; Milne and McCrea, 1934; Heckmann, 1942). It is especially interesting to note that homogeneity may be described in Newtonian gravitational fields so that examples of most if not all of the Bianchi types of cosmological models are possible in this description (Hibler, 1971). Cosmologies employing the Brans-Dicke theory of gravitation have also been described. Still more esoteric theories of space and time which either incorporate quantum field theory explicitly or which envisage a lattice or foam structure, or indeed some more strange structure still, have not found much useful application in cosmology.

From the opposite point of view observational cosmology has certainly been used to suggest a search for theories of spacetime and gravitation. There are three especially important examples of this interaction of cosmology and gravitational theory. The first is Einstein's inclusion of a cosmological constant to allow the construction of a static model. The second is the theory of spacetime which gives rise to the steady state cosmological model of Hoyle (1948, 1949), Bondi and Gold (1948), and their group. The third example is the use of the numerology of Eddington (1959) to suggest to Jordan (1952) a whole class of gravitational theories. The most important members of this class are the Brans-Dicke theories.

Brans and Dicke (1961) found the equivalent of the FRW models and recently Matzner, Ryan and Toton (1973) and Nariai (1972) have computed the anisotropic cosmologies. This competing theory of gravitation envisages a scalar field in addition to the metric tensor as being the cause of the gravitational force governing cosmic expansion. The cosmological models differ in some details from the general relativity cosmologies and in particular predict a slightly different helium abundance from that predicted by general relativity. The observed helium abundance is consistent with general relativity, but is sufficiently inaccurate that a Brans-Dicke model is not ruled out.

The other important example of a non-relativity cosmology is the steady state universe. From a philosophical point of view this universe is most attractive, for it avoids any singularity. It was proposed for this reason and led to a modification in the law of conservation of mass and energy. It basically uses the DeSitter line element (DeSitter, 1917), which satisfies Einstein's field equations only with a cosmological constant. By including continuous creation of matter the cosmological constant is eliminated. Although this model was quite popular in the 1950's and early 1960's, only a few die-hards are left who now still try to fit this model to observations. The death blow was the discovery of the 3K black body radiation causing most researchers to believe that the universe was hot and dense at one time.

In our discussion of cosmological models we concentrated our efforts on those which are homogeneous. Our concentration was based on two reasons: The first is that homogeneous models are much easier to handle than inhomogeneous ones. The second reason, however, is that the real universe apparently is homogeneous, in that matter appears to be smoothly spread provided large enough volume averages are taken. Although matter is concentrated in planets, stars, galaxies and clusters of galaxies, there does not seem to be any strong clustering on a larger scale. There are, however, weak indications that such large scale clustering may exist. DeVaucouleurs (1970) has put forward some interesting observational indications that clustering may exist at all orders, superclusters being clustered themselves and so forth. Needless to say, a hierarchical model as envisaged by deVaucouleurs could not be handled by the techniques of this book. Many of the techniques we discussed, however, could handle the models which are only weakly inhomogeneous, for example, using a high order perturbation theory. Strong large scale inhomogeneities of a particular type may also be treated (Gowdy, 1971; Ryan, 1972b).

Unconventional cosmology can also mean a conventional metric in a manifold with unusual topology. For example, the FRW models involve a three-space of constant curvature, but the global topology of these $t =$ const. hypersurfaces is not specified. In fact, many different topologies may be consistent with a given metric form (Wolf, 1967). There may be interesting physical effects dependent on global structures as distinct from localized fields (Wheeler, 1962b; Ellis, 1971a).

Relativistic Astrology and Religion

Most modern astrologers seem unaware of the advances that Einsteinian relativity has brought about. Indeed, most modern astrologers seem unaware of modern astronomy. The Newtonian effects of the precession of the equinoxes have not been incorporated into most astrological computations (but see Schmidt, 1970). The corrections due to relativity, and in particular, effects due to the continual expansion of the horizon in FRW models have not even been remotely envisaged by researchers in this area.

Some fideists have taken a serious view of the relation between relativistic cosmology and religion (Milne, 1952). The necessity of the singularity in general relativity can be taken as an indication of the creation of the physical universe by theological agents. A modern point of view has been expressed by Misner (1969c) who, however, does not use theological terminology. A dedicated anti-religionist may even feel disposed to deny those models in which there is a singularity (compare Bobin, 1960). We ourselves dislike singularities but rather from the point of view that every effect must have had a cause which is physical in nature.

15.2. A Call to Arms

Foremost among the problems of cosmology are observational studies, but many theoretical questions also beckon. As orbiting observatories become more common we should expect a flood of answers to observational questions, provided only that sufficient observing time is made available to cosmologists. Theoretical advances will be in the direction of explaining physical phenomena as well as development of the formal theory.

Desirable Observations

While many observations of importance to cosmology may be performed from the surface of the earth, others of necessity should be performed from an orbiting observatory. The atmosphere is not transparent to certain portions of the electromagnetic spectrum, and these portions include wave lengths where especially important information may be available. The infrared, ultraviolet, and x-ray portions of the spectra of distant galaxies and of background radiation could yield vital clues as to the nature of the early cosmos. Because of the cosmic redshift, the infrared spectrum in particular would be most informative.

The observations which are perhaps of greatest urgency and which can be performed from the surface of the earth include the following: a good determination of the deceleration parameter, whether there is significant invisible intergalactic matter, the average cosmic abundance of helium, deuterium and other significant isotopes, an estimate of any large scale inhomogeneities, a limit on the anisotropy of the Hubble redshift parameter, and whether galaxy orientations are correlated. The above list is mainly optical research, and other types of observation give cosmic information. In particular, the measurement of the black body radio background should be refined to see whether absorption or emission lines of cosmic significance exist, and just how isotropic this background is.

Observations that are not concentrated in the electromagnetic spectrum include neutrino astronomy and its application to cosmology, finding out whether there is a gravitational radiation background and if there is, whether there is a black body spectrum of gravitons, and observations of high energy particles and photons which would come from Leidenfrost radiation at the boundaries of sections of the universe that are matter and anti-matter.

A list of desirable observations in order of their priority is:

1) The determination of the deceleration parameter and the amount of invisible intergalactic matter.
2) Observations in the infrared, x-ray, and non-visible radiation.
3) The amount of large scale inhomogeneity in the universe.
4) The relative abundances of various isotopes in the universe.

Of course, the ease of doing certain of these experiments may mean that the results will come in a different order. It is important to a number of these observations to make the use of orbiting observatories practicable as soon as possible.

Theoretical Studies

Although theoretical programs may seem to be of less urgency than observational studies, we must be prepared for unanticipated experimental results: Surprises continually occur in astronomy. For example, although the discovery of pulsars was completely unexpected, rapid understanding of these objects was facilitated by early work on gravitational collapse. In observational cosmology programs can be described with some degree of definiteness, but in talking about future theory only directions can be pointed out.

The most vague but vital program which will hopefully be carried through in coming years is the application of quantum principles to the universe. The quantum field theory of gravitation is still in dispute. Even the interpretation of basic quantum principles as applied to the structure of space and time is in doubt. These controversies must be resolved before any significant application to cosmological studies can be made. It is important to find quantum solutions which correspond to the known cosmological models, but it is the details of what goes on at the epoch where a classical model has a singularity that are the most significant and which are most effected by detailed interpretation of quantum principles.

Such problems as particle production and the formation of galaxies may need entirely new physical theories, while other advances will be applications of well-known gravitational theories and electromagnetism. Until a good quantum understanding of the early universe is achieved, it may be necessary to postulate in an ad hoc way initial conditions to be applied within these theories. Further, it is by no means impossible that the understanding of galaxies and the cosmic number density of baryons can be achieved within known theoretical constructs, and further study of these problems within general relativity is certainly called for.

An example of new applications of general relativistic cosmology would be the study of inhomogeneous solutions. Study of inhomogeneous models is needed to confirm or confute chaotic cosmology. Chaotic cosmology is the idea that any initial conditions at the singularity lead to the FRW universe we see today. Especially if this idea is not true, we need to study inhomogeneous and anisotropic exact solutions to outline the range of initial conditions which lead to our present universe. A program such as chaotic cosmology is a call for new cosmological solutions which do not necessarily match the present universe but which may be a model for an earlier stage of evolution.

There are many computational problems which should be carried through and which do not involve any major changes in existing theory. The most important of these problems is a study of the early stages of the universe thought of as a plasma. Relativistic magneto-hydrodynamics is a difficult theory to work with but has the potential of producing significant results in a moderately short period of time. Within general relativity itself further study of perturbations and especially perturbations of rotating cosmologies is called for.

Finally, in this list of theoretical problems we mention further study of the nature and structure of singularities within general relativity.

Although much has been accomplished in this study, too many people are forgetting that there is still much to accomplish. In particular, the relationship between mathematical and physical singularities as we have defined them here is only very poorly understood. It cannot be overemphasized that this question remains as vital as it has always been.

In order of decreasing priority we suggest the following list of theoretical programs:

1) Inhomogeneous models.
2) Quantum general relativity applied to cosmology.
3) Singularities.
4) Magneto-hydrodynamics, particle and element production, and galaxy formation.
5) Resolution of the mystery of quasars.

Interest in cosmology has fluctuated over recent years, and of course, there is no guarantee that the coming decade will find the cosmos fashionable. During the next ten years, however, the world of astrophysics will be flooded by new astronomical data, many of which will contain surprises on a cosmic scale. Let us hope that response to this flood of information will be a flood of answers not only to the questions we now consider important but to those questions that will come into prominence.

EXERCISES AND PROBLEMS

Now I tell what I knew in Texas in my early youth
– WALT WHITMAN

For each chapter we have given problems of varying difficulty:

(no asterisk) = exercise
* = difficult problem
** = difficult research problem requiring new insights

At the end of the list are several briefly stated questions, again graded according to the above chart. The * and ** problems are meant to be suitable for Masters' and Doctors' theses respectively, and so far as we can tell, none have yet been satisfactorily discussed.

Chapter 1

1.1. The Einstein Universe (Einstein, 1917) is a static cosmological model with compact (finite volume) $t = \text{const.}$ hypersurfaces. Because it has finite total matter content, one might expect a resolution of Olbers' paradox – that is, that the sky as seen by an observer would be expected to be dark. Show, however, that if the cosmos really was an Einstein universe of infinite age, the sky seen by any observer would have the brightness of a stellar surface.

1.2. Compute the time of fall T for a radially travelling projectile dropped at rest at distance R from a central point mass M assuming Newtonian gravity and mechanics. Determine the numerical values of T for three situations: a) $R = 10^{11}$cm, $M = 10^{33}$gm. T is roughly the time scale involved in stellar gravitational collapse. b) $R = 10^{22}$cm, $M = 10^{44}$gm. T is characteristic of the formation time of a galaxy. c) $R = 10^{28}$cm, $M = 10^{55}$gm. T is the age of the universe.

1.3. Consider a sphere of uniform density ρ rotating with angular velocity ω. What is the minimum radius to which it can collapse if it always remains a sphere? Assume conservation of mass and angular momentum, and use Newtonian mechanics and gravitation.

*1.4. Study the evolution of a mass of fluid whose initial state of motion is that of the sphere of Problem 3. Can this study be done analytically or must it be carried through numerically?

Chapter 2

2.1. Give a coordinate system on the 2-sphere S^2 which entirely covers it except for a single point. Find explicitly a vector field (everywhere differentiable) on S^2 which vanishes only at one point. Show that any vector field on S^2 must vanish at least one point and therefore S^2 cannot have a globally defined basis of vector fields.

2.2. For physical reasons it is sometimes desirable to consider a metric which is "piecewise C^2," that is, the metric g is C^2 except on a set of hypersurfaces, and C^1 everywhere. To illustrate, let $f: M \to R$, and assume $df \neq 0$ at a point p where $f(p) = 0$. Prove that in a neighborhood of p the locus H of points such that $f = 0$ has the structure of a manifold. Let H be so small that it is coverable by a single coordinate patch and extend this patch on H to a coordinate patch in M by using the coordinates in H plus the function f. Find the general form of a metric g which is piecewise C^2, being C^2 except possibly on H, by finding the general functional form of its components in the frame $\{df, dx^i\}$ where x^i are coordinates on H.

2.3. A C^{17} manifold allows the existence of functions and coordinates which may be C^{17} but not of higher differentiability. Let V be an operation on these functions which is linear

$$V(f(p)+g(p)) = Vf(p)+Vg(p)$$

and a differentiation operator:

$$V(fg) = gVf + fVg\ .$$

Find such a V which is not expressible in terms of any basis of the form ∂_i. For simplicity, assume the manifold is one dimensional.

2.4. Let $u_1, \cdots, u_r$ be linearly independent vector fields $(r \le n = \dim M)$ which commute: $[u_i, u_j] = 0$ for all i, j. Let p be any point in M. Prove there is some coordinate patch about p in which $u_i = \partial_i$.

2.5. Prove that the commutator satisfies the Jacobi identity in a general basis of vector fields (coordinated or non-coordinated).

2.6. Let $T = t^i{}_{jk}\,\partial_i \otimes dx^j \otimes dx^k$ in a coordinate neighborhood N, coordinates $\{x^i\}$. Consider a second coordinate neighborhood $\overline{N}$, coordinates $\{\overline{x}^i\}$ and let $T = \overline{t}^i{}_{jk}\,\overline{\partial}_i \otimes \overline{dx}^j \otimes \overline{dx}^k$ in $N \cap \overline{N}$. Since the x^i are functions of $\overline{x}^i$ in $N \cap \overline{N}$, the matrix of functions $(\partial x^i/\partial \overline{x}^j)$ is invertible there. Find $\overline{t}^i{}_{jk}$ in terms of $t^i{}_{jk}$, $(\partial x^i/\partial \overline{x}^j)$, and the inverse matrix $(\partial \overline{x}^i/\partial x^j)$.

2.7. Suppose $\{X_i\}$ is a basis of vector fields and $C^i{}_{jk}$ is defined by the equation $[X_i, X_j] = C^s{}_{ij} X_s$. Let $\{\omega^i\}$ be the one-form basis dual to $\{X_i\}$. Prove that $d\omega^i = -\frac{1}{2} C^i{}_{st}\,\omega^s \wedge \omega^t$. Suppose $A = a^i{}_{jk} X_i \otimes \omega^j \otimes \omega^k$ and $U = u^s X_s$. Define $a^i{}_{jk;\ell}$ by the equation $\nabla_U A = a^i{}_{jk;\ell}\, u^\ell X_i \otimes \omega^j \otimes \omega^k \otimes \omega^\ell$, and compute $a^i{}_{jk;\ell}$ in terms of $X_\ell a^i{}_{jk}$ and the connection coefficients $\Gamma^i{}_{jk}$ defined in the text. Also show that $C^i{}_{jk} = \Gamma^i{}_{kj} - \Gamma^i{}_{jk}$, provided torsion vanishes, and complete the proof leading to the First Cartan equation.

2.8. Prove that $R(U,V,W,\omega)$ is linear in each term. By linear we mean linear over the set of functions, so that $R(f_1U_1 + f_2U_2, V, W, \omega) = f_1 R(U_1, V, W, \omega) + f_2 R(U_2, V, W, \omega)$ where f_1, f_2 are C^∞ functions.

If the Riemann curvature tensor field R has components $R^i{}_{jk\ell}$ defined in a general basis by $R = R^i{}_{jk\ell} X_i \otimes \omega^j \otimes \omega^k \otimes \omega^\ell$, find an expression for $R^i{}_{jk\ell}$ and thereby prove the Second Cartan Equation.

Chapter 3

**3.1. Relativistic kinetic theory (Taub, 1967) has been used to show that $0 \le p \le \frac{1}{3} w$ where p = pressure, w = energy density. Which of the assumptions may be modified in a physically reasonable way to produce in certain circumstances a strongly negative pressure or a pressure greater than $\frac{1}{3}$ w?

3.2. Show that the conservation law for a perfect fluid $[(w+p)u^\mu u^\nu + pg^{\mu\nu}]_{;\mu} = 0$, the continuity law $(\rho u^\sigma)_{;\sigma} = 0$, and the entropy law $\theta dS = d\varepsilon + pd(1/\rho)$ are not independent of $S_{,\mu}u^\mu = 0$. Note that $w = \rho(1+\varepsilon)$.

3.3. The full Bianchi identity $R^\mu{}_{\nu\alpha\beta;\gamma} + R^\mu{}_{\nu\beta\gamma;\alpha} + R^\mu{}_{\nu\gamma\alpha;\beta} = 0$ holds in any manifold. Its first contracted form is $R_{\nu\beta;\gamma} - R_{\nu\gamma;\beta} + R^\sigma{}_{\nu\beta\gamma;\sigma} = 0$. Its second contracted form is intimately connected to the conservation law $T^{\mu\nu}{}_{;\nu} = 0$ as described in the text. Show that if the dimension of M is 4, the full Bianchi identity and its first contracted form are in general the same.

3.4. We derived equations for $\theta,_\mu u^\mu$ and $w,_\mu u^\mu$ in the text in terms of rotation, shear, expansion, and $R_{\mu\nu}$. To derive equations for $\omega^{\alpha\beta}{}_{,\mu}u^\mu$ and $\sigma^{\alpha\beta}{}_{,\mu}u^\mu$ (that is, for the evolution of rotation and shear as seen by an observer with velocity u^μ) requires use of the full Bianchi equation, in first contracted form as described in the previous problem. Derive these equations. Part of the Riemann curvature tensor is determined by matter variables through Einstein's field equations, but the remaining curvature tensor components represent in some sense gravitational wave energy, which interacts with rotation and shear through these equations.

3.5. What is the analog of Raychaudhuri's equation in Newtonian cosmology? What does it say about the effect of rotation in a collapsing Newtonian cosmological model? (See Ellis, 1971b.)

Chapter 4

4.1. Consider a spherically symmetric star with radius $R(t)$, using Newtonian mechanics and gravitation. With the assumption that the density ρ is a function only of t, and that the velocity has the form $v_1(t)v_2(r)\hat{r}$ ($\hat{r}$ = unit radial vector), show that the outer radius R does not affect the dynamics. This fact is the basis for Newtonian cosmology, and therefore find the general solution for $\rho(t)$ if the pressure p is assumed identically zero. (Also see Problem 3.5.)

4.2. Solve the geodesic equation for the general path of a freely falling test particle in the closed, open, and flat FRW models. Null geodesics are particularly important, and you should decide whether any observers can ever see completely around a closed FRW model. See the section on horizons in Type I models, and especially see Schrodinger (1956). We chose particularly simple spaces of constant curvature for the t = const. hypersurfaces in these models, but Wolf (1967) shows that others should be considered, too. What physical effects could be seen if an unusual global topology actually existed in the t = const. hypersurface which best fits the real universe?

4.3. Although we illustrated the solution to (4.20) in Figure 4.3, we did not exhibit the exact solution. Find it. The functional form of $R(t)$ in the dust case is that of a cycloid if $k = 1$, also known as the solution to the brachistochrone problem. Discuss the significance of this coincidence in terms of a variational principle.

*4.4. Find a general cosmological model filled with n fluids, but having the overall symmetry of an FRW model. The velocity of any one fluid should not be taken as hypersurface orthogonal, and the fluids may or may not be postulated to interact. Notice that care must be taken to avoid anisotropic terms in the total stress-energy tensor, for although the perfect fluid criteria deny anisotropic stresses in $T^{\mu\nu}$ for a single fluid, the sum of two such fluids may produce an anisotropic total.

*4.5. Ni (1972) has produced a rather general scheme for generating theories to compete with general relativity. Which allow cosmological models of the FRW type, and can cosmological observations rule out any of Ni's theories?

**4.6. What modifications of an FRW model are inescapable when $R(t)$ becomes smaller than the Compton radius of a spinning elementary particle? Are modifications necessary at larger R? Is it really true that the gravitational field itself need not be quantized if R is greater than $(\hbar G/c^3)^{\frac{1}{2}} = 2 \cdot 10^{-33}$ cm? (See Parker, 1972b.)

Chapter 5

5.1. In a Riemannian (positive definite) metric space the metric topology has certain open sets defined as follows: 0 is open if 0 consists of all points Q such that $d(P,Q) < \varepsilon$ for ε a fixed number, p a fixed point, if 0 is entirely contained in a coordinate patch. Any open set M is a union of these open sets in the metric topology. The manifold topology consists of all unions of neighborhoods of the form $e_i < x^i < f_i$ where e_i, f_i are numbers and the x^i are coordinates in a patch in which allowed coordinates include all the e_i and f_i. Prove that all open sets in the manifold topology are open in the metric topology and all open sets in the metric topology are open in the manifold topology.

**5.2. Miller (1973) and Miller and Kruskal (1973) have considered extensions of manifolds in which the Hausdorff property is dropped. What other weakening of topological axioms can lead to interesting extensions either of incomplete but mathematically non-singular manifolds, of physically singular manifolds, or of complete manifolds with incomplete timelike paths of bounded acceleration?

5.3. Show that flat Minkowski space M^4 is complete and all timelike paths of bounded acceleration may be extended to infinite values of proper time.

**5.4. The Sachs method of assigning boundary points to a manifold is reminiscent of techniques used in relativistic kinetic theory (Ehlers, Geren, and Sachs, 1968). Suppose a collapsing stellar model is built up by use of relativistic kinetic theory. Use the Sachs' method to find the structure of the singularity – for example, can it be shown, that "black holes have no hair?"

*5.5. Computation of a boundary using the Schmidt or Geroch techniques is difficult, and in most cases hardly worth the effort. In cosmology, the diagonal Type V models (Chapter 9) are the simplest of the models which are poorly understood. Therefore compute the G- and B-boundaries of these models and interpret the results physically.

Chapter 6

6.1. Prove that a Killing vector with covariant components a_μ satisfies the equation $a_{\mu;\nu} + a_{\nu;\mu} = 0$.

6.2. Use the Jacobi equation (2.4) on ξ_i, X_μ, X_ν to show that $\xi_i D^\sigma_{\mu\nu} = 0$ as assumed in the discussion leading to (6.18).

**6.3. A Killing tensor K is a completely symmetric tensor field whose components $k_{\alpha\beta\cdots\gamma}$ satisfy the equation $k_{(\alpha\beta\cdots\gamma;\sigma)} = 0$ where

( ) signifies symmetrizing the indices (Walker and Penrose, 1970). A Killing tensor is irreducible if it cannot be expressed as the sum of symmetrized tensor products of lower rank Killing tensors. Note that the metric tensor is a Killing tensor. A Killing tensor provides constants of the motion for geodesics, just as a Killing vector does. However, no geometrical interpretation exists for a Killing tensor in the sense that a Killing vector is a generator of an infinitesimal isometry. Find such an interpretation, and also use the algebraic structure of the set of Killing tensors (Geroch, 1970b) to classify solutions of Einstein's field equations.

6.4. Find all the Killing vectors and Killing tensors (Problem 6.3) of Minkowski space. Are any of the Killing tensors reducible?

**6.5. MacCallum and Ellis (1970) have listed the spatially homogeneous cosmologies which also allow a discrete isotropy group. Find or classify all inhomogeneous cosmological models which allow a discrete symmetry group.

6.6. Explicitly find all 2 and 4 dimensional Lie algebras. To do so, find structure constants $C^i{}_{jk}$ which are antisymmetric in the lower indices and which satisfy the Jacobi identities. Since the structure constants appear in the commutator equations $[X_j, X_k] = C^i{}_{jk} X_i$, two sets of structure constants are equivalent if the X_i's of one may be transformed into the X_i's of the other by a real linear transformation. Thus $C^i{}_{jk}$ and $\bar{C}^i{}_{jk}$ are equivalent if $C^i{}_{jk} L_i{}^s = \bar{C}^s{}_{tu} L_j{}^t L_k{}^u$ where $(L_i{}^s)$ is a non-singular matrix of real numbers.

Chapter 7

**7.1. Find all geodesics in the various ST-homogeneous models. Which, if any, of the complete models have incomplete timelike paths of bounded acceleration?

**7.2. What are all the global topologies compatible with the various ST-homogeneous models listed?

7.3. Find coordinate patches to cover the charged dust model given. What is the minimum number of patches needed?

7.4. Does the Gödel model have irreducible Killing tensors of rank 2 or higher (other than the metric – see Problem 6.3)?

Chapter 8

8.1. Explicitly give the transformation of Taub space into itself which leads to the other analytic extension through a Misner bridge into a NUT region. The transformation can be given as a coordinate transformation which involves an infinite winding of coordinates near the boundaries of Taub space. Geodesics which are inextendible in one version of T-NUT-M space become extendible in the other version – except for certain geodesics which are inextendible in both versions.

8.2. From perturbations of Taub space, show that the Misner bridge cannot work in a matter-filled model with the symmetry of Taub space – that instead a singularity occurs.

8.3. Is Taub space at all an accurate description of the real universe? What observations bear on this question, and do existing observations definitely rule out Taub space as a model of the cosmos?

*8.4. Apply the Schmidt-Sachs ideas to Taub space and find all possible analytic extensions.

**8.5. Using the ideas of twistor theory (Penrose, 1968b), find solutions for the spin s massless field equations in T-NUT-M space. Which are regular across the Misner bridge?

Chapter 9

9.1. What would an observer see in a spatially-homogeneous cosmology with hypersurface orthogonal velocity? Assume the observer to be stationary with respect to the cosmic matter, and compute the relative intensity and redshift of light from stars at given proper distance from the observer as a function of angle.

9.2. Show that (9.15) can be solved uniquely for M, given β and L.

9.3. Find general forms which guarantee zero rotation in spatially homogeneous dust models. For example, all Type I models are rotationless, but only those Type IX models with diagonal metric are. Zero rotation in a Type IX model requires the velocity field to be orthogonal to the $t = \text{const.}$ hypersurfaces, but this requirement may not be necessary in other types.

**9.4. In a dust-filled FRW model, a spherical ball of matter may be extracted at each time t and the metric therein replaced by a Schwarzschild solution. The radius of the ball, $R(t)$, is chosen so that the metric is C^1 across the boundary – the mass of the Schwarzschild metric then agrees with the mass of extracted matter. Further, an arbitrary number of such holes may be created (if they don't overlap), the result being the "Swiss cheese" model for stars in an expanding universe. Find a Swiss cheese model for a rotating or anisotropic dust-filled cosmology. What vacuum solutions can be used to replace extracted sections?

9.5. In a perfect-fluid, spatially homogeneous cosmology, find explicit expressions by which acceleration, rotation, shear, and expansion may be calculated from the various matter and metric variables.

Chapter 10

**10.1. Are there any incomplete but mathematically non-singular fluid models? What is the physical interpretation of incompleteness in such a model?

10.2. What do the field equations for a rotating spatially homogeneous model look like if it is required that the velocity u be X_0, so that the timelike direction is no longer required to be curl-free?

*10.3. What is the effect of including an electromagnetic field in a collapsing fluid model? The nature of the singularity is known to be drastically changed if a vacuum model is changed to a model with magnetic field. Is a similar change apparent when such a field is added to a fluid model?

*10.4. Near the singularity in a collapsing cosmological model, the temperature presumably is well above the point where matter reaches the plasma state. What are the equations which govern the dynamics of a plasma in a rapidly changing universe? How are various plasma instabilities affected?

Chapter 11

11.1. Consider the model one-dimensional Hamiltonian given by $H^2 = p_x^2 + e^{-\alpha t} \cosh x$. Using the wall approximation discuss the behavior of the particle. For what value of α does the wall velocity w equal the free particle velocity v? Discuss the full theory in all three situations $w < v$, $w = v$, $w > v$.

**11.2. The normal Hamiltonian techniques will not work for the general Type V model. Find a variational principle for such a model, and use it to discuss its properties. In fact, it may be that the best approach to finding a Hamiltonian for Type V models starts with detailed knowledge about the solutions themselves. (Also see Problem 13.1.)

11.3. As mentioned in the text, a Dirac-type equation may be found in the quantization of the Kasner model. Find the equation, which involves a 2-spinor, and solve it. The best interpretation of the spinor components remains an open question.

11.4. The metric of a Type I fluid model may always be put into diagonal form. Show this fact within the Hamiltonian formalism.

Chapter 12

12.1. What effect does anisotropy in a Type I model have on element production in the early stages of the cosmos (Thorne, 1967)?

*12.2. Calculate the detailed behavior of the general symmetric Type IX model near the turnaround (time of maximum expansion). Note that the turnaround time is poorly represented with some of the techniques we developed, for Ω-time breaks down then.

12.3. A Type I FRW model is just marginally not going to recollapse. It represents both a limiting case of Type IX models which do reach a maximum point followed by recollapse and a limiting case of Type V models which have more than enough expansion to be far from eventual recollapse. However, does the inclusion of anisotropy in a Type I model make this model more or less likely to recollapse?

**12.4. Although a superposition principle is possible to a limited extent in FRW models, more generally it is unknown how to obtain new cosmologies from old. Find such a procedure in general Type IX models whereby a solution may be obtained from two already known models or whereby a new solution may be generated from one known one.

**12.5. Develop a good quantized theory of the general Type IX cosmology.

Chapter 13

**13.1. Find all the geodesics in a general Type V model. This problem is of more than routine importance due to the lack of a Hamiltonian technique for these models (see Problem 11.2).

*13.2. Derive numerical techniques for looking at the effects of quantizing matter in an anisotropic cosmological model of Type I. Periodic boundary conditions (because of identifications made in t = const. hypersurfaces) are especially appropriate here.

*13.3. Study the detailed behavior of a model with randomly fluctuating pressure. If the pressure is allowed to become sufficiently negative, a singularity may be avoided. Although the physical basis for the fluctuating pressure hypothesis comes from quantum field theory, a wide range of types of fluctuations should be studied.

13.4. Devise a program which will numerically find and display graphically the geodesics of a dust-filled FRW model (see Problem 4.2).

Chapter 14

*14.1. The magnetohydrodynamics of a plasma show that in general a magnetic field is "frozen in" to the matter. What affect does this association have on the structure of the singularity in a Type I model?

**14.2. How do galaxies form in a symmetric Type IX model? When would protogalactic perturbations be expected to start to form, and how would gravitational field energy enhance or retard their growth? Would rotational or shape parameters of galaxies be affected by those of the model so that we might observationally expect correlations in angular momentum or orientation among galaxies now?

**14.3. Develop a Hamiltonian technique along the lines of Taub (1969) for studying perturbations in a spatially homogeneous model which itself has a well-defined Hamiltonian.

**14.4. Because of the slowness of growth of perturbations in an FRW model, it is certainly likely that presently observed density fluctuations (galaxies, clusters of galaxies, etc.) are due to large-size perturbations existing at the very beginning of the present epoch of isotropic expansion. Rather than postulating these perturbations in an arbitrary manner, develop a reasonable theory which would explain their existence.

14.5. Suppose the universe as a whole were not electrically neutral. What would be the effects of a preponderance of one sign of charge near the singularity in a cosmological model?

Chapter 15

15.1. How could a Texas Congressman be convinced that scarce research funds should be appropriated for cosmology?

**15.2. With the insights provided by the current state of cosmology, astrophysics, and relativity, develop a good, strong metaphysics for cosmology. Apply this philosophy in a practical sense by examining and possibly changing the priorities we have listed.

*15.3. Catalog thoroughly all schemes of cosmology which have been devised. A classification scheme should be thought up and applied to this problem.

*15.4. In many cases a symmetry of the material content of a model does not imply a symmetry in the metric. For example, a perfect fluid, with isotropic pressure, may exist in an anisotropic model. This discrepancy leads to the questions: What is the best physical definition of homogeneity to be applied to a relativistic cosmological model? Are the models we've studied here all the ones appropriate to this definition?

15.5. In the next several decades interstellar probes may be launched. What cosmological questions could be answered better with such a probe than with observations made within the solar system?

Extra Questions

1. Is it necessary or even desirable that the manifold used in a relativistic cosmological model be orientable?

**2. What physical situations could be best described by a geometrical model using both a metric and a non-vanishing torsion tensor? [Possible Hint: Is there a pictorial representation of torsion similar to the arrows and screws with which we illustrated vectors and differential forms?]

3. What is the detailed singularity structure of the "Swiss cheese" model (Problem 9.4) taking into account both the cosmological singularity and the Schwarzschild singularity?

*4. Why should a cosmological model involve a differentiable manifold rather than a) a continuous manifold, b) a topological set or c) merely some abstract point set?

5. Find the magnitude in cgs units of the components of the Riemann curvature tensor on the surface of the earth in an appropriate coordinate system due to a) the mass of the earth and b) the expanding universe.

6. Determine the algebraic classification (Penrose, 1960) of the conformal tensor in various Bianchi-type cosmological models, and discuss the physical significance of the results.

**7. Does the minisuperspace quantum cosmology method provide reliable approximations to results which would be obtained from a full quantum theory of general relativity?

BIBLIOGRAPHY

This bibliography is as complete as possible, with some limitations. We have concentrated on papers on theoretical cosmology. Many observational aspects are not included. We have not included many papers we consider unimportant, our decisions being especially harsh if the paper appeared before about 1960. Decisions about the importance of a paper are necessarily idiosyncratic, so we must apologize in advance for omissions that will seem inexcusable to some readers. We did try to make the bibliography all-inclusive, but when the number of references exceeded 1,000, we realized that some discretion had to be exercised. Finally, we have not included references on alternative cosmological theories to general relativity.

With these limitations the bibliography is good through the first part of 1973. Since it concentrates on cosmology, rather than general relativity, and since it concentrates on recent references, earlier bibliographies should also be consulted. An especially useful bibliography on general relativity is given in Misner, Thorne, and Wheeler (1973), where other general relativity bibliographies are also listed. Heckmann (1942) is an important reference source for early cosmology. Two resource letters of the American Association of Physics Teachers should also be cited: "Resource Letter GR-1 on General Relativity" by Brill and Perisho (1968) and "Resource Letter C-1 on Cosmology" by Ryan and Shepley (1974). These letters list selected books and articles in their respective areas, with comments on content and usefulness.

As shown in the figure, interest in cosmology has unsteadily increased to its present high. The number of cosmology articles is presently holding rather steady at 0.3% of the total number of physics articles, having been

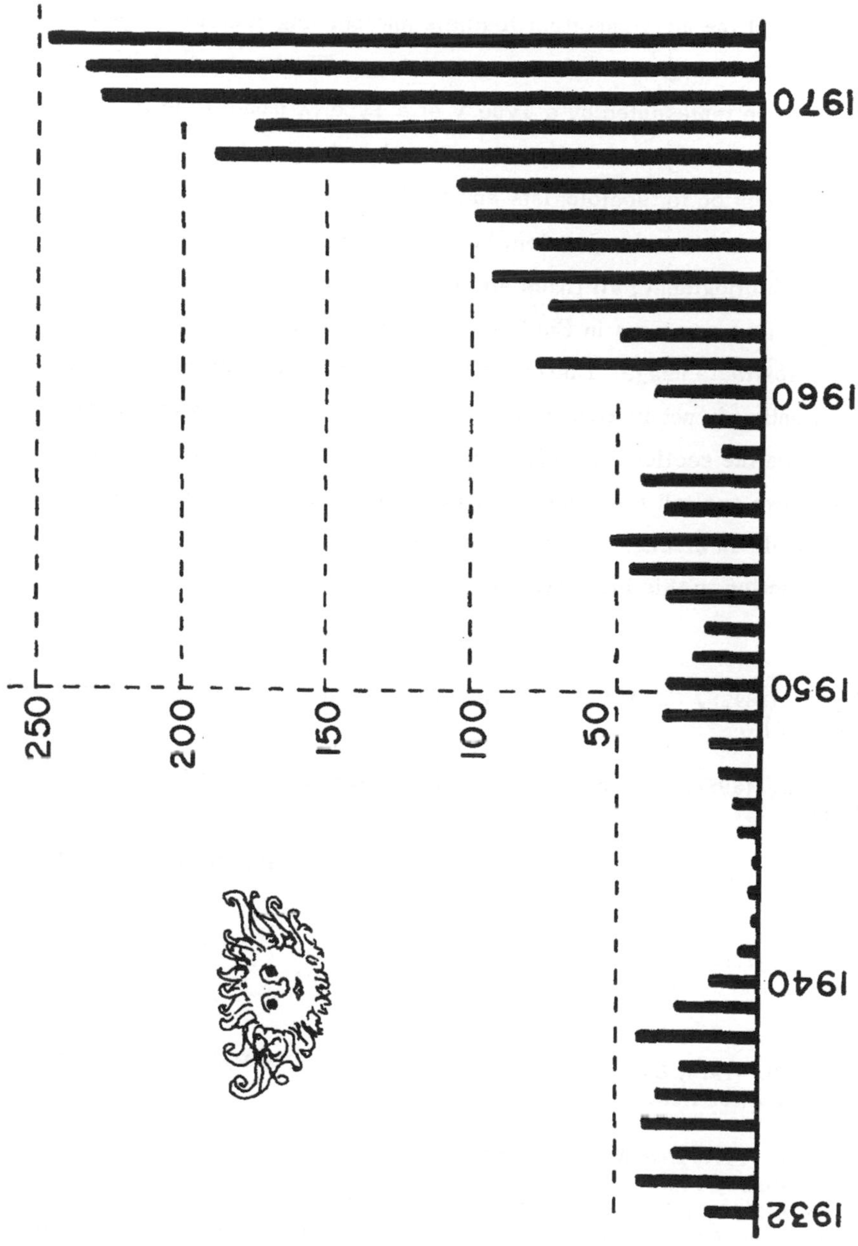

Fig. B1. Histogram of Cosmology Articles. The numbers represent the number of articles listed each year under the subject headings "Cosmology" and "Cosmogony" in *Physics Abstracts*.

about 0.7% of this total during the 1930's. Ryan and Shepley (1974) include a list of the most important journals and ongoing conferences which should be watched by researchers in cosmology. Whether interest in cosmology may be represented by a Type V or a Type IX model (that is, whether interest will continually expand or slow down and contract) is an important question for sociologists and funding agencies. Observation tends to favor continual expansion, but the evidence is not conclusive.

In this bibliography, all titles are given in English, and the vast majority of the papers are in English. Journal abbreviations conform to *Physics Abstracts* usage. Listing is by the first author's name, even when the first author is not in alphabetical sequence. The numbers in square brackets are the sections in which the entry is cited; a repeated number indicates two explicit references; but off-hand references to an entry are not included. In brackets, E indicates the exercise set, and B indicates the introduction to this bibliography.

1892

Killing, W.; (1892) *J.F.D.R.U.A. Math.* **109**, 121. "On the Foundations of Geometry" [6.2]

1895

Seeliger, H.; (1895) *Astron. Nachr.* **137**, 129. "On Newton's Law of Gravitation" [1.1]

1896

Neumann, C.; (1896) *On the Newtonian Principle of Action at a Distance* (Leipzig) [1.1]

1897

Bianchi, L.; (1897) *Mem. Soc. It. Della. Sc. (Dei. XL)(3)* **11**, 267. "On Three-Dimensional Spaces Which Admit a Group of Motions"

1914

Slipher, V. M.; (1914) *Lowell (Flagstaff) Obs. Bull.* **58**. "The Radial Velocity of the Andromeda Nebula"

1917

DeSitter, W.; (1917) *Mon. Not. Roy. Astron. Soc.* **78**, 3. "On Einstein's Theory of Gravitation and Its Astronomical Consequences" [5.2, 15.1]

Einstein, A., (1917) *S.-B. Preuss. Akad. Wiss.* p. 142. "Cosmological Considerations on the General Theory of Relativity" (in Lorentz, et al., 1923) [4.1, 7.1, 7.2, E]

1919

Palatini, A.; (1919) *Rend. Circ. Mat. Palermo* **43**, 203. "Invariant Derivation of the Gravitational Equations from Hamilton's Principles" [11.2]

1921

Kasner, E.; (1921) *Am. J. Math.* **43**, 217. "Geometrical Theorems on Einstein's Cosmological Equations" [9.5]

1922

Friedmann, A. A.; (1922) *Z. Phys.* **10**, 377. "On the Curvature of Space" [4.1, 4.2]

Weyl, H.; (1922) *Space, Time, Matter* (Methuen, London)

1923

Lorentz, H. A.; Einstein, A.; Minkowski, H.; Weyl, H.; (1923) *The Principle of Relativity* (Dover)

1924

Eddington, A. S.; (1924) *The Mathematical Theory of Relativity* (2nd Ed.) (Cambridge Univ. Press, Cambridge)

Eisenhart, L. P.; (1924) *Trans. Am. Math. Soc.* **26**, 205. "Space-Time Continua of Perfect Fluids in General Relativity" [3.1, 3.1]

Friedmann, A. A.; (1924) *Z. Phys.* 21, 326. "On the Possibility of a World With Constant Negative Curvature of Space" [4.2]

1926

Eisenhart, L. P.; (1926) *Riemannian Geometry* (Princeton Univ. Press, Princeton) [5.4, 6.4, 9.2, 10.1]

Gordon, W.; (1926) *Z. Phys.* **40**, 117. "The Compton Effect in Schrodinger's Theory" [11.4]

Schrödinger, E.; (1926) *Ann. Physik* **81**, 109. "Quantization as an Eigenvalue Problem" [11.4]

1927

Klein, O.; (1927) *Z. Phys.* **41**, 407. "Electrodynamics and Wave Mechanics from the Standpoint of the Correspondence Principle" [11.4]

1928

Robertson, H. P.; (1928) *Phil. Mag.* **5**, 835 "On Relativistic Cosmology"

1929

Jeans, J. H.; (1929) *Astronomy and Cosmogony* (2nd Ed.) (Cambridge Univ. Press, Cambridge) [14.3]

Robertson, H. P.; (1929) *Proc. Nat. Acad. Sci.* **15**, 822. "Foundations of Relativistic Cosmology" [1.1, 4.1, 4.2]

1931

Einstein, A.; (1931) *S.-B. Preuss. Acad. Wiss.* **12**, 235. "On The Cosmological Problem of General Relativity"

Lemaitre, G.; (1931) *Mon. Not. Roy. Astron. Soc.* **91**, 483. "A Homogeneous Universe of Constant Mass and Increasing Radius Accounting for the Radial Velocity of Extra-Galactic Nebulae"

1932

Veblen, O.; Whitehead, J. H. C.; (1932) Cambr. Tracts No. 29. "The Foundations of Differential Geometry" [2.1]

1933

Dingle, H.; (1933) *Proc. Nat. Acad. Sci.* **19**, 559. "Values of $T_\mu^{\ \nu}$ and the Christoffel Symbols for a Line Element of Considerable Generality"

Eddington, A. S.; (1933) *The Expanding Universe* (Cambridge Univ. Press, Cambridge)

Robertson, H. P.; (1933) *Rev. Mod. Phys.* **5**, 62. "Relativistic Cosmology" [1.1, 4.2]

1934

Milne, E. A.; (1934) *Quart. J. Math. Oxford. Ser.* **5**, 64. "A Newtonian Expanding Universe" [1.1, 15.1]

Milne, E. A.; McCrea, W. H.; (1934) *Quart. J. Math. Oxford Ser.* **5**, 73. "Newtonian Universes and the Curvature of Space" [1.1, 15.1]

Tolman, R. C.; (1934a) *Relativity, Thermodynamics, and Cosmology* (Clarendon Press, Oxford) [5.4]

————; (1934b) *Proc. Nat. Acad. Sci.* **20**, 169. "Effect of Inhomogeneity on Cosmological Models"

1935

Milne, E. A.; (1935) *Relativity, Gravitation, and World Structure* (Clarendon Press, Oxford)

Robertson, H. P.; (1935) *Astrophys. J.* **82**, 284. "Kinematics and World Structure [1.1, 4.2]

Walker, A. G.; (1935) *Quart. J. Math. Oxford Ser.* **6**, 81. "On Riemannian Spaces With Spherical Symmetry About a Line and the Conditions for Isotropy in General Relativity" [1.1, 4.1, 4.2]

1936

Hubble, E.; (1936) *Realm of the Nebulae* (Yale Univ. Press, 1936. Dover, 1958) [15.1]

Robertson, H. P.; (1936) *Astrophys. J.* **83**, 187. "Kinematics and World Structure" [1.1, 4.2]

————; (1936) *Astrophys. J.* **83**, 257. "Kinematics and World Structure" [1.1, 4.2]

Walker, A. G.; (1936) *Proc. Lond. Math. Soc.* **42**, 90. "On Milne's Theory of World-Structure"

1937

McVittie, G. C.; (1937) *Cosmological Theory* (Methuen, London)

Synge, J. L.; (1937) *Proc. Lond. Math. Soc.* **43**, 376. "Relativistic Hydrodynamics"

Taub, A. H.; (1937) *Phys. Rev.* **51**, 512. "Quantum Equations in Cosmological Spaces"

1942

Bergmann, P. G.; (1942) *Introduction to the Theory of Relativity* (Prentice-Hall, Englewood Cliffs) [3.2]

Heckmann, O.; (1942) *Theories of Cosmology* (Springer-Verlag, Heidelberg, reissued 1968) [15.1]

1946

Einstein, A.; Straus, E. G.; (1946) *Rev. Mod. Phys.* **18**, 148. Also *Rev. Mod. Phys.* **17**, 120 (1945). "The Influence of the Expansion of Space on the Gravitational Fields Surrounding the Individual Stars"

Gamow, G.; (1946a) *Nature* **158**, 549. "Rotating Universes?"
——————; (1946b) *Phys. Rev.* **70**, 572. "Expanding Universe and the Origin of Elements"
Lifshitz, E.; (1946) *J. Phys. USSR* **10**, 116. "On the Gravitational Stability of the Expanding Universe"

1947

Bondi, H.; (1947) *Mon. Not. Roy. Astron. Soc.* **107**, 410. "Spherically Symmetrical Models in General Relativity"
Dirac, P. A. M.; (1947) *The Principles of Quantum Mechanics* (3rd Ed.) (Clarendon Press, Oxford) [11.4]

1948

Alpher, R. A.; Bethe, H. A.; Gamow, G.; (1948) *Phys. Rev.* **73**, 803. "The Origin of Chemical Elements"
Bondi, H.; Gold, T.; (1948) *Mon. Not. Roy. Astron. Soc.* **108**, 252. "The Steady-State Theory of the Expanding Universe" [15.1]
Courant, R.; Friedrichs, K. O.; (1948) *Supersonic Flow and Shock Waves* (Interscience, New York) [3.2]
Hoyle, F.; (1948) *Mon. Not. Roy. Astron. Soc.* **108**, 372. "A New Model for the Expanding Universe" [15.1]
Milne, E. A.; (1948) *Kinematic Relativity* (Clarendon Press, Oxford)
Teller, E.; (1948) *Phys. Rev.* **73**, 801. "On the Change of Physical Constants"

1949

Eddington, A. S.; (1949) *Fundamental Theory* (Cambridge Univ. Press, Cambridge)
Gödel, K.; (1949) *Rev. Mod. Phys.* **21**, 447. "An Example of a New Type of Cosmological Solutions to Einstein's Field Equations of Gravitation" [7.1, 7.3]
Hoyle, F.; (1949) *Mon. Not. Roy. Astron. Soc.* **109**, 365. "On the Cosmological Problem" [15.1]
Weyl, H.; (1949) *Philosophy of Mathematics and Natural Science* (Princeton Univ. Press, Princeton) [2.1]

1950

Gödel, K.; (1950) *Proc. of the 1950 Int. Cong. of Math.*, Vol. I, p. 175. "Rotating Universes in General Relativity Theory" [3.4, 12.2]
Lemaitre, G.; (1950) *The Primeval Atom* (Van Nostrand, New York)
Schrödinger, E.; (1950) *Space-Time Structure* (Cambridge Univ. Press, Cambridge)

1951

Cartan, E.; (1951) *Lessons on the Geometry of Riemann Spaces* (Gauthier-Villars, Paris)
Taub, A. H.; (1951) *Ann. Math.* **53**, 472. "Empty Space-Times Admitting a Three Parameter Group of Motions" [6.4, Table 6.1, 8.1, 8.3, 9.4, 9.4, 9.5, 9.5]

1952

Jordan, P.; (1952) *Gravitation and Universe – Foundations of Theoretical Cosmology* (Vieweg & Sohn, Branschweig) [15.1]
Milne, E. A.; (1952) *Modern Cosmology and the Christian Idea of God* (Clarendon Press, Oxford) [15.1]
Møller, C.; (1952) *The Theory of Relativity* (Clarendon Press, Oxford)
Raychaudhuri, A.; (1952) *Phys. Rev.* **86**, 90. "Condensations in Expanding Cosmologic Models"

1953

Yano, K.; Bochner, S.; (1953) *Curvature and Betti Numbers* (Princeton Univ. Press, Princeton) [6.1, 6.2]

1954

Layzer, D.; (1954) *Astron. J.* **59**, 268. "On the Significance of Newtonian Cosmology"

McVittie, G. C.; (1954) *Astron. J.* **59**, 173. "Relativistic and Newtonian Cosmology"

Schouten, J.; (1954) *Ricci-Calculus* (Springer-Verlag, Berlin) [2.1, 7.1]

1955

Einstein, A.; (1955) *The Meaning of Relativity* (5th Ed.) (Princeton Univ. Press, Princeton) [2.5]

Koyre, A.; (1955) *Am. Phil. Soc. Trans.* **45**, Part 4. "A Documentary History of the Problem of Fall from Kepler to Newton" [1.1]

Lichnerowicz, A.; (1955) *Relativistic Theories of Gravitation and Electromagnetism* (Masson, Paris)

McCrea, W. H.; (1955) *Nature* **175**, 466. "Newtonian Cosmology"

Raychaudhuri, A.; (1955a) *Z. Astrophys.* **37**, 103. "Perturbed Cosmological Models"

______________ ; (1955b) *Phys. Rev.* **98**, 1123. "Relativistic Cosmology. I" [3.5]

Schweber, S.; Bethe, H.; DeHoffmann, F.; (1955) *Mesons and Fields*, Vol. 1 (Row, Peterson & Co., Evanston) [11.4]

1956

Heckmann, O.; Schücking, E.; (1956) in Mercier and Kervaire (1956) p. 114. "A World Model of Newtonian Cosmology with Expansion and Rotation"

Komar, A.; (1956) *Phys. Rev.* **104**, 544. "Necessity of Singularities in the Solution of the Field Equations of General Relativity"

Mercier, A.; Kervaire, M.; (Eds.) (1956) *Jubilee of Relativity Theory* (Birkhauser, Basel)

Milnor, J.; (1956) *Ann. Math.* **64**, 399. "On Manifolds Homeomorphic to the 7-Sphere" [2.1]

Rindler, W.; (1956) *Mon. Not. Roy. Astron. Soc.* **116**, 662. "Visual Horizons in World-Models"

Schrodinger, E.; (1956) *Expanding Universes* (Cambridge Univ. Press, Cambridge) [E]

1957

Bass, R. W.; Witten, L.; (1957) *Rev. Mod. Phys.* **29**, 452. "Remark on Cosmological Models"

Bonnor, W. B.; (1957) *Mon. Not. Roy. Astron. Soc.* **117**, 104. "Jeans' Formula for Gravitational Instability"

Burbidge, E. M.; Burbidge, G. R.; Fowler, W. A.; Hoyle, F.; (1957) *Rev. Mod. Phys.* **29**, 547. "Synthesis of the Elements in Stars"

Munitz, M. K.; (1957) *Theories of the Universe from Babylonian Myth to Modern Science* (The Free Press, Glencoe)

Raychaudhuri, A.; (1957) *Phys. Rev.* **106**, 172. "Singular State in Relativistic Cosmology"

Schücking, E.; (1957) *Naturwiss.* **44**, 507. "Homogeneous Shear-Free World Models in Relativistic Cosmology"

Synge, J. L.; (1957) *The Relativistic Gas* (North-Holland, Amsterdam)

Taub, A. H.; (1957) *Ill. J. Math.* **1**, 370. "Singular Hypersurfaces in General Relativity" [10.2]

Vaidya, P. C.; Shah, K. B.; (1957) *Proc. Nat. Inst. Sci. India* **A23**, 534. "A Radiating Mass Particle in an Expanding Universe"

Zwicky, F.; (1957) *Morphological Astronomy* (Springer-Verlag, Berlin)

1958

Adams, J. B.; Mjolsness, R.; Wheeler, J. A.; (1958) in Solvay (1958).

Choquet-Bruhat, Y.; (1958) *Bull. Soc. Math. France.* **86**, 155. "Existence Theorems in Relativistic Fluid Mechanics"

Dirac, P.; (1958a) *Proc. Roy. Soc. A.* **246**, 326. "Generalized Hamiltonian Dynamics"

———; (1958b) *Proc. Roy. Soc. A.* **246**, 333. "Theory of Gravitation in Hamiltonian Form"

Heckmann, O.; Schücking, E.; (1958) in Solvay (1958). "World Models"

Koyré, A.; (1958) *From the Closed World to the Infinite Universe* (Harper, New York)

Oort, J. H.; (1958) in Solvay (1958) "Distribution of Galaxies and the Density of the Universe"

Pauli, W.; (1958) *Theory of Relativity* (Pergamon, London)

Raychaudhuri, A.; (1958) *Proc. Phys. Soc.* **72**, 263. "An Anisotropic Cosmological Solution in General Relativity"

Ryle, M.; (1958) *Proc. Roy. Soc. A.* **248**, 289. "The Nature of the Cosmic Radio Sources"

Solvay; (1958)(11th Solvay Congress) *The Structure and Evolution of the Universe* (Stoops, Brussels)

1959

Dirac, P.; (1959) *Phys. Rev.* **114**, 924. "Fixation of Coordinates in the Hamiltonian Theory of Gravitation"

Eddington, A. S.; (1959) *Space, Time and Gravitation* (Harper, New York) [15.1]

Lichnerowicz, A.; (1959) *C. R. Acad. Sci. (Paris)* **249**, 2287. "On the Quantization of the Gravitational Field for a Space-Time of Constant Curvature"

Papapetrou, A.; Treder, H.; (1959) *Ann. Physik.* **3**, 360. "On the Existence of Singularity-Free Solutions to the Field Equations of the General Theory of Relativity, Which Can Represent Partial Models"

Rauch, H. E.; (1959) *Geodesics and Curvature in Differential Geometry in the Large* (Yeshiva Univ.)

Sciama, D. W.; (1959) *The Unity of the Universe* (Doubleday, Garden City)

Taub, A. H.; (1959) *Arch. Rat. Mech. Anal.* **3**, 312. "On Circulation in Relativistic Hydrodynamics" [3.2, 3.4]

1960

Birkhoff, G.; (1960) *Hydrodynamics, A Study in Logic, Fact and Similitude* (Princeton Univ. Press, Princeton) [3.1, 3.2]

Bobin, A.; (1960) in *Philosophical Encyclopedia* (F. Konstantinov, Ed.)(Soviet Encyclopedia Press, Moscow) "Infinity"

Bondi, H.; (1960) *Cosmology* (2nd Ed.)(Cambridge Univ. Press, Cambridge)

Graves, J. C.; Brill, D. R.; (1960) *Phys. Rev.* **120**, 1507. "Oscillatory Character of Reissner-Nordstrom Metric for an Ideal Charged Wormhole"

Infeld, L.; Plebanski, J.; (1960) *Motion and Relativity* (Pergamon, New York)

Israel, W.; (1960) *Nuovo Cimento* **18**, 397. "Evolution in General Relativity"

Jordan, P.; Ehlers, J.; Kundt, W.; (1960) *Akad. Wiss. Lit. Mainz., Abhandl. Mat.-Nat. Kl.* (Germany), No. 2. "Rigorous Solutions of the Field Equations of General Relativity"

Kruskal, M. D.; (1960) *Phys. Rev.* **119**, 1743. "Maximal Extension of Schwarzschild Metric" [5.4]

Papapetrou, A.; Treder, H.; (1960) *Ann. Physik* **6**, 311. "The Existence of Singularity-Free Solutions of the General Relativity Field Equations Which Can Represent Particle Models. II"

Penrose, R.; (1960) *Ann. Physics* **10**, 171. "A Spinor Approach to General Relativity" [E]

Synge, J. L.; (1960) *Relativity – The General Theory* (North-Holland, Amsterdam) [3.4, 3.4]

Szekeres, G.; (1960) *Pub. Math. Debrecen* **7**, 285. "On the Singularities of a Riemannian Manifold" [5.4]

1961

Brans, C.; Dicke, R. H.; (1961) *Phys. Rev.* **124**, 925. "Mach's Principle and a Relativistic Theory of Gravitation" [15.1]

Chandrasekhar, S.; Wright, J. P.; (1961) *Proc. Nat. Acad. Sci.* **47**, 341. "The Geodesics in Gödel's Universe" [7.3]

Ehlers, J.; (1961) *Akad. Wiss. Lit. Mainz., Abhandl. Mat.-Nat. Kl.* 11, 793. "Contributions to the Relativistic Mechanics of Continuous Media" [3.4, 3.5, 3.5]

Heckmann, O.; (1961) *Astron. J.* **66**, 599. "On the Possible Influence of a General Rotation on the Expansion of the Universe"

Khalatnikov, I. M.; Lifshitz, E. M.; Sudakov, V. V.; (1961) *Phys. Rev. Lett.* **6**, 311. "Singularities of the Cosmological Solutions of Gravitational Equations"

Lifshitz, E. M.; Sudakov, V. V.; Khalatnikov, I. M.; (1961) *Zh. Eksp. Teor. Fiz.* **40**, 1847 [Eng. trans. *Sov. Phys.–JETP* 13, 1298 (1961)]. "Singularities of Cosmological Solutions of Gravitational Equations. III"

Pachner, J.; (1961) *Z. Phys.* **164**, 574. "The Gravitational Field of a Point Mass in the Expanding Universe"

Robinson, B. B.; (1961) *Proc. Nat. Acad. Sci.* **47**, 1852. "Relativistic Universes with Shear"

Sandage, A.; (1961) *Astrophys. J.* 133, 355. "The Ability of the 200-Inch Telescope to Discriminate Between Selected World Models"

van den Bergh, S.; (1961) *Z. Astrophys.* **53**, 219. "The Luminosity Function of Galaxies

Weber, J.; (1961) *General Relativity and Gravitational Waves* (Interscience, New York) [3.3, 4.1, 7.2, 10.2]

Zel'dovich, Ya. B.; (1961) *Zh. Eksp. Teor. Fiz.* **41**, 1609 [Eng. trans. Sov. Phys.-JETP **14**, 1143]. "The Equation of State at Ultrahigh Densities and its Relativistic Limitations" [7.3]

1962

Abraham, R.; (1962) *J. Math. Mech.* 11, 553. "Piecewise Differentiable Manifolds and the Space-Time of General Relativity"

Arnowitt, R.; Deser, S.; Misner, C. W.; (1962) in Witten (1962). "The Dynamics of General Relativity" [11.2]

Behr, C.; (1962) *Z. Astrophys.* **54**, 268. "Generalization of the Friedmann World Model with Positive Space Curvature"

Brans, C. H.; (1962) *Phys. Rev.* **125**, 2194. "Mach's Principle and a Relativistic Theory of Gravitation. II"

Calabi, E.; Markus, L.; (1962) *Ann. Math.* **75**, 63. "Relativistic Space Forms" [3.2, 5.2, 6.4]

Choquet-Bruhat, Y.; (1962) in Witten (1962). "The Cauchy Problem" [9.2]

Dicke, R. H.; (1962) in Møller (1962). "Mach's Principle and Equivalence"

Edelen, D. G.; (1962) *The Structure of Field Space* (Univ. of California Press, Berkeley) [2.5]

Ehlers, J.; (1962) in Infeld (1962). "Relativistic Hydrodynamics and its Relation to Interior Solutions of the Gravitational Field Equations"

Fuller, R. W.; Wheeler, J. A.; (1962) *Phys. Rev.* **128**, 919. "Causality and Multiply Connected Space-Time"

Heckmann, O.; Shücking, E.; (1962) in Witten (1962). "Relativistic Cosmology"

Helgason, S.; (1962) *Differential Geometry and Symmetric Spaces* (Academic Press, New York) [2.1, 2.2, 2.4, 2.5, 5.1, 5.1, 5.1, 6.1, 7.1, 7.1]

Infeld, L.; (Mem. Vol.)(1962) *Recent Developments in General Relativity* (Pergamon, New York)

Jacobson, N.; (1962) *Lie Algebras* (Wiley, New York) [6.2]

McVittie, G. C.; (Ed.)(1962) *Problems of Extra-Galactic Research* (MacMillan, New York)

Møller, C.; (1962) *Evidence for Gravitational Theories* (Academic Press, New York)

Narlikar, J. V.; (1962) in Møller (1962). "Rotating Universes"

Ozsvath, I.; (1962) *Acad. Wiss. Lit. Mainz., Abhandl. Mat.-Nat. Kl.* (Germany), No. 13. "Solutions of the Einstein Field Equations with Simply Transitive Groups of Motions"

Ozsvath, I.; Schucking, E.; (1962) *Nature* **193**, 1168. "Finite Rotating Universes"

Pachner, J.; (1962) *Acta. Phys. Polon.* **22**, Suppl. 45. "On the Geodesics in an Expanding Universe"

Papapetrou, A.; (1962) *Ann. Physik.* **9**, 97. "The Question of the Existence of Non-Singular Solutions of the Field Equations of General Relativity. III"

Raychaudhuri, A. K.; Som, M. M.; (1962) *Proc. Cambridge Phil. Soc.* **58**, 338. "Stationary Cylindrically Symmetric Clusters of Particles in General Relativity"

Thomas, T. Y.; (1962) *Proc. Nat. Acad. Sci.* **48**, 611. "The Role of Gravitation in the Creation of Matter"

Weinberg, S.; (1962) *Nuovo Cimento* **25**, 15. "The Neutrino Problem in Cosmology"

Wheeler, J. A.; (1962a) *The Monist* **47**, 40. "The Universe in the Light of General Relativity"

__________; (1962b) *Geometrodynamics* (Academic Press, New York) [2.1, 2.3, 9.2, 9.4, 10.2, 15.1]

__________; (1962c) *Rev. Mod. Phys.* **34**, 873. "Problems on the Frontiers Between General Relativity and Differential Geometry"

Witten, L.; (Ed.)(1962) *Gravitation—An Introduction to Current Research* (John Wiley, New York)

1963

Dicke, R. H.; (1963) *Am. J. Phys.* **31**, 500. "Cosmology, Mach's Principle, and Relativity"

Flanders, H.; (1963) *Differential Forms with Applications to the Physical Sciences* (Academic Press, New York)

Gürsey, F.; (1963) *Ann. Physics* **24**, 211. "Reformulation of General Relativity in Accordance with Mach's Principle"

Kerr, R. P.; *Phys. Rev. Letters* **11**, 237. "Gravitational Field of a Spinning Mass as an Example of Algebraically Special Metrics"

Kobayashi, S.; Nomizu, K.; (1963) *Foundations of Differential Geometry*, Vol. 1 (Interscience, New York)

Kundt, W.; (1963) *Z. Phys.* **172**, 488. "Note on the Completeness of Spacetimes" [5.2, 5.2]

Lifshitz, E. M.; Khalatnikov, I. M.; (1963) *Adv. Phys.* **12**, 185. "Investigations in Relativistic Cosmology" [9.1, 9.5, 9.5]

Manasse, F. K.; Misner, C. W.; (1963) *J. Math. Phys.* **4**, 735. "Fermi Normal Coordinates and Some Basic Concepts in Differential Geometry"

Misner, C. W.; (1963) *J. Math. Phys.* **4**, 924. "The Flatter Regions of Newman, Unti, and Tamburino's Generalized Schwarzschild Space" [5.2, 5.2, 5.4, 8.1, 8.1]

Munkres, J.; (1963) *Elementary Differential Topology* (Princeton Univ. Press, Princeton) [2.1]

Newman, E.; Tamburino, L.; Unti, T.; (1963) *J. Math. Phys.* **4**, 915. "Empty-Space Generalization of the Schwarzschild Metric" [8.1]

Zel'dovich, Ya. B.; (1963) *Atomnaya Energiya* **14**, 92 [Eng. trans. *Sov. Atomic Energy* **14**, 83 (1963)]. "The Initial Stages of the Evolution of the Universe"

1964

Brill, D. R.; (1964) *Phys. Rev.* **133**, B845. "Electromagnetic Fields in a Homogeneous, Nonisotropic Universe" [10.2]

Cahen, M.; (1964) *Bull. Acad. Roy. Belgique* **50**, 972. "On a Class of Homogeneous Spaces in General Relativity"

Callan, C. G.; (1964) *Spherically Symmetric Cosmological Models* (Thesis, Princeton Univ.)

Chiu, H.; Hoffman, W.; (Eds.) (1964) *Gravitation and Relativity* (Benjamin, New York)

DeWitt, C.; DeWitt, B.; (1964) *Relativity Groups and Topology* (Gordon & Breach, New York)

Ellis, G. F. R.; (1964) *On General Relativistic Fluids and Cosmological Models* (Dissertation, Univ. of Cambridge)

Fock, V.; (1964) *The Theory of Space, Time, and Gravitation* (MacMillan, New York)

Hermann, R.; (1964) *J. Math. Mech.* **13**, 497. "An Incomplete Compact Homogeneous Lorentz Metric" [7.1]

Misner, C. W.; (1964) in DeWitt and DeWitt (1964). "Differential Geometry and Differential Topology" [6.1]

Novikov, I. D.; (1964a) *Zh. Eksp. Teor. Fiz.* **46**, 686 [Eng. Trans. *Sov. Phys.-JETP* **19**, 467 (1964)]. "The Possibility of Large Scale Inhomogenieties in the Expanding Universe"

__________; (1964b) *Astron. Zh.* **41**, 1075 [Eng. Trans. *Sov. Astron.-AJ* **8**, 857 (1965)]. "Delay of the Expansion of a Part of the Friedmann Universe and Superstars"

Shepley, L. C.; (1964) *Proc. Nat. Acad. Sci.* **52**, 1403. "Singularities in Spatially Homogeneous Dust-Filled Cosmological Models"

Swiatak, H.; (1964) *Acta. Phys. Polon.* **25**, 161. "On the Algebraic Structure of Gravitational Fields Admitting of 5- and 6-Parameter Groups of Motions"

Wheeler, J. A.; (1964) in DeWitt and DeWitt (1964). "Geometrodynamics and the Issue of the Final State"

Zel'dovich, Ya. B.; (1964) *Astron. Zh.* **41**, 873 [Eng. trans. *Sov. Ast.-AJ* **8**, 700 (1965)]. "Newtonian and Einsteinian Motion of Homogeneous Matter"

1965

Adler, R.; Bazin, M.; Schiffer, M.; (1965) *Introduction to General Relativity* (McGraw-Hill, New York)

Behr, C.; (1965a) *Astron. Abhandl. Hbg. Sternwarte* **7**, 245. "Analytical and Numerical Examination of Singularities of Homogeneous Rotating World Models" [13.1]

Behr, C.; (1965b) *Z. Astrophys.* **60**, 286. "A Singularity of Homogeneous Non-Isotropic Relativistic World Models"

Chernin, A. D.; (1965) *Astron. Zh.* **42**, 1124 [Eng. Trans. *Sov. Ast.-AJ* **9**, 871 (1966)]. "A Model of a Universe Filled by Radiation and Dust-Like Matter"

Dicke, R. H.; Peebles, P. J. E.; Roll, P. G.; Wilkinson, D. T.; (1965) *Astrophys. J.* **142**, 414. "Cosmic Black Body Radiation" [4.3, 15.1]

Harrison, B. K.; Thorne, K. S.; Wakano, M.; Wheeler, J. A.; (1965) *Gravitation Theory and Gravitational Collapse* (Univ. of Chicago Press, Chicago) [3.2]

Hawking, S. W.; (1965) *Phys. Rev. Letters* **15**, 689. "Occurence of Singularities in Open Universes"

Hawking, S. W.; Ellis, G. F. R.; (1965) *Phys. Letters* **17**, 246. "Singularities in Homogeneous World Models" [10.2, 10.2]

Hicks, N. J.; (1965) *Notes on Differential Geometry* (Van Nostrand, Princeton) [2.1, 2.2, 2.2]

Just, K.; (1965) *Nuovo Cimento* **36**, 1386. "Quantum Theory and the Singular State of Cosmology"

Khalatnikov, I. M.; (1965) *Zh. Eksp. Teor. Fiz.* **48**, 261 [Eng. Trans. *Sov. Phys.-JETP* **21**, 172 (1965)]. "Some Comments on the Singularities of the Cosmological Solutions of the Gravitation Equations"

McVittie, G. C. (1965) *General Relativity and Cosmology* (2nd Ed.) (Univ. of Ill. Press, Urbana)

North, J. D.; (1965) *Measure of the Universe* (Clarendon Press, Oxford)

Ozsvath, I.; (1965a) *J. Math. Phys.* **6**, 590. "New Homogeneous Solutions of Einstein's Field Equations with Incoherent Matter Obtained by a Spinor Technique" [7.1, 7.2, Table 7.1]

————; (1965b) *J. Math. Phys.* **6**, 1255. "Homogeneous Solutions of the Einstein-Maxwell Equations" [7.2]

————; (1965c) *J. Math. Phys.* **6**, 1265. "All Homogeneous Solutions of Einstein's Field Equations with Incoherent Matter and Electromagnetic Radiation"

Peebles, P. J. E.; (1965) *Astrophys. J.* **142**, 1317. "The Black-Body Radiation Content of the Universe and the Formation of Galaxies"

Penrose, R.; (1965) *Phys. Rev. Letters.* **14**, 57. "Gravitational Collapse and Space-Time Singularities"

Penzias, A. A.; Wilson, R. W.; (1965) *Astrophys. J.* **142**, 419. "A Measurement of Excess Antenna Temperature at 4080 Mc/s" [15.1]

Robinson, I.; Schild, A.; Schücking, E. L.; (Eds.) (1965) *Quasi-Stellar Sources and Gravitational Collapse* (Univ. of Chicago Press, Chicago)

Sakharov, A. D.; (1965) *Zh. Eksp. Teor. Fiz.* **49**, 345 [Eng. Trans. *Sov. Phys.-JETP* **22**, 241 (1965)]. "Initial Stage of an Expanding Universe and the Appearance of a Nonuniform Distribution of Matter"

Shepley, L. C.; (1965) *SO(3,R)-Homogeneous Cosmologies* (Thesis, Princeton Univ.)

Spivak, M.; (1965) *Calculus on Manifolds* (Benjamin, New York) [2.4]

Trautman, A.; Pirani, F. A. E.; Bondi, H.; (1965) *Lectures on General Relativity* (Prentice-Hall, Englewood Cliffs)

Wagoner, R. V.; (1965) *Phys. Rev.* **138**, B1583. "Rotation and Gravitational Collapse"

Wright, J. P.; (1965) *J. Math. Phys.* **6**, 103. "Solution of Einstein's Field Equations for a Rotating, Stationary, and Dust-Filled Universe"

1966

Alfven, H.; (1966) *Worlds-Antiworlds* (Freeman, San Francisco)

Cartan, E.; (1966) *The Theory of Spinors* (M.I.T. Press, Cambridge)

Davidson, W.; Narlikar, J. V.; (1966) *Rep. Prog. Phys.* **29**, 539. "Cosmological Models and their Observational Validation"

Doroshkevich, A. G.; (1966) *Astrofizika* **2**, 37. "The Gravitational Instability of Anisotropic Homogeneous Solutions" [14.3]

Farnsworth, D. L.; Kerr, R. P.; (1966) *J. Math. Phys.* **7**, 1625. "Homogeneous, Dust-Filled Cosmological Solutions" [7.1, 7.2, Table 7.1]

Geroch, R. P.; (1966) *Phys. Rev. Letters* **17**, 445. "Singularities in Closed Universes"

Gilbert, I. H.; (1966) *Astrophys. J.* **144**, 233. "An Integral Equation for the Development of Irregularities in an Expanding Universe"

Grischuk, L. P.; (1966) *Zh. Eksp. Teor. Fiz.* **51**, 475 [Eng. Trans. *Sov. Phys.-JETP* **24**, 320 (1967)]. "Some Remarks on the Singularities of the Cosmological Solutions of the Gravitational Equations" [10.2]

Hawking, S. W.; (1966a) *Phys. Rev. Letters* **17**, 444. "Singularities in the Universe"

——————; (1966b) *Proc. Roy. Soc. A* **294**, 511. "The Occurrence of Singularities in Cosmology"

——————; (1966c) *Proc. Roy. Soc. A* **295**, 490. "The Occurence of Singularities in Cosmology. II"

——————; (1966d) Essay Submitted for the Adams Prize. "Singularities and the Geometry of Space-Time"

——————; (1966e) *Astrophys. J.* **145**, 544. "Perturbations of an Expanding Universe"

Hoffmann, B.; (Ed.)(1966) *Perspectives in Geometry and Relativity* (Indiana Univ. Press, Bloomington)

Kantowski, R.; (1966) *Some Relativistic Cosmological Models* (Dissertation, Univ. Texas, Austin) [6.4, 10.1, 10.1]

Kantowski, R.; Sachs, R. K.; (1966) *J. Math. Phys.* **7**, 443. "Some Spatially Homogeneous Anisotropic Relativistic Cosmological Models" [10.1]

Kristian, J.; Sachs, R. K.; (1966) *Astrophys. J.* **143**, 379. "Observations in Cosmology" [3.3]

Maitra, S. C.; (1966) *J. Math. Phys.* **7**, 1025. "Stationary Dust-Filled Cosmological Solution with $\Lambda = 0$ and Without Closed Timelike Lines" [3.5, 10.2]

Narlikar, J. V.; (1966) *Mon. Not. Roy. Astron. Soc.* **131**, 501. "On Newtonian Universes"

Ozsvath, I.; (1966) in Hoffmann (1966). "Two Rotating Universes with Dust and Electromagnetic Field"

Peebles, P. J. E.; (1966) *Phys. Rev. Letters* **16**, 410. "Primeval Helium Abundance and the Primeval Fireball" [15.1]

Penrose, R.; (1966) Adams Prize Essay. "An Analysis of the Structure of Space-time" [2.1]

Zel'dovich, Ya. B.; (1966) *Uspekhi Fiz. Nauk.* **89**, 647 [Eng. Trans. *Sov. Phys.-Uspekni* **9**, 602 (1966)]. "The 'Hot' Model of the Universe"

1967

Abraham, R.; (1967) *Foundations of Mechanics* (Benjamin, New York)

Anderson, J. L.; (1967) *Principles of Relativity Physics* (Academic Press, New York)

Bichteler, K.; (1967) *Comm. Math. Phys.* **4**, 352. "On the Cauchy Problem of the Relativistic Boltzmann Equation" [3.1]

Burbidge, G. R.; Burbidge, E. M.; (1967) *Quasi-Stellar Objects* (Freeman, San Francisco)

Cahen, M.; Debever, R.; Defrise, L.; (1967) *J. Math. Mech.* **16**, 761. "A Complex Vectorial Formalism in General Relativity" [7.1, 7.2]

Conklin, E. K.; Bracewell, R. N.; (1967) *Phys. Rev. Letters* **18**, 614. "Isotropy of Cosmic Background Radiation at 10690 MHz"

DeWitt, B.; (1967a) *Phys. Rev.* **160**, 1113. "Quantum Theory of Gravity. I" [11.4, 11.4, 11.4]

————; (1967b) *Phys. Rev.* **162**, 1195. "Quantum Theory of Gravity. II" [11.4, 11.4, 11.4]

————; (1967c) *Phys. Rev.* **162**, 1239. "Quantum Theory of Gravity. III" [11.4, 11.4, 11.4]

Doroshkevich, A. G.; Zel'dovich, Ya. B.; Novikov, I. D.; (1967a) *Zh. Eksp. Teor. Fiz. Pis'ma* **5**, 119 [Eng. Trans. *JETP Letters* **5**, 96 (1967)]. "Neutrinos and Gravitons in the Anisotropic Model of the Universe" [12.3]

————————; (1967b) *Zh. Eksp. Teor. Fiz.* **53**, 644 [Eng. Trans. *Sov. Phys.-JETP* **26**, 408 (1968)]. "Weakly Interacting Particles in an Anisotropic Cosmological Model"

Ehlers, J.; (Ed.)(1967) *Relativity Theory and Astrophysics* (American Math. Soc. Providence)

Ellis, G. F. R.; (1967) *J. Math. Phys.* **8**, 1171. "Dynamics of Pressure-Free Matter in General Relativity" [3.5]

Geroch, R. P.; (1967) *J. Math. Phys.* **8**, 782. "Topology in General Relativity" [5.2, 8.3]

Grishchuk, L. P.; (1967) *Astron. Zh.* **44**, 1097 [Eng. Trans. *Sov. Astron.-AJ* **11**, 881 (1968)]. "Cosmological Models and Spatial Homogeneity Criteria"

Harrison, E. R.; (1967a) *Rev. Mod. Phys.* **39**, 862. "Normal Modes of Vibrations of the Universe"

————; (1967b) *Nature* **215**, 151. "Quantum Cosmology"

Hawking, S. W.; (1967) *Proc. Roy. Soc. A* **300**, 187. "The Occurrence of Singularities in Cosmology. III. Causality and Singularities" [8.3, 10.2]

Jacobs, K. C.; (1967) *Nature* **215**, 1156. "Friedmann Cosmological Model with Both Radiation and Matter"

Kundt, W.; (1967) *Comm. Math. Phys.* **4**, 143. "Non-Existence of Trouser-Worlds"

Lichnerowicz, A.; (1967) *Relativistic Hydrodynamics and Magnetohydrodynamics* (Benjamin, New York) [7.2]

Misner, C. W.; (1967a) Essay for Gravity Research Foundation, New Boston. "The Isotropy of the Universe" [12.3]

————; (1967b) in Ehlers (1967). "Taub-NUT Space as a Counterexample to Almost Anything"

————; (1967c) *Phys. Rev. Letters* **19**, 533. "Neutrino Viscosity and the Isotropy of Primordial Blackbody Radiation" [3.1]

————; (1967d) *Nature* **214**, 40. "Transport Processes in the Primordial Fireball" [3.1, 9.5]

Nelson, E.; (1967) *Tensor Analysis* (Mathematical Notes, Princeton Univ. Press, Princeton)

Ozsvath, I.; (1967) *J. Math. Phys.* **8**, 326. "Homogeneous Lichnerowicz Universes"

Partridge, R.B.; Wilkinson, D. T.; (1967) *Phys. Rev. Letters* **18**, 557. "Isotropy and Homogeneity of the Universe from Measurements of the Cosmic Microwave Background" [15.1]

Peebles, P. J. E.; (1967) *Astrophys. J.* **147**, 859. "The Gravitational Instability of the Universe"

Reinhardt, M. V.; (1967) *Z. Astrophys.* **66**, 292. "Quasars and Cosmological Models"

Sachs, R. K.; Wolfe, A. M.; (1967) *Astrophys. J.* **147**, 73. "Perturbations of a Cosmological Model and Angular Variations of the Microwave Background"

Saslaw, W. C., (1967) *Mon. Not. Roy. Astron. Soc.* **136**, 39 "Thermodynamic Instabilities in an Expanding Universe"

Taub, A. H.; (1967) in Ehlers (1967). "Relativistic Hydrodynamics" [3.1]

Thorne, K. S.; (1967) *Astrophys. J.* **148**, 51. "Primordial Element Formation, Primordial Magnetic Fields, and the Isotropy of the Universe" [E]

Wagoner, R. V.; Fowler, W. A.; Hoyle, F.; (1967) *Astrophys. J.* **148**, 3. "On the Synthesis of Elements at Very High Temperatures"

Wilkinson, D. T.; Partridge, R. B.; (1967) *Nature* **215**, 719. "Large Scale Density Inhomogeneities in the Universe"

Wolf, J.; (1967) *Spaces of Constant Curvature* (McGraw-Hill, New York) [5.4, 15.1, E]

Zel'dovich, Ya. B.; Novikov, I. D.; (1967) *Zh. Eksp. Teor. Fiz. Pis'ma* **6**, 772 [Eng. Trans. *JETP Letters* **6**, 236 (1967)]. "Physical Limitations on the Topology of the Universe"

1968

Banerji, S.; (1968) *Prog. Theor. Phys.* **39**, 365. "Homogeneous Cosmological Models Without Shear"

Brill, D. R.; Perisho, R. C.; (1968) *Am. J. Phys.* **36**, 85. "Resource Letter GR-1 on General Relativity"

Cahen, M.; McLenaghan, R.; (1968) *C. R. Acad. Sci. Paris* **266**, A1125. "Metrics of Symmetric Lorentz Spaces in Four Dimensions" [7.2]

Chernin, A. D.; (1968) *Nature* **220**, 250. "Radiation and Matter in an Open Cosmological Model"

DeWitt, C.; Wheeler, J. A.; (Eds.) (1968) *Battelle Rencontres* (Benjamin, New York)

Dicke, R. H.; (1968) *Astrophys. J.* **152**, 1. "Scalar-Tensor Gravitation and the Cosmic Fireball"

Dickson, F. P.; (1968) *The Bowl of Night* (M.I.T. Press, Cambridge)

Edelen, D. G.; (1968) *Nuovo Cimento* **55B**, 155. "Conformally Homogeneous Model Universes. I"

Ehlers, J.; Geren, P.; Sachs, R.; (1968) *J. Math. Phys.* **9**, 1344. "Isotropic Solutions of the Einstein-Liouville Equations" [E]

Estabrook, F. B.; Wahlquist, H. D.; Behr, C. G.; (1968) *J. Math. Phys.* **9**, 497. "Dyadic Analysis of Spatially Homogeneous World Models"

Field, G. B.; Shepley, L. C.; (1968) *Astrophys. Space Sci.* **1**, 309. "Density Perturbations in Cosmological Models"

Fowler, W. A.; Stephens, W. E.; (1968) *Am. J. Phys.* **36**, 289. "Resource Letter OE-1 on Origin of the Elements"

Geroch, R. P.; (1968a) *Ann. Phys.* **48**, 526. "What is a Singularity in General Relativity?"

__________; (1968b) *J. Math. Phys.* **9**, 450. "Local Characterization of Singularities in General Relativity" [5.2, 5.2, 5.3, 5.3]

Grishchuk, L. P.; Doroshkevich, A. G.; Novikov, I. D.; (1968) *Zh. Eksp. Teor. Fiz.* **55**, 2281 [Eng. Trans. *Sov. Phys.-JETP* **28**, 1210 (1969)]. "Anisotropy of the Early Stages of Cosmological Expansion and Observation"

Hawking, S. W.; (1968) *Proc. Roy. Soc. A* **308**, 433. "The Existence of Cosmic Time Functions"

Hawking, S. W.; Ellis, G. F. R.; (1968) *Astrophys. J.* **152**, 25. "The Cosmic Black-Body Radiation and the Existence of Singularities in our Universe" [10.1, 10.2, 10.2]

Jacobs, K. C.; (1968) *Astrophys. J.* **153**, 661. "Spatially Homogeneous and Euclidean Cosmological Models with Shear"

Krause-Astorga, J.; (1968) *Stability of an Homogeneous Anisotropic Relativistic Cosmological Model* (Dissertation, Univ. Texas Austin)

Kuchowicz, B.; (1968) *Nuclear and Relativistic Astrophysics and Nuclidic Cosmochemistry: 1963-1967* (Nuclear Energy Info. Center, Warsaw)

Longair, M. S.; (1968) *Nuovo Cimento Suppl.* **6**, 990. "Evolutionary Cosmologies, A Survey"

Matzner, R. A.; (1968) *J. Math. Phys.* **9**, 1063. "3-Sphere 'Backgrounds' for the Space Sections of the Taub Cosmological Solution"

Misner, C. W.; (1968) *Astrophys. J.* **151**, 431. "The Isotropy of the Universe" [3.1, 11.2, 12.1, 12.3, 12.3, 14.2, 15.1]

Misner, C. W.; Taub, A. H.; (1968) *Zh. Eksp. Teor. Fiz.* **55**, 233 [Eng. Trans. *Sov. Phys.-JETP* **28**, 122 (1969)]. "A Singularity-Free Empty Universe" [5.2, 5.2, 8.1, 8.1, 8.3]

Misner, C. W.; Thorne, K. S.; Wheeler, J. A.; (1968) *An Open Letter to Relativity Theorists*, August 19, 1968. "Sign Conventions"

Novikov, I. D.; (1968) *Astron. Zh.* **45**, 538 [Eng. Trans. *Sov. Astron.-AJ* **12**, 427 (1968)]. "An Expected Anisotropy of the Cosmological Radioradiation in Homogeneous Anisotropic Models"

Parker, L.; (1968) *Phys. Rev. Letters* **21**, 562. "Particle Creation in Expanding Universes"

Penrose, R.; (1968a) in DeWitt and Wheeler (1968). "Structure of Space-Time"

———; (1968b) *Int. J. Theor. Phys.* **1**, 61. "Twistor Quantisation and Curved Space-Time" [E]

Rees, M. J.; Sciama, D. W.; (1968) *Nature* **217**, 511. "Large-Scale Density Inhomogeneities in the Universe"

Robertson, H. P.; Noonan, T. W.; (1968) *Relativity and Cosmology* (Saunders, Philadelphia)

Ruban, V. A.; (1968) *Zh. Eksp. Teor. Fiz. Pis'ma* **8**, 669 [Eng. Trans. *JETP Letters* **8**, 414 (1968)]. "T-Models of 'Sphere' in General Relativity Theory"

Sandage, A.; (1968) *Observatory* **88**, 91. "Observational Cosmology"

Sato, H.; (1968) *Prog. Theor. Phys.* **40**, 781. "Observable Horizons in the Expanding Universe"

Stewart, J.; (1968) *Astrophys. Letters* **2**, 133. "Neutrino Viscosity in Cosmological Models" [12.3, 14.2]

Stewart, J.; Ellis, G. F. R.; (1968) *J. Math. Phys.* **9**, 1072. "Solutions of Einstein's Equations for a Fluid Which Exhibit Local Rotational Symmetry"

Wheeler, J. A.; (1968) *Am. Scientist* **56**, 1. "Our Universe: The Known and Unknown" [11.4, 11.4]

1969

Belinskii, V.; Khalatnikov, I. M.; (1969a) *Zh. Eksp. Teor. Fiz.* **56**, 1701 [Eng. Trans. *Sov. Phys.-JETP* **29**, 911 (1969)]. "On the Nature of the Singularities in the General Solution of the Gravitational Equations"

———; (1969b) *Zh. Eksp. Teor. Fiz.* **57**, 2163 [Eng. Trans. *Sov. Phys.-JETP* **30**, 1174 (1970)]. "General Solution to the Gravitational Equations with a Physical Singularity"

Carter, B.; (1969) *J. Math. Phys.* **10**, 70. "Killing Horizons and Orthogonally Transitive Groups in Space-Time"

Conklin, E. K.; (1969) *Nature* **222**, 971. "Velocity of the Earth with Respect to the Cosmic Background Radiation"

Doroshkevich, A. G.; Zel'dovich, Ya. B., Novikov, I. D.; (1969) *Astrofizika* **5**, 539. "The Kinetic Theory of Neutrinos in the Anisotropic Cosmological Models"

Douglas, J. N.; Robinson, I.; Schild, A.; Schucking, E. L.; Wheeler, J. A.; Woolf, N. J.; (Eds.) (1969) *Quasars and High Energy Astronomy* (Gordon & Breach, New York)

Ellis, G. F. R.; MacCallum, M. A. H.; (1969) *Comm. Math. Phys.* **12**, 108. "A Class of Homogeneous Cosmological Models" [6.4, Table 6.2]

Hawking, S. W.; (1969) *Mon. Not. Roy. Astron. Soc.* **142**, 129. "On the Rotation of the Universe"

Hawking, S. W.; Penrose, R.; (1969) *Proc. Roy. Soc. A* **314**, 529. "The Singularities of Gravitational Collapse and Cosmology"

Hughston, L. P.; (1969) *Astrophys. J.* **158**, 987. "Multifluid Cosmologies" [4.2, 4.2]

Iben, I.; Rood, R.; (1969) *Nature* **223**, 933. "Age and Initial Helium Abundance of Globular Cluster Stars" [15.1]

Jacobs, K. C.; (1969) *Astrophys. J.* **155**, 379. "Cosmologies of Bianchi Type I with a Uniform Magnetic Field"

Jaki, S. L.; (1969) *The Paradox of Oblers' Paradox* (Herder & Herder, New York) [1.1]

Johri, V. B.; (1969) *Astrophys. Letters* **3**, 65. "Perturbation of Cosmological Models"

Longair, M. S.; (1969) *Usphekhi Phys. Nauk.* **99**, 229 [Eng. Trans. *Sov. Phys.-Usphekhi,* **12**, 673 (1969)]. "The Counts of Radio Sources"

Matzner, R. A.; (1969) *Astrophys. J.* **157**, 1085. "The Evolution of Anisotropy in Nonrotating Bianchi Type V Cosmologies" [11.1]

Misner, C. W.; (1969a) *Phys. Rev. Letters* **22**, 1071. "Mix-Master Universe" [8.2, 12.2, 12.3, 12.3, 12.3]

__________ ; (1969b) *Phys. Rev.* **186**, 1319. "Quantum Cosmology. I" [4.2, 11.2, 11.3, 11.4, 12.2, 12.2, 12.3, 12.4, 12.4, 12.4]

__________ ; (1969c) *Phys. Rev.* **186**, 1328. "Absolute Zero of Time" [15.1]

Mjolsness, R. C.; (1969) *Phys. Rev.* **187**, 1753. "Finite Density Nonhomogeneous Newtonian Cosmologies"

Nariai, H.; (1969) *Prog. Theor. Phys.* **41**, 686. "The Lagrangian Approach to the Gravitational Instability in an Expanding Universe"

Noerdlinger, P. D.; (1969) *Phys. Rev.* **186**, 1347. "Geodesics of Robertson-Walker Universes"

Okerson, D.; (1969) *The Mixmaster Universe and Particle Formation* (B.A. Thesis, Princeton Univ.) [13.2]

Omnes, R.; (1969) *Phys. Rev. Letters* **23**, 38. "Possibility of Matter-Antimatter Separation at High Temperature" [15.1]

Ozsvath, I.; Schücking, E.; (1969) *Ann. Phys.* **55**, 166. "The Finite Rotating Universe"

Parker, L.; (1969) *Phys. Rev.* **183**, 1057. "Quantized Fields and Particle Creation in Expanding Universes. I"

Petrov, A. Z.; (1969) *Einstein Spaces* (Pergamon, Oxford) [5.2, 6.4, 6.4, 7.1]

Rees, M. J.; (1969) *Observatory* **89**, 193. "The Collapse of the Universe – An Eschatological Study"

Ruban, V. A.; (1969) *Zh. Eksp. Teor. Fiz.* **56**, 1914 [Eng. Trans. *Sov. Phys.-JETP* **29**, 1027 (1969)]. "Spherically Symmetric T-Models in the General Theory of Relativity"

Ryan, M. P.; (1969) *J. Math. Phys.* **10**, 1724. "General Form of the Einstein Equations for a Bianchi Type IX Universe"

Schmidt, B. G.; (1969) *Comm. Math. Phys.* **15**, 329. "Discrete Isotropies in a Class of Cosmological Models"

Sexl, R. U.; Urbantke, H. K.; (1969) *Phys. Rev.* **179**, 1247. "Production of Particles by Gravitational Fields" [15.1]

Shepley, L. C.; (1969a) *Phys. Letters* **28A**, 695. "A Cosmological Model in Which 'Singularity' Does Not Require a 'Matter Singularity'" [10.2]

————— ; (1969b) in Douglas, et al. (1969). "Can a Dust-Filled Cosmology Bounce?"

Stewart, J. M.; (1969) *Mon. Not. Roy. Astron. Soc.* **145**, 347. "Non-Equilibrium Processes in the Early Universe"

Taub, A. H.; (1969) *Comm. Math. Phys.* **15**, 235. "Stability of General Relativistic Gaseous Masses and Variational Principles" [E]

Treder, H. J.; (1969) *Monatsber. Deutschen Akad. Wiss. Berlin* **11**, 226. "Symmetry and Cosmology"

Vajk, J. P.; (1969) *J. Math. Phys.* **10**, 1145. "Exact Robertson-Walker Cosmological Solutions Containing Relativistic Fluids"

Yu, J. T.; Peebles, P. J. E.; (1969) *Astrophys. J.* **158**, 103. "Superclusters of Galaxies?" [3.1]

Zel'dovich, Ya. B.; Novikov, I. D.; (1969) *Astron. Zh.* **46**, 960 [Eng. Trans. *Sov. Astron.-AJ* **13**, 754 (1970)]. "The Hypothesis About Initial Spectrum of Metric Perturbations in Friedmann Models"

1970

Belinskii, V.; Khalatnikov, I. M.; (1970) *Zh Eksp. Theor. Fiz.* **59**, 314 [Eng. Trans. *Sov. Phys.-JETP* **32**, 169 (1971)]. "On a General Solution of the Gravitational Equations with a Physical Singularity of an Oscillatory Nature"

Belinskii, V. A.; Khalatnikov, I. M.; Lifshitz, E. M.; (1970) *Adv. in Phys.* **19**, 525. "Oscillatory Approach to a Singular Point in the Relativistic Cosmology" [11.3]

Bisnovatyi-Kogan, G. S.; Zel'dovich, Ya. B.; (1970) *Astron. Zh.* **47**, 942 [Eng. Trans. *Sov. Astron.-AJ* **14**, 758 (1971)]. "The Growth of Perturbations in an Expanding Universe of Free Particles"

Carmeli, M.; Fickler, S. I.; Witten, L.; (Eds.) (1970) *Relativity* (Plenum Press, New York)

Danziger, I. J.; (1970) *Ann. Rev. Astron. Astrophys.* **8**, 161. "The Cosmic Abundance of Helium"

DeVaucouleurs, G.; (1970) *Science* **167**, 1203. "The Case for a Hierarchial Cosmology" [3.1, 4.3, 10.1, 15.1]

DeWitt, B.; (1970) in Carmeli, et al. (1970). "Spacetime as a Sheaf of Geodesics in Superspace"

Doroshkevich, A. G.; (1970) *Astrofizika* **6**, 581. "The Space Structure of Perturbation and the Origin of Rotation of Galaxies in the Theory of Fluctuation"

Doroshkevich, A. G.; Novikov, I. D.; (1970) *Astron. Zh.* **47**, 948 [Eng. Trans. *Sov. Astron.-AJ* **14**, 763 (1971)]. "Mixmaster Universes and the Cosmological Problem"

Feinblum, D. A.; (1970) *J. Math. Phys.* **11**, 2713. "Global Singularities and the Taub-NUT Metric"

Geroch, R. P.; (1970a) in Carmeli, et al. (1970). "Singularities"

—————; (1970b) *J. Math. Phys.* **11**, 1955. "Multipole Moments. I. Flat Space" [E]

Gowdy, R. H.; (1970) *Phys. Rev. D* **2**, 2774. "Action Functional of General Relativity as a Path Length. I. Closed Empty Universes" [11.4]

Hagedorn, R.; (1970) *Nucl. Phys. B* **B24**, 93. "Remarks on Thermodynamical Model of Strong Interactions" [15.1]

Haggerty, M. J.; (1970) *Bull. Am. Astron. Soc.* **2**, 245. "Influence of Binary Gravitational Interactions in Hierarchial Cosmologies"

Hoyle, F.; Narlikar, J. V.; (1970) *Nature* **228**, 544. "Effect of Quantum Conditions in a Friedmann Cosmology"

Huang, K.; Weinberg, S.; (1970) *Phys. Rev. Letters* **25**, 895. "Ultimate Temperature and the Early Universe"

Hughston, L.; Jacobs, K.; (1970) *Astrophys. J.* **160**, 147. "Homogeneous Electromagnetic and Massive-Vector Meson Fields in Bianchi Cosmologies" [11.2, 14.1]

Hughston, L. P.; Shepley, L. C.; (1970) *Astrophys. J.* **160**, 333. "Anisotropic Multifluid Cosmologies with Hypersurface Orthogonal Velocity Fields" [4.2]

Jacobs, K.; Hughston, L.; (1970) "More Quantum Cosmology" (unpublished) [11.3]

Lifshitz, E. M.; Khalatnikov, I. M.; (1970) *Zh. Eksp. Theor. Fiz. Pis'ma* **11**, 200 [Eng. Trans. *JETP Letters* **11**, 123 (1970)]. "Oscillatory Approach to Singular Point in the Open Cosmological Model"

Lifshitz, E.; Lifshitz, I.; Khalatnikov, I. M.; (1970) *Zh. Eksp. Teor. Fiz.* **59**, 322 [Eng. Trans. *Sov. Phys.-JETP* **32**, 173 (1971)]. "Asymptotic Analysis of Oscillatory Mode of Approach to a Singularity in Homogeneous Cosmological Models"

MacCallum, M. A. H.; Ellis, G. F. R.; (1970) *Comm. Math. Phys.* **19**, 31. "A Class of Homogeneous Cosmological Models. II. Observations" [E]

MacCallum, M. A. H.; Stewart, J. M.; Schmidt, B. G.; (1970) *Comm. Math. Phys.* **17**, 343. "Anisotropic Stresses in Homogeneous Cosmologies"

Matzner, R. A.; (1970) *J. Math. Phys.* **11**, 2432. "Rotation in Closed Perfect-Fluid Cosmologies"

Matzner, R. A.; Shepley, L. C.; Warren, J. B.; (1970) *Ann. Physics* **57**, 401. "Dynamics of SO(3,R)-Homogeneous Cosmologies" [12.2, 13.2]

Misner, C. W.; (1970) in Carmeli, et al. (1970). "Classical and Quantum Dynamics of a Closed Universe"

Oort, J. H.; (1970) *Astron. and Astrophys.* **7**, 405. "The Density of the Universe"

Ozsvath, I.; (1970a) *J. Math. Phys.* **11**, 2860. "Spatially Homogeneous World Models"

__________; (1970b) *J. Math. Phys.* **11**, 2871. "Dust-Filled Universes of Class II and Class III"

Peebles, P. J. E.; (1970) *Phys. Rev. D* **1**, 397. "Nonlinear Limit on Primeval Adiabatic Perturbations"

Peebles, P. J. E.; Yu, J. T.; (1970) *Astrophys. J.* **162**, 815. "Primeval Adiabatic Perturbation in an Expanding Universe"

Ryan, M. P.; (1970) *Qualitative Cosmology—Diagrammatic Solutions for Bianchi Type IX Universes with Expansion, Rotation, and Shear* (Ph.D. Thesis, Univ. of Maryland) [8.2]

Sandage, A. R.; (1970) *Physics Today* **23**, No. 2, p. 34. "Cosmology—A Search for Two Numbers"

Schmidt, S.; (1970) *Astrology 14—Your New Sun Sign* (Bobbs-Merrill, Indianapolis) [15.1]

Silk, J.; (1970) *Mon. Not. Roy. Astron. Soc.* **147**, 13. "The Instability of a Rotating Universe"

Sokolov, D. D.; (1970) *Dok. Akad. Nauk. SSSR* **195**, 1307 [Eng. Trans. *Sov. Phys. Dokl.* **15**, 1112]. "Topology of Models of the Universe"

Urankar, A.; (1970) *Ann. Physik* **24**, 110. "A Homogeneous Anisotropic Relativistic World Model with Negative Space Curvature"

Vajk, J. P.; Eltgroth, P. G.; (1970) *J. Math. Phys.* **11**, 2212. "Spatially Homogeneous, Anisotropic Cosmological Models Containing Relativistic Fluid and Magnetic Field"

Walker, M.; Penrose, R.; (1970) *Comm. Math. Phys.* **18**, 265. "On Quadratic First Integrals of the Geodesic Equations for Type {2,2} Spacetimes" [E]

Zapolosky, H.; (1970) Unpublished Notes. Reported on in Ryan (1972d) [12.4]

Zel'dovich, Ya. B.; (1970) *Zh. Eksp. Teor. Fiz. Pis'ma* **12**, 443 [Eng. Trans. *JETP Letters* **12**, 307 (1970)]. "Particle Production in Cosmology"

Zel'dovich, Ya. B.; Novikov, I. D.; (1970) *Astrofizika* **6**, 379. "Vortex Perturbations in the Friedmann Cosmological Model"

1971

Alfvén, H.; (1971) *Phys. Today* **24**, No. 2, p. 28. "Plasma Physics Applied to Cosmology"

Arens, R.; (1971) *Comm. Math. Phys.* **21**, 125. "Hamiltonian Structures for Homogeneous Spaces"

Bahcall, J. N.; Frautschi, S.; (1971) *Astrophys. J. Lett.* **170**, L81. "The Hadron Barrier in Cosmology and Gravitational Collapse" [15.1]

Belinskii, V.; Khalatnikov, I. M.; Ryan, M.; (1971) Preprint Landau Institute of Theoretical Physics [Appeared as Ryan (1972a)] [12.2]

Belinskii, V. A.; Lifshitz, E. M.; Khalatnikov, I. M.; (1971a) *Uspekhi Fiz. Nauk* **102** 463 [Eng. Trans. *Sov. Phys.-Uspekhi* **13**, 745 (1971)]. "Oscillatory Approach to the Singular Point in Relativistic Cosmology" [12.2]

———————— ; (1971b) *Zh. Eksp. Teor. Fiz.* **60**, 1969 [Eng. Trans. *Sov. Phys.-JETP* **33**, 1061 (1971)]. "Oscillatory Mode of Approach to Singularity in Homogeneous Cosmological Models with Rotating Axes"

Bonanos, S.; (1971) *Comm. Math. Phys.* **22**, 190. "On the Stability of the Taub Universe"

Burbidge, G.; (1971) *Nature* **233**, 36. "Was There Really a Big Bang?" [15.1]

Chernin, A. D.; (1971) *Astrophys. Lett.* **8**, 31. "Evolution of Structure in the Early Universe"

Collins, C. B.; (1971) *Comm. Math. Phys.* **23**, 137. "More Qualitative Cosmology" [11.3]

Collins, C. B.; Stewart, J. M.; (1971) *Mon. Not. Roy. Astron. Soc.* **153**, 419. "Qualitative Cosmology" [11.3, 12.3]

DeVaucouleurs, G.; Wertz, J. R.; (1971) *Nature* **231**, 109. "Hierarchial Big Bang Cosmology"

Doroshkevich, A. G.; Lukash, V.; Novikov, I. D.; (1971) *Zh. Eksp. Teor. Fiz.* **60**, 1201 [Eng. Trans. *Sov. Phys.-JETP* **33**, 649 (1971)]. "Impossibility of Mixing in a Cosmological Model of the Bianchi IX Type"

Doroshkevich, A. G.; Zel'dovich, Ya. B.; Novikov, I. D.; (1971) *Zh. Eksp. Teor. Fiz.* **60**, 3 [Eng. Trans. *Sov. Phys.-JETP* **33**, 1 (1971)]. "Perturbations in an Anisotropic Homogeneous Universe"

Ellis, G. F. R.; (1971a) *Gen. Relativ. and Gravitation* **2**, 7. "Topology and Cosmology" [15.1]

———— ; (1971b) in Sachs (1971). "Relativistic Cosmology" [E]

Ernst, F. J.; (1971) *J. Math. Phys.* **12**, 2395. "Exterior-Algebraic Derivation of Einstein Field Equations Employing a Generalized Basis"

Geroch, R. P.; (1971a) in Sachs (1971). "Spacetime Structure From a Global Viewpoint"

———— ; (1971b) *Gen. Rel. Grav.* **2**, 61. "General Relativitv in the Large"

Ginzburg, V. L.; (1971) *Comments. Astrophys. Space Phys.* **3**, 7. "About the Singularities in General Relativity and Cosmology"

Ginzburg, V. L.; Kirzhnits, D. A.; Lyubushin, A. A.; (1971) *Zh. Eksp. Teor. Fiz.* **60**, 451 [Eng. Trans. *Sov. Phys.-JETP* **33**, 242 (1971)]. "On the Role of Quantum Fluctuations of a Gravitational Field in General Relativity and Cosmology"

Gowdy, R.; (1971) *Phys. Rev. Letters* **27**, 826. "Gravitational Waves in Closed Universes" [15.1]

Grishchuk, L. P.; Doroshkevich, A. G.; Lukash, V.; (1971) *Zh. Eksp. Teor. Fiz.* **61**, 3 [Eng. Trans. *Sov. Phys.-JETP* **34**, 1 (1971)]. "The Model of Mixmaster Universe with Arbitrarily Moving Matter"

Gunn, J. E.; (1971) *Astrophys. J.* **164**, L113. "On the Distances of the Quasi-Stellar Objects"

Haggerty, M. J.; (1971) *Astrophys. J.* **166**, 257. "A Newtonian Model for the Creation of a Hierarchial Cosmology"

Hawking, S. W.; (1971) *Mon. Not. Roy. Astron. Soc.* **152**, 75. "Gravitationally Collapsed Objects of Very Low Mass"

Hibler, D. L.; (1971) Univ. of Texas (Unpublished Ph.D. work; Communicated by J. Ehlers) [15.1]

Jacobs, K. C.; (1971) *Bull. Am. Astron. Soc.* **3**, 480. "General Relativity Theory"

Kundt, W.; (1971) in Sachs (1971). "Cosmological Density Fluctuations During Hadron Stage"

Kuper, G. C.; Peres, A.; (Eds.)(1971) *Relativity and Gravitation* (Gordon & Breach, New York)

Landau, L. D.; Lifshitz, E. M.; (1971) *The Classical Theory of Fields* (3rd Ed.) (Addison-Wesley, Reading, Mass.)

Liang, E. P. T.; (1971) *Velocity Dominated Singularities in General Relativistic Cosmologies* (Ph.D. Thesis, Univ. California, Berkeley)

Longair, M. S.; (1971) *Rep. Prog. Phys.* **34**, 1125. "Observational Cosmology"

MacCallum, M. A. H.; (1971a) *Nature Phys. Sci.* **230**, 112. "Problems of the Mixmaster Universe" [12.3]

__________; (1971b) *Comm. Math. Phys.* **20**, 57. "A Class of Homogeneous Cosmological Models III, Asymptotic Behavior"

Matzner, R. A.; (1971a) *Ann. Physics* **65**, 482. "Anisotropic Calculations; Application to Observations"

__________; (1971b) *Comm. Math. Phys.* **20**, 1. "Collisionless Radiation in Closed Cosmologies"

Matzner, R. A.; Chitre, D. M.; (1971) *Comm. Math. Phys.* **22**, 173. "Rotation Does Not Enhance Mixing in the Mixmaster Universe"

Matzner, R. A.; Perko, T. E.; Shepley, L. C.; (1971) *Phys. Letters* **35A**, 467. "Perturbations in Anisotropic Euclidean-Homogeneous Cosmologies"

Nariai, H.; Tomita, K.; (1971) *Prog. Theor. Phys. Suppl.* **49**, 83. "Formation of Proto-Galaxies in the Expanding Universe – Gravitational Instability"

O'Connell, D. J. K.; (Ed.) (1971) *Nuclei of Galaxies* (North-Holland, Amsterdam)

Ozsvath, I.; (1971) *J. Math. Phys.* **12**, 1078. "Spatially Homogeneous Rotating World Models"

Parker, L.; (1971) *Phys. Rev. D* **3**, 346. "Quantized Fields and Particle Creation in Expanding Universes. II" [15.1]

Peebles, P. J. E.; (1971a) *Astrophys. Space. Sci.* **11**, 443. "Primeval Turbulence" [14.3]

——————; (1971b) *Physical Cosmology* (Princeton Univ. Press, Princeton) [15.1, 15.1, 15.1]

Perko, T. E.; (1971) *Perturbations in Anisotropic Euclidean-Homogeneous Cosmologies* (Ph.D. Thesis, Univ. Texas, Austin)

Rees, M. J.; (1971) in Sachs (1971) "Some Current Ideas on Galaxy Formation"

Ryan, M. P.; (1971a) *Ann. Physics* **65**, 506. "Qualitative Cosmology – Diagramatic Solutions for Bianchi Type IX Universes with Expansion, Rotation and Shear. I. The Symmetric Case" [12.2, 12.2, 12.2, 12.2]

——————; (1971b) *Ann. Physics* **68**, 541. "Qualitative Cosmology – Diagramatic Solutions for Bianchi Type IX Universes with Expansion, Rotation and Shear. II. The General Case" [12.2]

Sachs, R. K.; (Ed.) (1971) *General Relativity and Cosmology* (Academic Press, New York)

Sandage, A.; (1971) in O'Connell (1971). "The Age of Galaxies and Globular Clusters – Problems of Finding the Hubble Constant and Deceleration Parameter"

Schmidt, B. G.; (1971) *Gen. Rel. Grav.* **1**, 269. "A New Definition of Singular Points in General Relativity" [5.3]

Schutz, B.; (1971) *Phys. Rev. D* **4**, 3559. "Hamiltonian Theory of a Relativistic Perfect Fluid" [11.2]

Sciami, D. W.; (1971) *Modern Cosmology* (Cambridge Univ. Press, Cambridge)

Shikin, I. S.; (1971) *Zh. Eksp. Teor. Fiz.* **61**, 445 [Eng. Trans. *Sov. Phys.-JETP* **34**, 236 (1972)]. "Analogues of Homogeneous Anisotropic Models of General Relativity in Newtonian Cosmology"

Smith, J. W.; (1971) *Proc. Camb. Phil. Soc.* **69**, 295. "On the Differential Topology of Space-Time"

Weinberg, S.; (1971) *Astrophys. J.* **168**, 175. "Entropy Generation and the Survival of Protogalaxies in an Expanding Universe"

Wertz, J. R.; (1971) *Astrophys. J.* **164**, 227. "A Newtonian Big-Bang Hierarchial Cosmological Model"

1972

Barnes, R.; Prondzinski, R.; (1972) *Astrophys. Space Sci.* **18**, 34. "A Comparative Study of Brans-Dicke and General Relativistic Cosmologies in Terms of Observationally Measurable Quantities"

Batakis, N.; Cohen, J. M.; (1972) *Ann. Physics* **73**, 578. "Closed Anisotropic Cosmological Models"

Belinskii, V. A.; Khalatnikov, I. M.; (1972) *Zh. Eksp. Teor. Fiz.* **63**, 1121 [Eng. Trans. *Sov. Phys.-JETP* **36**, 591 (1973)]. "Effect of Scalar and Vector Fields on the Nature of the Cosmological Singularity"

Belinskii, V. A.; Lifshitz, E. M.; Khalatnikov, I. M.; (1972) *Zh. Eksp. Teor. Fiz.* **62**, 1606 [Eng. Trans. *Sov. Phys.-JETP* **35**, 838 (1972)]. "Construction of a General Cosmological Solution of the Einstein Equation with a Time Singularity"

Bonanos, S.; (1972) *Comm. Math. Phys.* **26**, 259. "Stability of Homogeneous Universes"

Chernin, A. D.; (1972a) *Dokl. Akad. Nauk SSSR* **206**, 62 [Eng. Trans. *Sov. Phys.-Dokl.* **17**, 825 (1973)]. "Hydrodynamic Motions and the Vacuum Stage in an Anisotropic Cosmological Model"

——————; (1972b) *Astrophys. Letters* **10**, 125. "Dynamic Motions in the Early Universe"

Chernin, A. D.; Moros, B. S.; Vandakurov, Y. V.; (1972) *Phys. Letters A* **39 A**, 233, "Non-Homologous Expansion in Cosmology"

Chernin, A. D.; Shvarts, A. N.; (1972) *Dok. Akad. Nauk SSSR* **205**, 1057 [Eng. Trans. *Sov. Phys.-Dokl.* **17**, 733 (1973)]. "Anisotropic Cosmology and Cosmic Structure"

Chibisov, G. V.; (1972) *Astron. Zh.* **49** 74 [Eng. Trans. *Sov. Astron.-AJ* **16**, 56 (1972)]. "Damping of Adiabatic Perturbations in an Expanding Universe"

Chitre, D.; (1972a) *Investigation of Vanishing of a Horizon for Bianchi Type IX (The Mixmaster Universe)* (Thesis, Univ. Maryland) [12.2, 12.3]

———; (1972b) *Phys. Rev. D* **6**, 3390. "High Frequency Sound Waves to Eliminate a Horizon in the Mixmaster Universe"

Collins, C. B.; (1972) *Comm. Math. Phys.* **27**, 37. "Qualitative Magnetic Cosmology"

Dallaporta, N.; Lucchin, F.; (1972) *Astron. Astrophys.* **19**, 123. "On Galaxy Formation from Primeval Universal Turbulence"

Davies, P. C. W.; (1972) *J. Phys. A* **5**, 1722. "Is the Universe Transparent or Opaque?"

DeGraaf, T.; (1972) in *Proc. Neutrino '72 Europhysics Conf.* Vol. I (OMKD Technoinform, 1972, Budapest). "The Lepton Era of the Big Bang"

Dehnen, H.; Obregon, O.; (1972) *Astrophys. Space. Sci.* **17**, 338. "Cosmological Vacuum Solutions in Brans and Dicke's Scalar-Tensor Theory"

Demianski, M.; Grishchuk, L. P.; (1972) *Comm. Math. Phys.* **25**, 233. "Homogeneous Rotating Universe with Flat Space"

Eardley, D.; Liang, E. P. T.; Sachs, R.; (1972) *J. Math. Phys.* **13**, 99. "Velocity Dominated Singularities in Irrotational Dust Cosmologies" [5.3]

Ellis, G. F. R.; Sciama, D. W.; (1972) in O'Raifeartaigh (1972). "Global and Non-Global Problems in Cosmology"

Evans, D. S.; Wills, D.; Wills, B. J.; (Eds.) (1972) *External Galaxies and Quasi-Stellar Objects* (Reidel, Dordrecht)

Field, G. B.; (1972) *Ann. Rev. Astron. Astrophys.* **10**, 227. "Intergalactic Matter"

Harrison, E. R.; (1972) *Phys. Today* **25**, No. 12, p. 30. "The Cosmic Numbers"

Hu, B. L.; Regge, T.; (1972) *Phys. Rev. Letters* **29**, 1616. "Perturbations on the Mixmaster Universe" [14.3]

Isaacson, R. A.; (1972) *Nature* **239**, 447. "Cosmic Effects of Gravitational Waves"

Jones, B. T. J.; Peebles, P. J. E.; (1972) *Comments. Astrophys. Space Phys.* **4**, 121. "Chaos in Cosmology"

Klauder, J.; (Ed.) (1972) *Magic Without Magic* (Freeman, San Francisco)

Kubo, M.; (1972) *Sci. Reports Tohoku. Univ. Ser. 1*, **55**, No. 1. "Classification of Uniform Cosmological Models"

Liang, E. P. T.; (1972a) *J. Math. Phys.* **13**, 386. "Velocity Dominated Singularities in Irrotational Hydrodynamic Cosmological Models"

———; (1972b) *Phys. Rev. D* **5**, 2458. "Quantum Models for the Lowest-Order Velocity-Dominated Solutions of Irrotational Dust Cosmologies"

MacCallum, M. A. H.; (1972a) *Phys. Letters A* **40 A**, 325. "On Criteria of Cosmological Spatial Homogeneity"

———; (1972b) *Phys. Letters A* **40 A**, 385. "On 'Diagonal' Bianchi Cosmologies"

———; (1972c) *Comm. Math. Phys.* **25**, 173. "Variational Principles and Spatially Homogeneous Universes Including Rotation"

MacCallum, M. A. H.; Taub, A. H.; (1972) *Comm. Math. Phys.* **25**, 173. "Variational Principles and Spatially Homogeneous Universes Including Rotation" [11.3]

Matzner, R. A.; (1972) *Astrophys. J.* **171**, 433. "Dissipative Effects in the Expansion of the Universe. II. A Multicomponent Model for Neutrino Dissipation of Anisotropy in the Early Universe"

Matzner, R. A.; Misner, C. W.; (1972) *Astrophys. J.* **171**, 415. "Dissipative Effects in the Expansion of the Universe. I"

Misner, C. W.; (1972) in Klauder (1972). "Minisuperspace" [11.4, 11.4, 11.4]

Nariai, H.; (1972) *Prog. Theor. Phys.* **47**, 1824. "Hamiltonian Approach to the Dynamics of Expanding Homogeneous Universes in the Brans-Dicke Cosmology" [15.1]

Ni, W.-T.; (1972) *Astrophys. J.* **176**, 769. "Theoretical Frameworks for Testing Relativistic Gravity, IV: A Compendium of Metric Theories of Gravity and Their Post-Newtonian Limits" [E]

Novikov, S. P.; (1972) *Zh. Eksp. Teor. Fiz.* **62**, 1977 [Eng. Trans. *Sov. Phys.-JETP* **35**, 1031 (1972)]. "Some Properties of Cosmological Models"

Omnes, R.; (1972) *Phys. Reports* **3C**, 1. "The Possible Role of Elementary Particle Physics in Cosmology"

O'Raifeartaigh, L.; (1972) *General Relativity* (Clarendon Press, Oxford)

Parker, L.; (1972a) *Phys. Rev. Letters* **28**, 705. "Particle Creation in Isotropic Cosmologies" [15.1]

———— ; (1972b) *Phys. Rev. Letters* **28**, 1497. "Particle Creation in Isotropic Cosmologies" [15.1, E]

Peach, J. V.; (1972) in Evans, et al. (1972). "Cosmological Information From Galaxies and Radio Galaxies"

Penrose, R.; (1972) *Reg. Conf. Ser. Appl. Math.* No. 1, p. 1. "Techniques of Differential Topology in Relativity"

Perko, T. E. ; Matzner, R. A.; Shepley, L. C.; (1972) *Phys. Rev. D* **6**, 969. "Galaxy Formation in Anisotropic Cosmologies" [14.3]

Rees, M. J.; (1972) *Comm. Astrophys. Space Phys.* **4**, 179. "Cosmological Significance of e^2/GM^2 and Related 'Large Numbers'"

Reeves, H.; (1972) *Phys. Rev. D* **6**, 3363. "Densities of Baryons and Neutrinos in the Universe From an Analysis of Big-Bang Nucleosynthesis"

Reines, F.; (Ed.) (1972) *Cosmology Fusion and Other Matters* (Colorado Assoc. Univ. Press, Boulder)

Reinhardt, M.; (1972) *Astrophys. Letters* **12**, 135. "Cosmological Inferences From the Angular Diameters of Quasars"

Rood, H.; Page, T. L.; Kintner, E.; King, I. R.; (1972) *Astrophys. J.* **175**, 627. "The Structure of the Coma Cluster of Galaxies"

Rowan-Robinson, M.; (1972) *Astrophys. J. Letters* **178**, L81. "The Cosmological Implications of Counts of Galaxies"

Ruban, V. A.; (1972) *Dok. Akad. Nauk. SSSR* **204**, 1086 [Eng. Trans. *Sov. Phys.-Dokl.* **17**, 568 (1972)]. "Non-Singular Metrics of the Taub-Nut Type With Electromagnetic Fields"

Ruban, V. A ; Chernin, A. D., (1972) *Astron. Zh.* **49**, 447 [Eng. Trans. *Sov. Astron.-AJ* **16**, 363 (1972)]. "Rotational Perturbations in the Anisotropic Cosmology"

Ruban, V. A.; Finkelstein, A. M.; (1972) *Lett. Nuovo Cimento* **5**, 289. "Generalization of the Taub-Kasner Cosmological Metric in the Scalar-Tensor Gravitation Theory"

Ryan, M. P.; (1972a) *Ann. Physics* **70**, 301. "The Oscillatory Regime Near the Singularity in Bianchi-Type IX Universes"

———— ; (1972b) *Ann. Physics* **72**, 584. "Surface-of-Revolution Cosmology" [15.1]

Ryan, M. P.; (1972c) *Astrophys. J. Letters* **177**, L79. "Is the Existence of a Galaxy Evidence for a Black Hole at Its Center?"

———— ; (1972d) *Hamiltonian Cosmology* (Springer-Verlag, Heidelberg) [11.2, 11.2, 11.3, 11.3, 11.4]

Sandage, A.; (1972) "The Red-Shift-Distance Relation." This has three parts: I. *Astrophys. J.* **173**, 485; II. *Astrophys. J.* **178**, 1; III. *Astrophys. J.* **178**, 25. See also Sandage (1973). [4.3, 15.1]

Sarfatt, J.; (1972) *Nature (Phys. Sci.)* **240**, 101. "Gravitation, Strong Interactions and the Creation of the Universe"

Saslaw, W. C.; (1972) *Astrophys. J.* **173**, 1. "Conditions for the Rapid Growth of Perturbations in an Expanding Universe"

Saslaw, W. C.; Jacobs, K. C.; (Eds.) (1972) *The Emerging Universe* (Univ. Press of Virginia, Charlottesville)

Segal, I.; (1972) *Astron. Astrophys.* **18**, 143. "Covariant Chronogeometry and Extreme Distances. I" [15.1]

Stauffer, D.; (1972) *Phys. Rev. D* **6**, 1797. "Temperature Maxima of the Early (Hadron) Universe"

Stecker, F. W.; Puget, J. L.; (1972) *Astrophys. J.* **178**, 57. "Galaxy Formation from Annihilation – Generated Supersonic Turbulence in the Baryon-Symmetric Big-Bang Cosmology and the Gamma-Ray Background Spectrum"

Stewart, J.; (1972) *Astrophys. J.* **176**, 323. "Perturbations in an Expanding Universe of Free Particles"

Syunyaev, R. A.; Zel'dovich, Ya. B.; (1972) *Astron. Astrophys.* **20**, 189. "Formation of Clusters of Galaxies, Protocluster Fragmentation and Intergalactic Gas Heating"

Thaddeus, P.; (1972) *Ann. Rev. Astron. Astrophys.* **10**, 305. "The Short-Wavelength Spectrum of the Microwave Background"

Tomita, K.; (1972) *Prog. Theor. Phys.* **48**, 1503. "Primordial Irregularities in the Early Universe"

Weinberg, S.; (1972) *Gravitation and Cosmology* (John Wiley & Sons, New York)

Wyler, J. A.; (1972) Preprint, Princeton Univ. "Rasputin, Science and the Transmogrification of Destiny"

Zel'dovich, Ya. B.; (1972) *Mon. Not. Roy. Astron. Soc.* **160**, 1p. "A Hypothesis Unifying the Structure and the Entropy of the Universe"

1973

Alpher, R. A.; (1973) *Am. Sci.* **61**, 52. "Large Numbers, Cosmology and Gamow"

Barnes, A.; (1973) *Gen. Rel. Grav.* **4**, 105. "On Shear-Free Normal Flows of a Perfect Fluid"

Berezdivin, R.; Sachs, R. K.; (1973) *J. Math. Phys.* **14**, 1254. "Matter Symmetries in General Relativistic Kinetic Theory"

Bowers, R.; Zimmerman, R.; (1973) *Phys. Rev. D* **7**, 296. "Relativistic Quantum Many-Body Theory in Riemannian Space-Time" [15.1]

Burbidge, G. R.; O'Dell, S. L.; (1973) *Astrophys. J.* **183**, 759. "The Redshift-Magnitude Relation for Quasi-Stellar Objects"

Carlitz, R.; Frautschi, S.; Nahm, W.; (1973) *Astron. Astrophys.* **26**, 171. "Implications of the Statistical Bootstrap Model for Cosmology and Galaxy Formation"

Carpenter, R. L.; Gulkis, S.; Sato, T.; (1973) *Astrophys. J. Letters* **182**, L61. "Search for Small-Scale Anisotropy in the 2.7°K Cosmic Background Radiation at a Wavelength of 3.56 Centimeters"

Clarke, C. J. S.; (1973) *Comm. Math. Phys.* **32**, 205. "Local Extensions in Singular Space-Times"

Collins, C. B.; Hawking, S.; (1973a) *Astrophys. J.* **180**, 317. "Why is the Universe Isotropic?" [12.3]

—————; (1973b) *Mon. Not. Roy. Astron. Soc.* **162**, 307. "The Rotation and Distortion of the Universe" [15.1]

Collins, P. A.; Williams, R. M.; (1973) *Phys. Rev. D* **7**, 965. "Dynamics of the Friedmann Universe Using Regge Calculus"

Dallaporta, N.; Lucchin, F.; (1973) *Astron. Astrophys.* **26**, 325. "On Galaxy Formation from Primeval Universal Turbulence. II."

Date, T. H.; (1973) *Curr. Sci. (India)* **42**, 15. "A Universe Filled with Magnetofluid"

Davies, P. C. W.; (1973) *Mon. Not. Roy. Astron. Soc.* **161**, 1. "The Thermal Future of the Universe"

Dirac, P. A. M.; (1973) *Proc. Roy. Soc. A* **333**, 403. "Long Range Forces and Broken Symmetries"

Duncan, D.; (1973) *Some Aspects of the Theory of Singularities in General Relativity* (Dissertation, Univ. Texas, Austin) [5.13]

Eardley, D.; Sachs, R. K.; (1973) *J. Math. Phys.* **14**, 209. "Spacetimes with a Future Projective Infinity"

Ellis, G. F. R.; (1973) in Schatzman (1973a). "Relativistic Cosmology" [10.1]

Evans, A. B.; (1973) *Phys. Letters A* **44A**, 211. "Newtonian Universes with Non-Linear Velocity"

Fanaroff, B. L.; Longair, M. S.; (1973) *Mon. Not. Roy. Astron. Soc.* **161**, 393. "Cosmological Information from Surveys of Radio Source Spectra"

Gold, R.; (1973) *Nature* **242**, 24. "Multiple Universes"

Gurevich, L. E.; Finkelstein, A. M.; Ruban, V. A.; (1973) *Astrophys. Space Sci.* **22**, 231. "On the Problem of the Initial State in the Isotropic Scalar-Tensor Cosmology of Brans-Dicke"

Hagedorn, R.; (1973) in Schatzman (1973a) "Thermodynamics of Strong Interactions"

Hakim, R.; Vilain, C.; (1973) *Astron. Astrophys.* **25**, 211. "The Applications of Relativistic Kinetic Theory to Cosmological Models – Some Observational Consequences"

Harrison, E. R.; (1973a) *Phys. Rev. Letters* **30**, 188. "Origin of Magnetic Fields in the Early Universe"

—————; (1973b) *Ann. Rev. Astron. Astrophys.* **11**, 155. "Standard Model of the Early Universe"

Hauser, M. G.; Peebles, P. J. E.; (1973) "Statistical Analysis of Catalogs of Extragalactic Objects." This has two parts: I. *Astrophys. J.* **185**, 413. II. *Astrophys. J.* **185**, 757.

Hawking, S. W.; Ellis, G. F. R.; (1973) *The Large Scale Structure of Space-Time* (Cambridge Univ. Press, Cambridge)

Hecht, H. F.; (1973) *Astron. Astrophys.* **26**, 123. "The Role of the Electron Neutrino Interaction in the Primordial Gas"

Hegyi, D. J.; (Ed.) (1973) *Ann. New York Acad. Sci.* **224**, 1. "The Sixth Texas Symposium on Relativistic Astrophysics"

Heller, M.; Klimek, Z.; Suszycki, L.; (1973) *Astrophys. Space Sci.* **20**, 205. "Imperfect Fluid Friedmannian Cosmology"

Horedt, G.; (1973) *Astrophys. J.* **183**, 383. "On the Expanding-Universe Postulate"

Hu, B. L.; (1973) *Phys. Rev. D* **8**, 1048. "Scalar Waves in the Mixmaster Universe. I. The Helmholtz Equation in a Fixed Background"

Hughston, L. P.; Sommers, P.; (1973) *Comm. Math. Phys.* **32**, 147. "Spacetimes with Killing Tensors"

Isaacson, R. A.; Winicour, J.; (1973) *Astrophys. J.* **184**, 49. "Gravitational Radiation and Observational Cosmology"

Jones, B. J. T.; (1973) *Astrophys. J.* **181**, 269. "Cosmic Turbulence and the Origin of Galaxies"

King, A. R.; Ellis, G. F. R.; (1973) *Comm. Math. Phys.* **31**, 209. "Tilted Homogeneous Cosmological Models"

Klimek, Z.; (1973) *Postepy. Astron.* **21**, 115. "Singularities in General Relativity and Cosmology"

Krieger, C. J.; Solheim, J.-E.; (1973) *Astrophys. Space Sci.* **23**, 325. "On Redshifts and Number Counts of Extragalactic Objects in Two Cosmological Models"

Layzer, D.; Hively, R.; (1973) *Astrophys. J.* **179**, 361. "Origin of the Microwave Background"

Lerner, D. E.; (1973) *Comm. Math. Phys.* **32**, 19. "The Space of Lorentz Metrics"

Longair, M. S.; Rees, M. J.; (1973) in Schatzman (1973a). "Lecture Notes on Observational Cosmology"

MacCallum, M. A. H.; (1973) in Schatzman (1973a). "Cosmological Models from a Geometric Point of View"

Matzner, R. A.; Ryan, M. P.; Toton, E.; (1973) *Nuovo Cimento B* **14B**, 161. "Brans-Dicke Theory and Anisotropic Cosmologies" [15.1]

Miller, J. G.; (1973) *J. Math. Phys.* **14**, 486. "Global Analysis of the Kerr-Taub-Nut Metric" [8.3, E]

Miller, J. G.; Kruskal, M. D.; (1973) *J. Math. Phys.* **14**, 484. "Extension of a Compact Lorentz Manifold" [5.4, 8.3, E]

Misner, C. W.; Thorne, K. S.; Wheeler, J. A.; (1973) *Gravitation* (Freeman & Co., San Francisco)

Moser, A. R.; Matzner, R. A.; Ryan, M. P.; (1973) *Ann. Physics* **79**, 558. "Numerical Solutions for Symmetric Bianchi Type IX Universes" [13.2]

Narlikar, J. V.; (1973) *Nature (Phys. Sci.)* **242**, 135. "Singularity and Matter Creation in Cosmological Models"

Osinovsky, M. E.; (1973) *Comm. Math. Phys.* **32**, 39. "Highly Mobile Einstein Spaces in the Large"

Parijskij, Y. N.; (1973) *Astrophys. J. Letters* **180**, L47. "New Limit on Small-Scale Irregularities of 'Black Body' Radiation"

Parker, L.; Fulling, S. A.; (1973) *Phys. Rev. D* **7**, 2357. "Quantized Matter Fields and the Avoidance of Singularities in General Relativity"

Partridge, R. B.; (1973) *Nature* **244**, 263. "Absorber Theory of Radiation and the Future of the Universe"

Plazowski, J.; (1973) *Acta. Phys. Pol. B.* **B4**, 49. "The Imbedding Method of Finding the Maximal Extensions of Solutions of Einstein Field Equations"

Reeves, H.; Audouze, J.; Fowler, W. A.; Schramm, D. N.; (1973) *Astrophys. J.* **179**, 909. "On the Origin of the Light Elements" [15.1]

Reinhardt, M.; (1973) *Z. Naturforsch. A* **28A**, 529. "Mach's Principle – A Critical Review"

Rogerson, J. B.; York, D. G.; (1973) *Astrophys. J. Letters* **186**, L95. "Inter-Stellar Deuterium Abundance in the Direction of Beta Centauri" [15.1]

Sandage, A.; (1973) "The Redshift-Distance Relation" This has four parts: IV. *Astrophys. J.* **180**, 687; V. *Astrophys. J.* **183**, 711; VI *Astrophys. J.* **183**, 731; VII. (with E. Hardy) *Astrophys. J.* **183**, 743 [4.3, 15.1]

Schatzman, E.; (Ed.)(1973a) Cargese *Lectures in Physics*, Vol. 6 (Gordon & Breach, New York)

———— ; (1973b) *Comm. Astrophys. Space Sci.* **5**, 23. "Nucleosynthesis in the Symmetric Universe"

Schmidt, B. G.; (1973) *Comm. Math. Phys.* **29**, 49. "The Local b-Completeness of Space-Times"

Schramm, D. N.; Arnett, W. D.; (Eds.)(1973) *Explosive Nucleosynthesis* (Univ. of Texas Press, Austin)

Schulman, L. S.; (1973) *Phys. Rev. D* **7**, 2868. "Correlating Arrows of Time"

Shepley, L. C.; (1973) in Schatzman (1973a). "Type I Cosmologies"

Silk, J.; (1973a) in Schatzman (1973a). "Primordial Turbulence and Intergalactic Matter"

———— ; (1973b) *Ann. Rev. Astron. Astrophys.* **11**, 269. "Diffuse X and Gamma Radiation"

———— ; (1973c) *Comm. Astrophys. Space Sci.* **5**, 9. "The Case for a Chaotic Cosmogony"

Silk, J.; Ames, S.; (1973) *Astrophys. J.* **178**, 77. "Primordial Turbulence and the Formation of Galaxies"

Silk, J.; Lea, S.; (1973) *Astrophys. J.* **180**, 669. "Primordial Random Motions and Angular Momenta of Galaxies and Galaxy Clusters"

Sistero, R. F.; (1973) *Astrophys. Space Sci.* **20**, 19. "Cosmological Hadronic Fireball"

Snyder, L. E.; Buhl, D.; Zuckerman, B.; (1973) *Nature* **242**, 33. "Origin of Elements"

Sommers, P.; (1973) *J. Math. Phys.* **14**, 787. "On Killing Tensors and Constants of Motion"

Steigman, G.; (1973) in Schatzman (1973a). "The Case Against Antimatter in the Universe" [15.1]

Strecker, F. W.; (1973) *Nature (Phys. Sci.)* **241**, 74. "Diffuse Cosmic Gamma Rays: Present Status of Theory and Observations"

Strom, R. G.; (1973) *Nature (Phys. Sci.)* **244**, 2. "Radio Source Depolarization, Size, and Cosmology"

Szekeres, P.; (1973) *Nuovo Cimento B* **17B**, 187. "Global Description of Spherical Collapsing and Expanding Dust Clouds"

Trautman, A.; (1973) *Nature (Phys. Sci.)* **242**, 7. "Spin and Torsion May Avert Gravitational Singularities"

Wagoner, R. V.; (1973) *Astrophys. J.* **179**, 343. "Big-Bang Nucleosynthesis Revisited"

White, P. C.; (1973) *J. Math. Phys.* **14**, 831. "C^{∞} Perturbations of a Cosmological Model"

1974

Collins, C. B.; (1974) (Preprint) "Tilting at Cosmological Singularities" [3.5, 10.2]

Ellis, G. F. R.; King, A. R.; (1974) (Preprint) "Was the Big Bang a Whimper?"

Epstein, R. I.; Arnett, W. D.; Schramm, D. N.; (1974) (Preprint) "Can Supernovae Produce Deuterium?"

Field, G. B.; (1974) in *Stars and Stellar Systems*, Vol. IX: Galaxies and the Universe, A. and M. Sandage (Eds.) (Univ. Chicago Press, Chicago) (to appear). "The Origin of Galaxies" [15.1]

Gott, J. R.; Gunn, J. E.; Schramm, D. N.; Tinsley, B. M.; (1974) (Preprint) "An Unbound Universe?"

Ryan, M. P.; (1974) *J. Math. Phys.* **15**, 812. "Hamiltonian Cosmology: Death and Transfiguration" [11.3, Table 11.1]

Ryan, M. P.; Shepley, L. C.; (1974) *Am. J. Phys.* (in preparation). "Resource Letter C–1 on Cosmology"

INDEX

Library of Congress Cataloging in Publication Data

Ryan, Michael P 1943-
Homogeneous relativistic cosmologies.

(Princeton series in physics)
Bibliography: p.
1. Cosmology. I. Shepley, Lawrence C., 1939-
joint author. II. Title.
QB981.R88 523.1'01 74-2976
ISBN 0-691-08146-8
ISBN 0-691-0153-0 (pbk.)

Ingram Content Group UK Ltd.
Milton Keynes UK
UKHW021819010623
422734UK00007B/518

9 780691 618197